Fiber Optics Engineering

For further volumes:
http://www.springer.com/series/6976

Optical Networks

Series Editor: Biswanath Mukherjee
University of California, Davis
Davis, CA

Mohammad Azadeh

Fiber Optics Engineering

Springer

Mohammad Azadeh
Source Photonics, Inc.
20550 Nordhoff St.
Chatsworth, CA 91311
USA
azadeh@sourcephotonics.com

Series Editor
Biswanath Mukherjee
University of California
Davis, CA
USA

ISSN 1935-3839
ISBN 978-1-4614-2933-3 e-ISBN 978-1-4419-0304-4
DOI 10.1007/978-1-4419-0304-4
Springer Dordrecht Heidelberg London New York

Printed on acid-free paper

Springer is part of Springer Science+Business Media (www.springer.com)

Preface

Within the past few decades, information technologies have been evolving at a tremendous rate, causing profound changes to our world and our ways of life. In particular, fiber optics has been playing an increasingly crucial role within the telecommunication revolution. Not only most long-distance links are fiber based, but optical fibers are increasingly approaching the individual end users, providing wide bandwidth links to support all kinds of data-intensive applications such as video, voice, and data services.

As an engineering discipline, fiber optics is both fascinating and challenging. Fiber optics is an area that incorporates elements from a wide range of technologies including optics, microelectronics, quantum electronics, semiconductors, and networking. As a result of rapid changes in almost all of these areas, fiber optics is a fast evolving field. Therefore, the need for up-to-date texts that address this growing field from an interdisciplinary perspective persists.

This book presents an overview of fiber optics from a practical, engineering perspective. Therefore, in addition to topics such as lasers, detectors, and optical fibers, several topics related to electronic circuits that generate, detect, and process the optical signals are covered. In other words, this book attempts to present fiber optics not so much in terms of a field of "optics" but more from the perspective of an engineering field within "optoelectronics." As a result, practicing professionals and engineers, with a general background in physics, electrical engineering, communication, and hardware should find this book a useful reference that provides a summary of the main topics in fiber optics. Moreover, this book should be a useful resource for students whose field of study is somehow related to the broad areas of optics, optical engineering, optoelectronics, and photonics.

Obviously, covering all aspects of fiber optics in any depth requires many volumes. Thus, an individual text must out of necessity be selective in the topics it covers and in the perspectives it offers. This book covers a range of subjects, starting from more abstract basic topics and proceeding towards more practical issues. In most cases, an overview of main results is given, and additional references are provided for those interested in more details. Moreover, because of the practical character of the book, mathematical equations are kept at a minimum, and only essential equations are provided. In a few instances where more mathematical details are given and equations are derived, an elementary knowledge of calculus is sufficient for following the discussion, and the inconvenience of having to go through the math is well rewarded by the deeper insights provided by the results.

The logical flow of the book is as follows. The first three chapters act as a foundation and a general background for the rest of the book. Chapter 1 covers basic physical concepts such as the nature of light, electromagnetic spectrum, and a brief overview of fiber optics. Chapter 2 provides an overview of important networking concepts and the role of fiber optics within the telecommunication infrastructure. Chapter 3 provides an introduction to fiber optics from a signal viewpoint. This includes some basic mathematical background, as well as characterization of physical signals in the electrical and optical domains.

Chapters 4–7 cover the main elements of a fiber optic link in more depth. Chapter 4 is dedicated to diode lasers which are the standard source in fiber optics. Chapter 5 deals with propagation of optical signals in fibers and signal degradation effects. PIN and APD detectors that convert photons back to electrons are the topic of Chapter 6. Thus, these three chapters deal with generation, propagation, and detection of optical signals. Chapter 7, on the other hand, deals with light coupling and passive components. Therefore, Chapter 7 examines ways of transferring optical signals between elements that generate, detect, and transport the optical signals.

The next two chapters, Chapters 8 and 9, essentially deal with electronic circuits that interface with diode lasers and optical detectors. In particular, Chapter 8 examines optical transmitter circuits and various electronic designs used in driving high-speed optical sources. Chapter 9 examines the main blocks in an optical receiver circuit as well as ways of characterizing the performance of a receiver. A feature of this book is that in addition to traditional CW transceivers, burst mode transmitter and receiver circuits, increasingly used in PON applications, are also discussed.

The final three chapters of the book cover areas that have to do with fiber optics as a viable industry. Chapter 10 presents an overview of reliability issues for optoelectronic devices and modules. A viable fiber optic system is expected to operate outside the laboratory and under real operating conditions for many years, and this requires paying attention to factors outside pure optics or electronics. Chapter 11 examines topics related to test and measurement. In an engineering environment, it is crucial not only to have a firm grasp on theoretical issues and design concepts, but also to design and conduct tests, measure signals, and use test instruments effectively. Finally, Chapter 12 presents a brief treatment of fiber optic related standards. Standards play a crucial rule in all industries, and fiber optics is no exception. Indeed, it is oftentimes adherence to standards that enables a device or system to go beyond a laboratory demonstration and fulfill a well-defined role in the jigsaw of a complex industry such as fiber optics.

* * *

I am greatly indebted to many individuals for this project. In particular, I would like to thank Dr. A. Nourbakhsh who inspired and encouraged me to take on this work. I would also like to acknowledge my past and present colleagues at Source Photonics for the enriching experience of many years of working together. In particular, I would like to thank Dr. Mark Heimbuch, Dr. Sheng Zheng, Dr. Near Margalit, Dr. Chris LaBounty, and Dr. Allen Panahi, for numerous enlightening discussions on a variety of technical subjects. Without that experience and those discussions, this book could not have been created. I would also like to thank Springer for accepting this project, and in particular Ms. Katelyn Stanne, whose guidance was essential in bringing the project to its conclusion.

Mohammad Azadeh

Contents

Chapter 1
Fiber Optic Communications: A Review

1.1 Introduction

There is no doubt that telecommunication has played a crucial role in the makeup of the modern world. Without the telecommunication revolution and the electronic foundations behind it, the modern life would be unimaginable. It is not hard to imagine why this is the case, after all, it is communication that shapes us as human beings and makes the world intelligible to us. Our daily life is intricately intertwined with telecommunication and its manifestations. While we are driving, we call our friend who is traveling on the other side of the world. We can watch events live as they are unfolding in another continent. We buy something in Australia and our credit card account is charged in United States. We send and receive emails with all kinds of attachments in a fraction of second. In short, under the effects of instant telecommunications, the world is shrinking from isolated lands separated by vast oceans to an interconnected global village.

If telecommunication is a product of modern technology, communication itself in the form of language has been with humans as long as there have been humans around. Whether we are talking about human language or electronic communications, there are certain common fundamental features at work. To begin with, let us consider the case of common spoken language. Let us assume I am talking to my friend who is sitting next to me in a restaurant.

Conceptually, we can break down the process to certain stages. In the first stage, the process starts in my mind. I have some thoughts, memories, or concepts that I like to share with my friend. We can call these initial stages the processing layers (Fig. 1.1). The term "processing layers" in its plural form is meant to represent the extremely complex set of functions that logically constitute some sort of hierarchy. Data processing takes place in these various layers until eventually in the last stage the intended message is converted to a serial stream of data, for instance in the form of a sentence. The next process involves converting this stream of data into a physical signal. I can achieve this task by using my speech organs, through which I can produce and modulate sound waves in a precise manner, representing the sentence that I intend to communicate to my friend. Effectively, my vocal system acts as an audio transmitter.

Next comes the transmission of the physical signal. Sound waves carrying energy travel through the medium of the air until they reach my friend. The next step is receiving the physical signal and converting it back to a format that can be processed by the brain. This is achieved by the ear, which acts as the receiver. The function of the receiver is to convert the physical signal back to a form that is suitable for further processing by the nervous system and ultimately the brain.

M. Azadeh, *Fiber Optics Engineering*, Optical Networks,
DOI 10.1007/978-1-4419-0304-4_1,

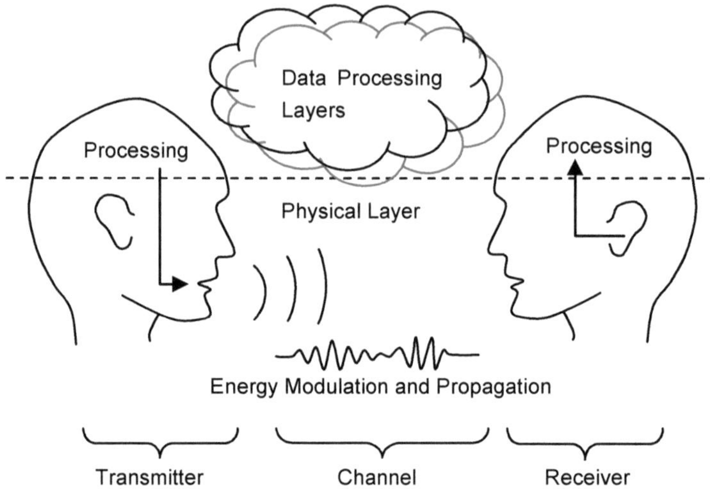

Fig. 1.1. Simple model depicting the essential elements in a communication link

Several crucial features of a communication link are evident from this example. Let us spend some time discussing these points, as they are directly relevant to all communications systems, including fiber optic links.

- First, note that the whole process can be divided into two separate domains which we have called "date processing layers" and "physical layer." This emphasizes the hierarchical nature of the process. More specifically, the physical layer does not care about the nature and details of the processing that takes place in the brain. The physical layer represents the process of converting data to a physical signal and sending that signal to the destination effectively without losing data.
- The information that needs to be transferred has to be processed and eventually converted into a representation that is understood by both sides. In the case of humans, this would be the common language. In a communication system, this would mean some form of formatting or coding, as well as a set of standards known by both the transmitter and the receiver that govern those formats and codes.
- Communication requires modulation of energy. Indeed, it is precisely the modulation of some form of energy, and not just the existence of energy (in this case sound waves), that can represent information. A constant monotonic sound cannot represent any information, even though it has energy and it travels from one point to another, unless some aspect of it is changed. For instance, it must be turned off and on. In every case, a transmitter must modulate some form of energy that carries the information from the sender to the receiver and thus enables communication. No modulation, no communication!

- Another feature is that the more information I intend to transmit to my friend in a given time, the faster I must modulate the sound waves. If I want to tell a long story about all the interesting things that I saw in my trip to South America, I need to talk much faster than if I just wanted to complain about the weather. This obvious fact lies behind much of the efforts to achieve higher modulation speeds in communication links.
- The modulated energy must travel through a medium or a channel. This channel should support the propagation and transfer of the modulated energy: obviously sound waves can propagate in the air. Electromagnetic waves do not need a physical medium, as they can propagate in free space. So in that case vacuum can also act as the channel. Moreover, anytime information is transferred in the form of modulated energy through any medium some form of degradation takes place. It may get weaker as it propagates through the medium, or it could be mixed with other unwanted noise signals, or the waveform of the signal itself may distort.
- In order to combat these degradations, either the transmitter or the receiver (or both) should somehow compensate them. If my friend is sitting further away from me, or if the restaurant is too busy and the noise level is high, I must talk louder. If he misses something that I say, he may ask me to repeat what I said or I may have to talk slower. If neither the transmitter nor the receiver is willing to modify itself to accommodate the signal degradation, the communication link may break down.

These observations, simple as they seem, are directly relevant to all practical telecommunication systems. We will revisit these concepts throughout this book on different occasions, especially as they apply to fiber optic links.

1.2 The nature of light

The main distinction of fiber optic links is that they use light as the form of energy that they modulate, and they use optical fibers to propagate that energy from the source to the destination. Indeed the main advantage of using light energy for communication is the ease with which light can be modulated with high-speed signals and transported over long distances in an optical fiber with minimal degradation. Thus, in order to understand the nature of optical communications we must start with a brief discussion about the nature of light.

In spite of the abundance of our various experiences and encounters with light, the actual nature of light remains elusive and mysterious. In ancient times, the interest in light was mainly expressed as a fascination with one of the most amazing optical instruments, i.e., the eye. Thus, there was debate between philosophers about the nature of vision, how it takes place, and how it results in the perception of shapes and colors. For instance, Aristotle, who exercised great influence on scientific thinking for centuries, explained light and vision in terms of his theoretical concepts like potentiality, actuality, and form and matter. He thought that the form of an object, as opposed to its matter, can somehow travel through space in the form of an image and be received by the viewer. Perception then takes place when this form is impressed upon the soul. Furthermore, transparency is a potentiality in some substances, and brightness (i.e., light) is the actualization of that potential [1]. The Atomists on the other hand, and chief among them Democritus, believed that everything consisted of atoms. Therefore they thought in terms of "atoms of light" [2]. There were also theories that regarded vision and light as rays emanating from the eye and reaching toward the objects. Plato, for instance, believed that light emanating in the form of rays from the eye combines with the light of day, and the result, in the form of a ray, will reach the object [3][1].

These views, alien as they may seem today, in some ways remind us of the modern views of light. For example, Aristotle's theories about an image traveling in the air have some resemblance to the modern theory of imaging. More notably, the belief in the particle nature of light has resurfaced in twentieth century quantum physics [4,5].

In the past few centuries, discussion on the nature of light had divided scientists in mainly two camps. On one side was the particle or corpuscular theory of light. One of the most prominent supporters of this view was Sir Isaac Newton. Partly due to Newton's influence, the corpuscular theory held sway for almost a century after him. On the opposite side was the wave theory, a main proponent of which was Christian Huygens. Eventually, however, Newton's name and prestige was insufficient to overcome the weight of experimental evidence favoring the wave theory, and this is how the wave theory became the first thoroughly scientific theory that was able to explain all the known phenomena at the time.

1.2.1 The wave nature of light

Numerous optical phenomena such as diffraction and interference provide strong evidence for the wave nature of light. One of the first scientists responsible for the wave theory in its modern form was Thomas Young who is famous for his experiments on interference [6]. Eventually, however, the wave theory of light found its most eloquent expression in nineteenth century by James Clerk Maxwell and his electromagnetic theory. Maxwell combined all known phenomena related to electricity and magnetism and summarized the results in his famous four equations, which in their differential form are as follows[2] [7]:

---

[1] Although it must be said that Plato always talks through other characters, most notably Socrates. So we should be careful in assigning a view to Plato directly.

[2] In fact, the four equations that are commonly known as Maxwell's equations are more properly called Maxwell–Heaviside, as their current formulation is due to Oliver Heaviside. Maxwell's own formulation was much more cumbersome [8].

$$\nabla \times \mathbf{E} = -\frac{\partial \mathbf{B}}{\partial t} \tag{1.1}$$

$$\nabla \times \mathbf{H} = J + \frac{\partial \mathbf{D}}{\partial t} \tag{1.2}$$

$$\nabla . \mathbf{D} = \rho \tag{1.3}$$

$$\nabla . \mathbf{B} = 0 \tag{1.4}$$

In these equations **E** is the electric field vector, **H** is the magnetic field vector, **D** is the electric flux density vector, **B** is the magnetic flux density vector, ρ is the charge density, and ∇ is the differential operator representing space derivatives. These equations are based on earlier works from Faraday, Ampere, Coulomb, and Gauss, among others. Notice that in these equations we have two magnetic variables and two electric variables. These variables are further related to each other according to the constitutive relations[3]:

$$\mathbf{D} = \varepsilon \mathbf{E} \tag{1.5}$$

$$\mathbf{B} = \mu \mathbf{H} \tag{1.6}$$

Here ε is the permittivity and μ is the permeability of the medium. It is through these constitutive relations that the dynamics of bound charges and currents in real materials come into play. Before moving on, we should make a few comments.

These equations summarize a body of experimental evidence, obtained earlier by scientists like Faraday, Ampere, and Gauss. More specifically, setting the time derivatives to zero, we get the familiar electrostatic and magnetostatic field equations. Moreover, there are certain symmetries between magnetic and electric fields, with one major exception, which is clearly evident from Eqs. (1.3) and (1.4). Equation (1.3) states that electric fields can "diverge" from a charge density. In other words, electric field lines can have beginnings and ends. However, Eq. (1.4) states that magnetic fields do not begin or end, which means they always have to be in the form of loops. Physically, this means that no magnetic charge or magnetic monopole exists. The lack of a "magnetic current" term similar to J in Eq. (1.1) is another aspect of this fact. This curious lack of symmetry has prompted an extensive search for magnetic monopoles, but with no results so far.

An elegant consequence of Maxwell's theory is that light is identified as a form of electromagnetic waves, and indeed all classical optics can be driven from Maxwell's equations. It can be seen from these equations that the field variables E and

[3] Technically these relationships hold only for linear, homogenous materials. More complex forms of constitutive relations must be used for more exotic materials.

H are coupled together. Thus, the equations can be combined to arrive at a single equation for a single variable. For example, in free space (or in a non-conducting dielectric) the current term J in Eq. (1.2) is zero. We can then take the curl of Eq. (1.1) and, after some manipulation including using the constitutive relations, arrive at [9]

$$\nabla^2 \mathbf{E} = \varepsilon\mu \frac{\partial^2 \mathbf{E}}{\partial t^2} \tag{1.7}$$

which is the well-known wave equation for electric field. The same wave equation can be obtained for the magnetic field. The general solution for wave equation can conveniently be expressed in the mathematical form using the complex notation

$$\mathbf{E}(\mathbf{r}, t) = \mathbf{A}(\mathbf{r}) e^{j(\mathbf{k}.\mathbf{r} - \omega t)} \tag{1.8}$$

where $\mathbf{r}$ is the space vector and non-harmonic space variations are lumped into the vector variable $\mathbf{A}(\mathbf{r})$. Harmonic time variations are represented by the angular frequency ω, and harmonic space variations are represented by the wave vector $\mathbf{k}$.

The solution represented by Eq. (1.8) describes a wave propagating in the direction of the vector $\mathbf{k}$, with a speed that can be shown to be

$$c = \frac{1}{\sqrt{\varepsilon\mu}} \tag{1.9}$$

Maxwell himself realized that the speed calculated from Eq. (1.9) was remarkably similar to the available experimental measurements of speed of light. He concluded that light waves are transverse electromagnetic waves which like any other wave can be described by amplitude, frequency, and wavelength. Therefore, electromagnetic waves in general and light in particular are in fact one class of solutions to Maxwell's equations.

Because electric and magnetic fields are vectors, in general they have components in x, y, and z directions (in the Cartesian coordinates). However, in fiber optic applications, we are generally interested in waves that propagate along a single direction, for instance along the fiber, or along the optical axis of a device. In such cases we can write the wave solution as

$$\mathbf{E}(\mathbf{r}, t) = \mathbf{A}(\mathbf{r}) e^{j(kz - \omega t)} \tag{1.10}$$

where now z is the direction of wave propagation, and k is the component of wave vector **k** in the z direction.[4] Equation (1.10) can describe a plane wave traveling in the z direction or a guided wave propagating along a wave guide such as an optical fiber. We will revisit this equation in Chapter 5 where we discuss light propagation in optical fibers. Note that in Eq. (1.10) the electric field **E** (and the field amplitude **A**) is still a vector and a function of three spatial coordinates.

Let us now analyze the term $(kz-\omega t)$ in Eq. (1.10). The angular frequency ω is given by $\omega = 2\pi f$, where f is the frequency in Hertz and is related to the oscillation period T as $\omega=2\pi/T$. Moreover, k is related to the wavelength λ as $k = 2\pi/\lambda$. Therefore, k represents the periodicity of the waves in space, and ω represents the periodicity in time. To better appreciate the meaning of these quantities, we should remember that a phase front is a point in the wave where the phase does not change. In terms of Eq. (1.10), this means all the points in space where the argument of the exponential function is constant. Setting the argument equal to a constant and differentiation with respect to time, we obtain

$$\omega - k\frac{dz}{dt} = 0 \tag{1.11}$$

The term dz/dt can be recognized as the speed of the wave along the z direction, which is in fact the speed of light. After substituting the wave vector and angular frequency with wavelength and frequency, we arrive at the simple but important relationship between the speed of light, its wavelength, and its frequency:

$$c = f\lambda \tag{1.12}$$

Equation (1.12) simply states that the speed of light is equal to the number of wavelengths that pass from a point in space in a second (frequency) times the length of one wavelength. Of the two quantities on the right-hand side of Eq. (1.12), f is determined by the source and, under normal circumstances, remains constant as the wave propagates through various media. In other words, we can think of frequency as an inherent property of the light, i.e., a property that is determined by the source and (usually) remains the same once the light is generated. On the other hand, λ changes as the wave goes from one medium to the next.

From an optical point of view a medium is characterized by its index of refraction, n. The index of refraction of vacuum is one, but other media have indices of refraction greater than unity. If we represent the vacuum speed of light and wave length by c_0 and λ_0, respectively, we obtain $c=c_0/n$ and $\lambda=\lambda_0/n$. In other words, both the speed of light and its wavelength decrease as it enters a medium with an

[4] Technically, we should add a subscript z to k and denote it as k_z as a reminder of the fact that it is the z component of the wave vector **k**, and that the wave vector in general has other components as well. However, to keep the notation simple, we will drop the z subscript whenever we are only interested in propagation along the z direction.

index of refraction of n. Therefore, the speed and wavelength are extrinsic properties of light, i.e., properties that are affected by the medium.

A harmonic wave described by Eq. (1.10) may seem too artificial. But we can think of it as a building block for constructing more complex waveforms. This is possible because of a very important property of Maxwell's equations: their linearity. This means that if we find two solutions for the equations, for example two plane waves with two different frequencies or amplitudes, any linear combination of those two plane waves is also a solution to the equations. In this way, complex waveforms with arbitrary profiles in space or time can be studied.

1.2.2 The particle nature of light

In spite of the success of Maxwell's theory in describing a wide range of optical phenomena, in twentieth century the picture once again changed, especially with the advent of quantum theory. Although a wide range of phenomena are best explained through the wave nature of light, there are other instances, such as the photoelectric effect, that are easier to explain by taking light to consist of individual packets of energy, called photons. Later in the twentieth century certain phenomena such as the Lamb shift and photon antibunching were discovered that, unlike the photoelectric effect, do not have any classical field explanation. As a result, although quantum field theories present fundamental challenges to our classical notions of reality, quantum optics is now a well-established and growing field [10–13]. In fact, quantum light states in which a precise and known number of photons are present can now be realized experimentally [14].

One of the first triumphs of quantum theory of light came when Max Planck realized that he could explain the problem of black body radiation spectrum by assuming that the electromagnetic energy could be radiated or absorbed only in multiples of a minimum amount of energy, or quantum of energy, whose value is directly proportional to the frequency of electromagnetic radiation [15]. The result is the well-known relationship:

$$E = hf \tag{1.13}$$

where E is the energy in Joules and h is the Plank's constant whose value in the international system is 6.623×10^{-34} Joules times seconds.

Note that by itself, all Eq. (1.13) claims is that the energy of a photon is proportional to its frequency. When it comes to interaction of light with atomic systems, the quantization of energy comes from another equation, called the Schrödinger equation, whose steady-state solutions in a potential field are discrete. One of the most important (and abundant!) examples of such a system is the atom. The nucleus provides the potential field, and the steady-state solutions of the Schrödinger equation result in the familiar discrete energy levels the electrons can occupy. According to Maxwell's equations an electron orbiting a nucleus is an accelerating charge which must radiate electromagnetic energy. If it were not for energy quantization and if the electron could change its energy continuously, it would have to radiate all its energy in a flash and crash into the atom's nucleolus.

However, the electron can move between the allowed discrete levels in an atom. If an electron moves from a higher energy level to a lower energy level, conservation of energy requires that energy be released in some other way. A radiative transition is one where the energy difference is released in the form of a photon whose frequency is related to the energy difference according to Eq. (1.13). That is why this equation is of primary importance to lasers. We can think of a laser as a system with two energy levels. When the electrons are somehow pushed into the higher level, a situation known as population inversion is created. When they jump back to the lower level in a coherent manner, they produce coherent electromagnetic waves, or laser light. The frequency of the light is determined by Eq. (1.13). In semiconductors, the two levels are typically the valence band and the conduction band, and the energy difference between the two levels can be as high as a few electron-Volts.[5] By knowing this energy difference, we can calculate the frequency or the wavelength of the light generated by that laser through Eq. (1.13).

1.2.3 The wave particle duality

So is light wave or particle? This is for sure an interesting and still challenging question for physicists. As a result of both theoretical and experimental evidence and in spite of persistent mysteries in interpretation of the evidence, it has now been accepted that light has a dual nature in that it can behave both as particle and as wave [16–18]. In fact, such a duality is not limited to light. According to quantum theory, not only light waves can have particle-like properties, particles of matter can and do have a wave-like nature. One way to appreciate this dual nature is through the well-known de Broglie equation:

$$\lambda = \frac{h}{p} \tag{1.14}$$

which postulates a wavelength λ for *any* particle with a momentum of p. Thus, an electron that is moving with a certain speed can show wavelike behavior, something indeed verified experimentally. The reason that cars and billiard balls do not act like waves is that due to their large masses, and because of the small value of Planck's constant, the wavelength associated with them is so exceedingly small that for all practical purposes it can be neglected.

---

[5] An electron-Volt (eV) is the energy needed to move an electron up a potential barrier of 1 volt. We remember that increasing the potential of a charge of 1 Coulomb by 1 Volt requires 1 Joule. Thus, 1 eV is 1.60×10^{-19} J, because the electron charge is 1.60×10^{-19} C.

The denominator of Eq. (1.14) represents the momentum of a particle. When it comes to electromagnetic waves, Maxwell's theory indeed predicts a momentum for the wave. For uniform plane waves, we have [6]

$$p = \frac{E}{c} \tag{1.15}$$

where E is energy (per square meter per second), p is the momentum (per square meter of cross section), and c is the speed of light. If we substitute p from Eq. (1.15) for p in Eq. (1.14) and use Eq. (1.12) to convert wavelength to frequency, we arrive at the familiar Planck equation, Eq. (1.13). This shows that the pieces of the puzzle fit together once we recognize that the wave-like and particle-like behaviors are not contradictory but complementary.

Equation (1.14) yields another important insight. When the energy of a wave increases, so does its momentum, and an increase in momentum means a decrease in wavelength. Thus, we can expect high-energy waves to show particle-like behavior more clearly. In fact, gamma rays, which are the shortest wavelength and highest energy form of the electromagnetic spectrum, behave not like waves, but like rays of high-energy particles. On the other hand radio waves, which are in the lower side of the electromagnetic spectrum, have such low energies that for all practical purposes their particle nature can be neglected.

What is the practical relevance of all these concepts to engineering applications? From a practical point of view, the light behaves more like particles when it is being generated or detected. On the other hand, when it comes to propagation of light, it behaves more like waves. This approach has resulted in a view known as semiclassical theory of light. If we want to study the generation or detection of light in such devices as semiconductor lasers and detectors, we use the quantum physical approach and think of light as photons. When it comes to propagation of light in free space or other media, we use classical field theory, as described by Maxwell's equations.

1.3 The electromagnetic spectrum

As mentioned in the previous section, when it comes to light propagation, we can safely treat it as an electromagnetic wave. The electromagnetic spectrum covers a wide range of frequencies, and visible light occupies only a small fraction of it [19]. It is imperative for engineers to gain an overall understating of this spectrum. In fact, a large portion of electrical engineering deals with frequencies that correspond to the low side of this very same spectrum. In these lower frequencies, we can manipulate signals through electronic devices such as transistors. The optical frequency is far too high to be handled directly by electronic devices. Therefore, in fiber optic applications we are in fact dealing with two separate bands of the spectrum: the optical frequency and the much lower modulation frequencies. Thus, it is doubly important for optical engineers to be familiar with the characteristics of electromagnetic waves at different regions of the spectrum.

An overview of the electromagnetic spectrum is shown in Fig. 1.2. Electromagnetic waves can be characterized according to their frequency or wavelength. Generally, at the low side of the spectrum working with frequency is more convenient. Thus, we have the AM radio band which functions in the range of hundreds of kilohertz or VHF and UHF TV signals that operate up to the range of several hundred megahertz. The microwave and millimeter range includes wavelengths in the range of roughly 1 cm–1 mm. The corresponding frequency varies roughly from a gigahertz to tens of gigahertz. Above these frequencies the ability of electronic circuits to modulate the electromagnetic waves starts to diminish. But the spectrum itself continues into the infrared region, where most fiber optic communication links operate. Here the frequencies start to get so large that it becomes more convenient to talk in terms of wavelength. That is why the higher regions, including the visible light, are usually characterized by wavelength. Beyond visible light and at shorter wavelengths, we have ultraviolet, X-rays, and finally gamma rays.

One way to gain a better insight into the behavior of electromagnetic waves in practical systems is to divide the spectrum into three regions, based on the ratio of the system's physical dimensions to the wavelength of signals of interest. Depending on this ratio, the behavior of signals can be studied by application of circuit theory, wave theory, or ray theory. If we denote the physical size of the system that we work with as D, the three regions are as follows:

- $D << \lambda \Rightarrow$ circuit theory
- $D \approx \lambda \Rightarrow$wave theory
- $D >> \lambda \Rightarrow$ ray theory

When the wavelength is much larger than the dimensions of our system, we are in the domain of circuit theory. In this regime we can neglect the wave nature of the electromagnetic energy, assume instantaneous energy propagation within the system (infinite wave speed), and use simplified lumped circuit analysis. This is where most of conventional electronic circuits operate. Thus, we can think of circuit theory as the low-frequency approximation to Maxwell's equation.

The other extreme is the case where the wavelength is much shorter than the dimensions of our system. In this case we can also neglect the wave nature of electromagnetic waves and treat them as rays. The best example of this approximation is geometrical optics. The mid range, however, is where we cannot utilize either of the above approximations. Here we have to use wave theory which in its most complete form is expressed by Maxwell's equations, although depending on the application oftentimes certain simplifying assumption are made here too.

The above categorization can also illuminate various modes of transmission in each region of spectrum, which is also shown in Fig. 1.2. Theoretically, all electromagnetic waves can propagate in free space. However, at low frequencies the most efficient mode of transferring electromagnetic energy is through conducting wires. This is the domain of circuits and lumped elements.

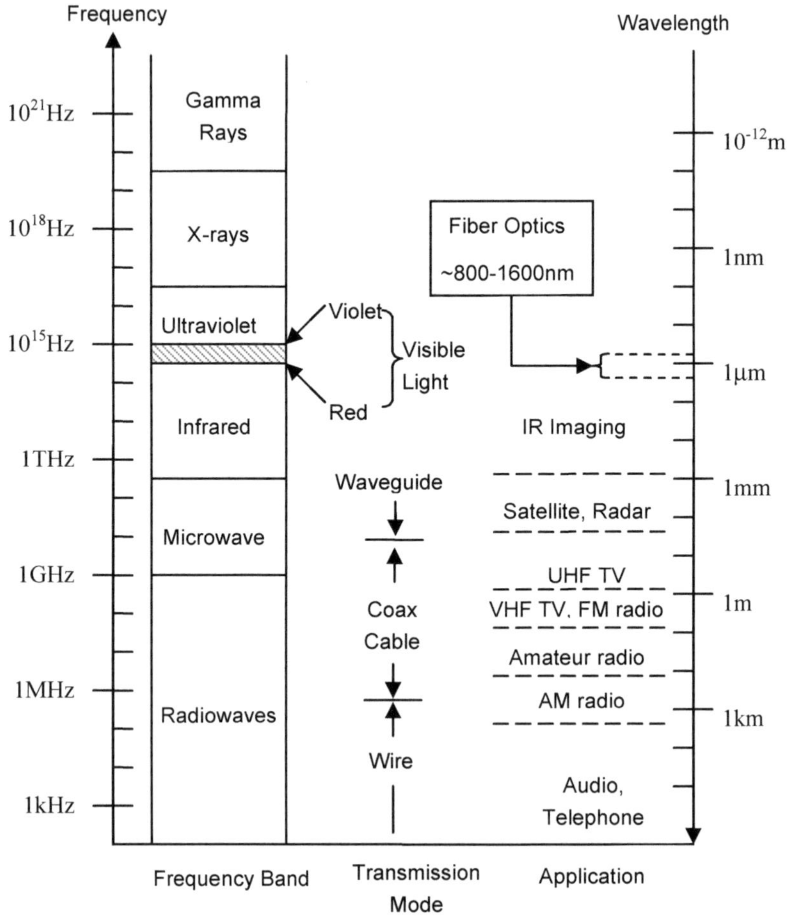

Fig. 1.2. The electromagnetic spectrum

But as the frequency is increased, the wavelength decreases, and the wave nature of the signals must be taken into account. That is why for transmission distances comparable to the wavelength we need to use a controlled impedance medium, such as a coaxial cable. Still at higher frequencies, a waveguide must be used. In a waveguide we have essentially electromagnetic waves subject to the boundary conditions forced by the geometry of the waveguide. This is the case for microwave frequencies where hollow pipes guide the electromagnetic energy in the desired direction. In fact, the optical fibers that are used to guide light waves at much higher optical frequencies are also waveguides that enforce boundary conditions on light waves and hence prevent them from scattering in space.

Figure 1.2 also shows the relatively small portion of the spectrum used in fiber optics. The most common wavelengths used in fiber optic communication range from 800 to 1600 nm, which happen to be mostly in the infrared range. The reasons these wavelengths are particularly attractive for optical communication have to do with both light sources and the fiber medium. Many useful semiconductor laser structures have bandgap energies that fall in this range, which makes them efficient light sources at these wavelengths. On the other hand, propagation losses in silica fibers reach their minimum values in this range. The availability of efficient light sources and suitable propagation properties of fibers make this range of wavelengths the optimal choice for fiber optic communications.

1.4 Elements of a fiber optic link

We started this chapter with a model of communication based on common language. With that background in mind, and with the insights into the electromagnetic spectrum, we can now narrow down the discussion further and go over the building blocks of a typical fiber optic link, as shown in Fig. 1.3.

First, notice that the figure is divided vertically into two domains: a physical layer and a set of data processing layer(s). This is meant to be a schematic reflection of the layered structure of most fiber optic systems. The processing layers are where complex signal processing functions such as multiplexing and demultiplexing, error detection, routing, and switching take place. This is equivalent to what happens in the brain in our language model of Fig. 1.1.

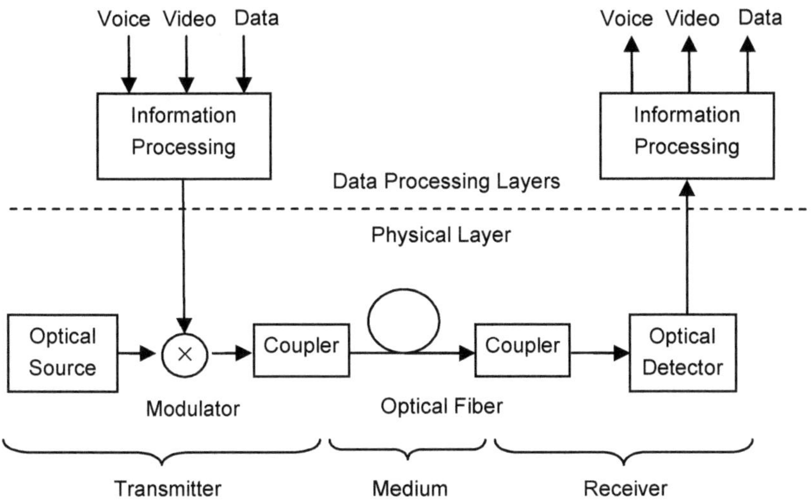

Fig. 1.3. Elements of a typical fiber optic link

The physical layer, on the other hand, does not care (at least directly) about the contents or formatting of the data. Its purpose is to convert the electrical data to optical data, send the optical signal over the fiber, receive it on the other side of the fiber, and convert it back to electrical data. Once this is done, the data will be sent back to the higher levels where further processing in the electrical domain will take place to recover the actual contents in their original format.

The exact architecture of the data processing layers varies and depends on the application. In a simple system, for instance, it may consist of a simple transducer that converts some physical quantity, for example, sound energy, into an electrical signal. That electrical signal then is converted to optical signal and sent over the fiber. On the other hand, in a complex system, the processing layers may include a complicated structure with many layers, each specialized at performing a specific task. Moreover, in such a system each layer makes the layers below itself "transparent" to layers above it. By doing so, each layer can focus on a specific task without having to worry about the details of what goes on below or above it.

As we move up and away from the physical layer, the functions become increasingly algorithmic and require complex data processing. Therefore, in general, moving down toward the physical layer means more hardware, and moving up toward processing layers means more software. In this book we are mainly concerned with the physical layer, although major concepts related to processing layers are reviewed in Chapter 2.

As can be seen from Fig. 1.3, the physical layer consists of an optical transmitter, an optical fiber or channel, and a receiver. The optical transmitter includes a light source along with a modulator. It provides the necessary optical energy that carries the information along the fiber. In practice, this source is either an LED or a semiconductor laser. The modulator's function is to modulate the light from this source with the serial sequence of data. Modulators can be divided into two broad categories. *Direct modulators* modulate the light by directly controlling the current that is running through the light source. As a result, in a direct modulation scheme the same device, say a semiconductor laser, both generates and modulates the light. Direct modulation schemes are very attractive because of simplicity and low cost and are preferred at lower modulation speeds. *External modulators*, on the other hand, do not generate the light themselves. Instead, they manipulate the constant or CW light that is generated by a separate optical source. As a result, external modulation is more complex and costly. However, external modulators can operate at higher speeds and provide better performance.

Once the optical signal with the desired modulation is generated, it must be coupled into the fiber. This is done through a coupler. The reason a separate block is dedicated to this function in Fig. 1.3 is to emphasize its non-trivial nature and the involved challenges. Unlike electrical signals that can easily be coupled from one conductor to the next through physical contact, transferring optical signals in fiber optic systems involves more mechanical challenges and requires, among other things, careful alignment. Even then, every time an optical signal is transferred from one medium to another a portion of it is lost.

Once the optical signal is coupled into the optical fiber, it can generally propagate for long distances with relatively little degradation. The exact nature and amount of these degradations is a function of the structure of the fiber, wavelength, and spectral width of the optical source. *Single-mode* fibers can generally support longer distances with much less degradation, while *multimode fibers* are suitable for shorter distances.

The final stage in the link is the receiver. Once the light reaches the other side of the link, it has to go through another coupler which directs the light to an optical detector. The detector converts the modulated optical signal to an electrical signal. However, the electrical signal coming out of a detector is generally too weak to be useful for further processing. As a result, the receiver must provide additional amplification in order to bring up the amplitude of the electrical signals within acceptable levels. This is typically done through adding a *preamplifier*, or a *transimpedance amplifier*, immediately after and at close physical proximity to the detector. The receiver may also perform further signal processing and conditioning after the preamplifier stage. At any rate, the receiver must provide a clean replica of the original signal at its output, something that can then be passed up to the processing layers, where various functions such as demultiplexing, error correction, and routing take place.

1.5 Light sources, detectors, and glass fibers

In the previous section we discussed some of the main characteristics of fiber optic links in general terms. In order to complete our general discussion, it is necessary to have a brief review of optical sources, optical detectors, and the fiber.

1.5.1 Optical sources

The role of the optical source is to generate light energy that once modulated carries the information across the fiber. Although there are a multitude of ways to generate light, in almost all fiber optic systems a semiconductor device is used for this purpose. These devices include diode lasers and light-emitting diodes (LEDs). There are many reasons for using these devices. For one thing, they are generally easy to operate and integrate with electronic circuits. From a circuit standpoint, diode lasers and LEDs all behave like a diode. In order to turn them on, a forward voltage must be applied to them, which in turn results in a current flow, which in turn will turn the device on. To increase the optical power, all we need to do is to increase the current. To turn the device off, we just need to turn off the diode by shutting down the current. Thus, in many applications the modulation of light is achieved by directly modulating the current flow through the device.

This method, called direct modulation, is very convenient and can be utilized up to modulation rates of several gigahertz. Figure 1.4 shows a simple circuit for converting an electric signal to an optical signal using a semiconductor laser.

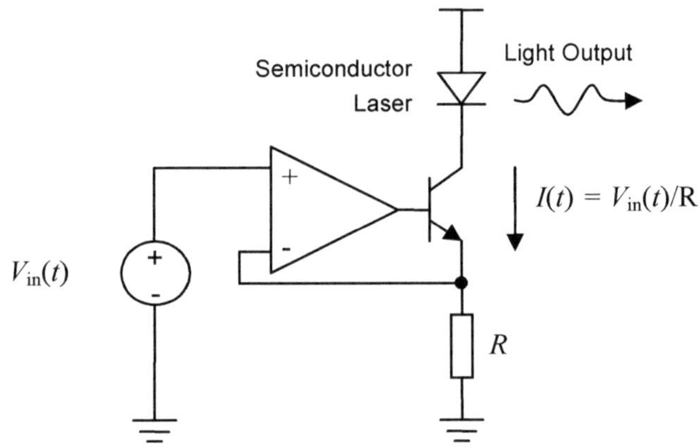

Fig. 1.4. Simplified laser driver circuit

The circuit converts the voltage of the source to a proportional current signal that flows through the laser. Because the laser behaves more or less linearly above its threshold current, the voltage signal is converted to a similar optical signal. Although this simple circuit has several shortcomings, it serves well to illustrate the basic idea behind a *laser driver*.

Another advantage of semiconductor lasers is that these devices are generally very efficient light sources. Semiconductor lasers are characterized by their *threshold current* and *slope efficacy*. A typical semiconductor laser optimized for high-speed communication applications may have a threshold current of 5 mA and a slope efficiency of 0.5 mW/mA at room temperature. This means that it will start generating light as soon as the current flowing through it exceeds 5 mA. After that, the optical output power increases by 0.5 mW for every milliampere increase in current. Thus, at 10mA above threshold, this device can generate 5 mW of optical power. If we assume a forward voltage of 2 V, this represents a power dissipation of 2 V×15 mA=30 mW. The overall efficiency of this device at this particular biasing point is 5/30=0.17, which is not bad at all.

Semiconductor devices also have an important advantage in terms of the optical wavelengths they produce. The wavelength generated or absorbed by a semiconductor is a function of its bandgap energy. The bandgap of semiconductors used in optoelectronics is in the range of 0.5–2 eV. This range corresponds to a wavelength range of approximately 500–4000 nm, which happens to include the wavelength windows of 1300 and 1500 nm, where the attenuation of glass fiber is minimum. Therefore, the light output of these devices can propagate in fibers for long distances with little attenuation. Semiconductor devices are also very small in size, generally cheap, and very reliable. They can be produced in large quantities through wafer processing techniques similar to other semiconductor devices. All these advantages make these devices ideal sources for numerous optical communication applications.

We mentioned that semiconductor light sources used in optical communications can be either LEDs or lasers. LEDs are cheaper, and thus they are mainly used in low data rates or short-reach applications. The main disadvantage of LEDs is that their light output has a wide spectrum width, which in turn causes high dispersion as the light propagates in the fiber. Dispersion causes the smearing of sharp edges in a signal as it propagates in the fiber and is directly proportional to the spectral width of the source. This is why LEDs cannot be used for long distance or high modulation rates in optical communication. Semiconductor lasers, on the other hand, have much narrower spectral widths, and therefore they are usually preferred in high-speed or long-reach links. In this book, we focus our discussions on lasers.

The properties of these lasers depend on the materials used in constructing them as well as the physical and geometrical structures used in their design. Generally speaking, a laser is an optical oscillator, and an oscillator is realized by applying feedback to an amplifier. Semiconductor lasers can be divided into two main categories depending on the nature of this feedback.

In a Fabry–Perot (FP) laser, the feedback is provided by the two facets on the two sides of the active region. The optical cavity in FP lasers generally supports multiple wavelengths. Therefore, the output spectrum, although much narrower compared to an LED, still consists of several closely located peaks, or optical modes. A distributed feedback (DFB) laser, on other hand, includes additional structures that greatly attenuate all but one of those modes, and therefore a DFB laser comes closest to producing an ideal single wavelength output. This is why DFB lasers can minimize dispersion and support the longest attainable reaches.

Both FP and DFB lasers are *edge-emitting* devices, i.e., the light propagates in parallel to the semiconductor junction and comes out from the sides. A different structure is the vertical cavity surface emitting laser, or VCSEL. In a VCSEL the light output is perpendicular to the surface of the semiconductor. VCSELs are very attractive because of low threshold current and high efficiency. Moreover, many VCSELs can be integrated in the form of one-or two-dimensional arrays, a feature not available for edge-emitting devices.

However, VCSELs usually work at short wavelengths around 850 nm, a wavelength unfortunately unusable for long haul communication. Although research into production of VCSELs at longer wavelengths of 1300 and 1500 nm is intense, such devices are still not available in an industrial scale [20]. As a result, VCSELs like LEDs are usually used for short-distance links. We will come back to these various laser structures in Chapter 4 where we discuss their principles of operation and properties as well as their advantages and disadvantages in more depth.

1.5.2 Optical detectors

Like optical sources, the optical detectors used in fiber optics are almost exclusively semiconductor devices in the form of PIN diodes and avalanche photo diode (APD) detectors. From a circuit perspective, these are again diodes that like any other diode (including laser diodes or LEDs) can be forward or reverse biased. However, for them to act as detector, they must be reverse biased. A PIN diode can operate with a very low reverse bias. This makes the PIN diode attractive because it can be operated as an element within a standard electronic circuit that runs at a low supply voltage of say 3.3 V. Figure 1.5 shows a simple detector circuit. The output of the detector is a current that is proportional to the light power received by the detector. Thus, the detector can be considered as a light-controlled current source. A transimpedance amplifier (TIA) converts the photocurrent generated by the detector to a voltage.

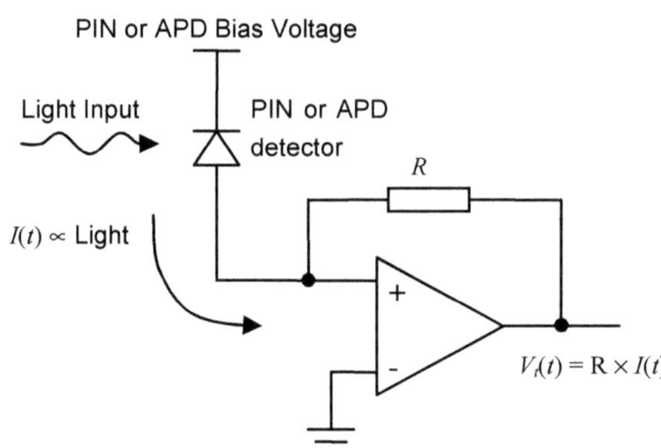

Fig. 1.5. Simplified detector using PIN diode or APD

APD detectors are not very different from a circuit point of view. However, they require a much higher reverse bias voltage to operate. Currently low-voltage APDs require at least 30–40 V of reverse bias. Moreover, the required optimal reverse bias voltage for an APD varies with temperature. As the temperature rises, the reverse bias applied to the APD must also increase to maintain the gain of the APD constant. Because of these complications, APDs are harder to use and more care must be taken in designing them. They are also more costly in terms of both the APD itself and the additional required support circuitry. However, APDs have a big advantage that makes them attractive for high-end receivers: unlike PIN detectors, an APD detector provides inherent current gain. This additional gain directly translates to improved performance. Typically an APD can work with a fraction of the optical power needed for a similar PIN diode, and this means longer links can be supported with an APD receiver.

1.5.3 The optical fiber

Perhaps the most critical factor that has made fiber optic communication possible is the development of low-loss silica fibers. The idea that light can be "guided" in a medium other than air is not inherently revolutionary. The physical principle behind such guiding is the refraction of light, as summarized by Snell's law:

$$n_1 \sin(\theta_1) = n_2 \sin(\theta_2) \tag{1.16}$$

where θ_1 and θ_2 are the angles of incidence and refraction, and n_1 and n_2 are the indices of refraction of the two media (Fig. 1.6). This means that when light enters from a medium with a lower index of refraction to a medium with a higher index of refraction, it bends toward the line normal to the two media. Going the opposite direction, the light bends away from the normal direction. In this case if the angle of incidence is more than what is known as the critical angle, θ_T, all the energy reflects back into the medium with the higher index of refraction. This phenomenon is called *total internal reflection*. This is what happens if you try to look outside from under the water. You can see things that are immediately above you, but as you look away, beyond a certain angle the surface of water acts like a mirror, reflecting light from under the water while blocking light from outside. The same principle is behind the light guiding properties of an optical fiber.[6] An optical fiber consists of a *core* with higher index of refraction and a *cladding* with a lower index of refraction. If light is launched into the fiber either in parallel with its axis or with a small angle, total internal reflection will trap all the energy inside the fiber and the optical wave will be guided along the axis of the fiber.

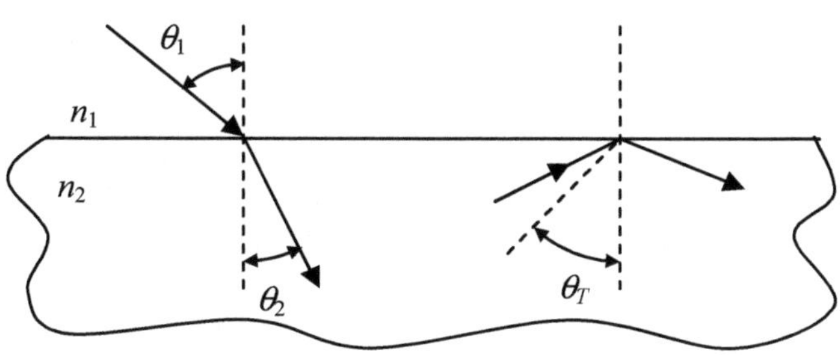

Fig. 1.6. Snell's law of refraction and total internal reflection

This would constitute a *guided mode* that the fiber can support. If the light enters with a large angle with respect to the axis of fiber, at each reflection a portion of energy is radiated out, and therefore after a few reflections the energy inside the

[6] To be more accurate, multi-mode transmission of light can be understood in terms of total internal reflection. However, single-mode transmission requires the wave theory of light. We will discuss this point in Chapter 5.

fiber will attenuate to undetectable levels. This would be an example of a *leaky mode*, i.e., a mode the fiber cannot support.

An optical fiber can normally support several optical modes depending on the thickness of its core. There are two ways to look at wave propagation in a fiber. Based on geometrical optics, we can think of each mode as being a certain path for light rays as they bounce back and forth between the sides of the fiber core while traveling forward. However, a more accurate description requires the wave theory of light, where each mode is like a resonant state in which Maxwell's equations along with the appropriate boundary conditions are satisfied. As the size of the core reduces, so does the number of modes that it can support. Of particular interest is the case where the size of the core is so small that only one mode can propagate in the fiber in parallel with the fiber's axis. This would be a *single-mode fiber*. In a single-mode fiber the loss can be as low as 0.25 dB/km.[7] This means after 1 km, more than 90% of the original optical power remains in the fiber. This is indeed a remarkable degree of transparency. Imagine a piece of glass 1 km thick that only absorbs 10% of the light going through it! In comparison, normal glass starts to become opaque only after a few centimeters. This is why an optical signal can travel as much as 100 km or more in a fiber without any additional amplification and still be detected.

1.6 Advantages of fiber optics

In the beginning of this chapter when we introduced the model of human communication through language, we mentioned the rather obvious fact that in order to transfer more data in a given time, one needs to talk faster. In communication systems, this is equivalent to increasing the modulation speed of signals. In an electrical link, the transmitter and the receiver are connected through a conductive path, for example a pair of twisted wires or a coax cable. However, copper connections suffer from increased attenuation of signals at higher frequencies. The sources of these losses include the skin effect in the conductors, radiation, and loss within the dielectrics and the shields [21].

Optical fibers, on the other hand, provide much higher bandwidths. The frequency bandwidth corresponding to the wavelength range of 850–1600 nm is roughly 100 THz. With advanced techniques, single channel modulation speeds in excess of 50 GHz have been demonstrated and components for 40 Gbps transmission are commercially available [22]. Much higher aggregate data rates can be achieved through wavelength division multiplexing (WDM). As long as the optical power remains low, the fiber acts as a linear medium, which means various wavelengths of light can travel simultaneously within the same fiber without affecting each other. It is this property that is used in WDM systems to increase the link capacity even further. In a WDM system, several wavelengths are launched

[7] Later in this chapter we will discuss the decibel unit of power.

into the fiber in parallel, each modulated with an independent stream of data. At the other side of the link, optical filters separate each wavelength, sending each signal to a separate receiver. In this way the available bandwidth of the fiber is multiplied by the number of wavelengths or channels. For instance, in a course WDM (CWDM) system, four to eight wavelengths are used. In a dense WDM (DWDM) system 80 or more channels may be used. Using advanced techniques, transmission with aggregate data rates well into terra bits per second (Tbps) have been demonstrated [23–25].

We should mention that regardless of the inherent wide bandwidth of the fiber and the multiplexing techniques that utilize that bandwidth, optical signals suffer degradation mainly due to attenuation and dispersion. However, and in spite of these degradations, the bandwidth of an optical fiber far exceeds that of a similar coax cable or electrical conductor. For instance, sending an optical signal at the rate of 10 Gbps in a single mode fiber over a distance of 10 km is well within the capabilities of commercially available parts. If we assume a voice channel takes up 100 Kbps of bandwidth, a single strand of fiber can support around 100,000 simultaneous voice channels. To establish this bandwidth capacity through copper cables requires massive amounts of parallel cabling and a lot of physical space and supporting infrastructure.

The benefits of fiber are not limited to higher bandwidth and lower volume of physical infrastructure. Because fibers carry information in the form of light, they are immune to external electromagnetic interference. A copper cable can act as an antenna, receiving electromagnetic radiations from other manmade or natural radiation sources. Typical manmade sources include the power grid and radio and TV stations. Typical natural sources of noise include lightening or microwave radiation from outer space. None of these sources can interfere with the optical signals in a fiber. This isolation works both ways, i.e., fibers do not radiate energy either. This makes them more suitable for applications where security is a matter of concern. Once a fiber link is established, it is very difficult, if not impossible, to "tap into it" without breaking it. Another advantage of fiber links is that they do not need to carry electrical signals, and more generally, they do not require a conductive path between the transmitter and the receiver. This makes them ideal for applications where the link must pass through an environment where presence of electrical signals poses safety risks or cases where it is desired to keep the two sides of the link electrically isolated.

1.7 Digital and analog systems

So far we have not distinguished between the various ways in which information can be represented. In the physical world, and at least in the macro-level which we deal with in everyday life, most variables are analog. This means that the quantities of interest vary continuously within a range of possible values. In fact we perceive most of the physical quantities that we sense, like light, heat, and pressure, in an analog manner. Likewise, the most fundamental electrical quantities that represent signals are analog. The voltage of a point in a circuit or the current that flows through an element is an analog variable. However, in spite of the analog nature of these physical quantities, most fiber optic communication links are categorized as digital and function in a digital manner. The reason behind this is that digital signals are more immune to noise, less sensitive to nonlinearities, and easier to produce and detect.

From a historic perspective, the information revolution is inherently tied with computers. Internet, after all, was invented to connect digital computers together. Likewise, all the applications that somehow are related to computers are naturally digital and deal with digital data. This is why when it comes to fiber optics, not only the physical layer, but also most of the backbone data-intensive links are digital, precisely because they have to handle digital data. In spite of this, analog links are still used in some applications, such as video transmission and RF feeds [26]. For this reason, throughout this book, we mostly focus on digital links, although whenever appropriate we also include discussions on analog links.

1.8 Characterization of fiber optic links

So far we have reviewed the building blocks that make up a fiber optic link. However, it is also important to characterize the performance of a particular link in a quantitative way. This is particularly important when a link is being designed. We want to know, for instance, what kind of components are needed in order to achieve our target performance characteristics. Alternatively, we may want to know what kind of *margin* an existing link is operating at. If a link is working with little or no margin, it means a small change in the performance of any component used in the link may cause problems. This is why we need the concept of *link budgeting* in order to quantify the performance of a link.

We should start by discussing the important notion of decibel or dB, which is a logarithmic measure of the ratio of two optical powers. The ratio of two optical powers P_1 and P_2 in decibels is defined as follows:

$$dB = 10\log_{10}\left(\frac{P_1}{P_2}\right) \tag{1.17}$$

For example, if P_1 is ten times larger than P_2, we can say P_1 is 10 dB higher than P_2. Equation (1.17) also provides a way of measuring power as long as we agree on a reference power level. By definition, a power level of 1 mW is used for such a reference. That allows for defining a logarithmic scale, referred to as dB-milliwatt or dBm, to measure optical power:

$$P_{\mathrm{dBm}} = 10\log_{10}(P_{\mathrm{mW}}) \tag{1.18}$$

where P_{dBm} is the power in dBm and P_{mW} is the power in units of mW. Note that Eq. (1.17) is the same as Eq. (1.18) when P_2 is set to 1 mW. It should be noted that in both these equations the logarithm is defined in base 10.

One may question the reason for using a logarithmic scale instead of the more familiar linear scale. In a linear scale, the optical power is usually expressed in units of milliwatt. Likewise, the ratio of two powers (or any other quantity) is just a number, for example 0.5 or 10. So why should we go to the trouble of calculating logarithmic powers and ratios? The answer lies in some of the very useful properties of the logarithm function.

Remember that logarithm is an operator that converts multiplication to addition, and exponents to multiplication. In other words: $\log(ab)=\log(a)+\log(b)$, and $\log(a^b)=b\times\log(a)$. These properties become handy in link budget calculations because the optical power generally decreases exponentially in optical fibers with distance. That is why the attenuation of a particular kind of fiber is usually given in units of decibel per kilometer. For example, a single mode fiber at the wavelength of 1310 nm has a typical attenuation of 0.5 dB/km. This means that for every kilometer of this fiber, we should expect 0.5 dB reduction in power. Another reason that using logarithmic scales is useful is the fact that gain and loss are inherently multiplicative. Therefore, whenever we want to calculate the effects of gains or losses, logarithmic scales allow us to add or subtract instead of multiply or divide. To summarize, once we are used to thinking in terms of decibels, it becomes much easier to understand and visualize the status of an optical signal or an optical link without going through more complex calculations. Table 1.1 summarizes approximate equivalences between linear and logarithmic powers and ratios.

Table 1.1. Linear ratios (powers) and their logarithmic equivalent in dB (dBm)

Amplification/power higher than 1 mW		Attenuation/power lower than 1 mW	
Ratio/mW	dB/dBm	Ratio/mW	dB/dBm
1000	30	1	0
100	20	0.99	−0.05
10	10	0.9	−0.5
5	7	0.8	−1
3	5	0.5	−3
2	3	0.3	−5
1.25	1	0.2	−7
1.1	0.5	0.1	−10
1.01	0.05	0.01	−20
1	0	0.001	−30

It is very useful to remember some of the more common ratios as rules of thumb. For example, doubling the power means adding 3 dB to it, while dividing it by two means subtracting 3 dB from it. Similarly, multiplying by a factor of 10 means adding 10 dB to the signal power, and attenuating a signal by a factor of 10 is equivalent to subtracting 10 dB from its power. Thus, the left side of the table represents amplification of a signal by the given ratios, and the right side represents attenuation of the signal by the given ratios.

This table can also be used to convert back and forth between linear power and decibel-milliwatt. For example, a laser with 2 mW of optical power has an output power of 3 dBm. A receiver with a *sensitivity* of −30 dBm is capable of working properly with optical power levels as low as 0.001 mW. We should note that in this table decibel equivalences for numbers other than exponents of 10 are approximate. However, exact conversions can easily be done through Eqs. (1.17) and (1.18).

We are now in a position to discuss the topic of link budgets. To illustrate the topic, let us consider Fig. 1.7 which illustrates an example of an optical link along with the power levels at each point in the link. The transmitter's output power is assumed to be at −3 dBm. The output of the transmitter has to go through a coupler with a loss factor of 0.5 dB before it enters the fiber. This means the power level at the beginning of the fiber is −3.5 dBm. The fiber has a length of 20 km and is assumed to have a loss of 0.5 dB/km at the operating wavelength. This means for each kilometer of fiber, the power is reduced by 0.5 dB. That adds up to a total loss of 10 dB, which means if the power is −3.5 dBm at the input of the fiber, at the output it is reduced to −13.5 dBm. Then there is another coupler that couples the output of the link to the receiver. Thus, the power level at the receiver is −14 dBm.

Let us now assume that the receiver has a sensitivity of −21 dBm at the data rate and wavelength of interest. In this case, the receiver can work with much lower power levels than it is actually receiving. To be exact, we can say that this link has 7 dB of margin. Referring to Table 1.1 we find that 7 dB corresponds to a factor of 5, and −7 dB corresponds to a factor of 0.2. In other words, in this example the receiver can work with 5 times less power than what it is now receiving. This is indeed a fair amount of margin.

What does it mean to have extra margin in a link? In reality, there are always reasons for component degradation over time. The power of the transmitter may degrade over time because of aging of the laser, just as the sensitivity of the receiver may get worst as the receiver ages. Other causes that influence the behavior of the link include environmental factors such as temperature. If the numbers given in the example are not guaranteed over the entire temperature range in which the link is designed to work, then temperature degradations must also be taken into account. Another cause for additional loss is accumulation of dust particles or lack of perfect mating in optical couplers which could cause further degradation. All these factors must be considered when a link is being designed. As a result, extra margin should always be set aside for these various degradations.

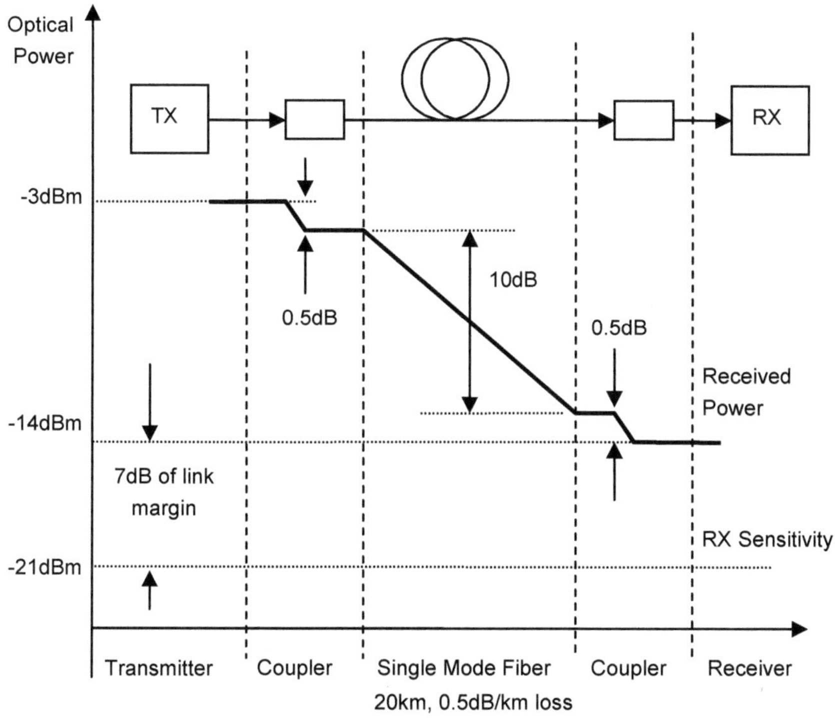

Fig. 1.7. Link budget in a simple optical link

On the other hand, having too much margin could indicate that the link is over-designed, i.e., it is likely that more expensive components have been used where cheaper parts could have worked just as well. Transmitters with higher power or receivers with better sensitivity are generally more expensive.

The choice of link margin is a tradeoff between many factors such as link specifications, cost, robustness, and link quality. Oftentimes standards define the link budget for a specific class of application. Moreover, reliability models based on statistical measures can provide a means of forecasting the performance of individual elements as they go through environmental stress and aging. We will discuss some of these topics as well as the relevant standards in the last chapters of this book.

1.9 Summary

This chapter provided an introduction to fiber optic communication. Through the example of language, we highlighted several basic concepts, including the fact that communication typically involves some hierarchy, and that at the lowest level, it requires modulation of energy.

The form of energy used in fiber optics is light, and therefore it is important to gain some insight into the nature of light. From the classical perspective, the most complete description of light (and more generally electromagnetic waves) is given by Maxwell's equations. From the viewpoint of modern physics, light is characterized in terms of discrete packets of energy or photons. These two perspectives are not contradictory, but complementary. In fact, as we move from the long wavelength side of the electromagnetic spectrum to the low wavelength extremes, the energy of waves/photons increases, and particle-like behavior becomes more dominant. From a practical and engineering perspective, it is usually sufficient to describe the generation and absorption of light in terms of photons and the propagation of light in terms of classical fields described by Maxwell's equations.

With this basic view of light in mind, we next turned to the description of a generic fiber optic link. Here again we encountered the hierarchical nature of the flow of data, from higher data processing layers to a physical layer and vice versa. In the next chapter we will examine the hierarchical structure of communication networks in more detail. However, in this book we are mainly concerned with the physical layer, which consists of optical sources, optical detectors, and optical fibers. These elements were briefly introduced in this chapter and are examined more carefully in the other chapters of this book.

Fiber optics has become an integral part of the modern telecommunication infrastructure, and for good reasons. We discussed some of the main advantages of fiber optics, which for the most part, can be summarized in terms of higher bandwidths and longer distances. These two elements directly address the bottlenecks that modern telecommunication systems have to deal with, in both digital and analog domains.

We also introduced the concept of link budgeting, which takes the concept of a fiber optic link one step further by quantifying it in terms of link margin. Of particular importance here are the units of decibels and decibel-milliwatt for optical power, which are ubiquitous in the analysis and design of fiber optic links. Getting used to these logarithmic units will be helpful in gaining intuitive insight into a wide range of problems having to do with link budgeting, range, gain, and loss.

References

[1] Aristotle, *On The Soul*, Book II

[2] R. L. Oldershaw, "Democritus-scientific wizard of the 5th century BC," Speculations in Science and Technology, Vol. 21, pp. 37–44, 1988

[3] Plato, *Timaeus*

[4] K. Seyrafi, *Light, The Mystery of Universe*. Electro-Optical Research Company Los Angeles, CA, 1986

[5] R. Loudon, *The Quantum Theory of Light*, 3rd Ed., Oxford University Press, Oxford, 2000

[6] E. Hecht, *Optics*, 4th Ed., Addison-Wesley, Menlo Park, CA, 2002

[7] W. C. Chew, M. S. Tong, and B. Hu, *Recent Advances in Integral Equation Solvers for Electromagnetics*, Morgan and Claypool, San Rafael, CA, 2009

[8] J. Z. Buchwald, "Oliver Heaviside, Maxwell's apostle and Maxwellian apostate," Centaurus, Vol. 28, pp. 288–330, 1985

[9] G. Keiser, *Optical Fiber Communications*, McGraw-Hill, New York 1999

[10] V. Jacques, "Experimental realization of Wheeler's delayed-choice gedanken experiment," Science, Vol. 315, pp. 966–968, 2007

[11] V. V. Dodonov, "Nonclassical states in quantum optics: a 'squeezed' review of the first 75 years," Journal of Optics. B, Quantum and Semiclassical Optics, Vol. 4, pp. R1–R33, 2002

[12] M. M. deSouza, "Classical electrodynamics and the quantum nature of light," Journal of Physics A-Mathematical and General, Vol. 30, pp. 6565–6585, 1997

[13] D. F. Walls, "Evidence for the quantum nature of light," Nature, Vol. 280, pp. 451–454, 1979

[14] B Lounis and M. Orrit, "Single-photon sources," Reports on Progress in Physics, Vol. 68, pp. 1129–1179, 2005

[15] S Gasiorowicz, *Quantum Physics*, John Wiley & Sons, Hoboken; NJ, 1974

[16] T. L. Dimitrova and A. Weis, "The wave-particle duality of light: a demonstration experiment," American Journal of Physics, Vol. 76, pp. 137–142, 2008

[17] K. Camilleri, "Heisenberg and the wave-particle duality," Studies in History and Philosophy of Modern Physics, Vol. 37, pp. 298–315, 2006

[18] S. S. Afshar et al., "Paradox in wave-particle duality," Foundations of Physics, Vol. 37, pp. 295–305, 2007

[19] D. W. Ball, "The electromagnetic spectrum: a history," Spectroscopy, Vol. 22, pp. 14–20, 2007

[20] D. Supper et al. "Long-wavelength VCSEL with integrated relief for control of singlemode emission," Optics Communications, Vol. 267, pp. 447–450, 2006

[21] C.C. Xu and S. A. Boggs, "High frequency loss from neutral wire-shield interaction of shielded power cable," IEEE Transactions on Power Delivery, Vol. 23, pp. 531–536, 2008

[22] Y. D. Chung et al., "A 60-GHz-band analog optical system-on-package transmitter for fiber-radio communications," Journal of Lightwave Technology, Vol. 25, pp. 3407–3412, 2007

[23] K. Gnauck et al., "25.6-Tb/s WDM transmission of polarization-multiplexed RZ-DQPSK Signals," Journal of Lightwave Technology, Vol. 26, pp. 79–84, 2008

[24] H. Suzuki et al., "12.5 GHz spaced 1.28 Tb/s (512-channel×2.5 Gb/s) super-dense WDM transmission over 320 km SMF using multiwavelength generation technique," IEEE Photonics Technology Letters, Vol. 14, pp. 405–407, 2002

[25] J. X. Cai et al., "Long-haul 40 Gb/s DWDM transmission with aggregate capacities exceeding 1 Tb/s," Journal of Lightwave Technology, Vol. 20, pp. 2247–2258, 2002

[26] C. H. Cox, *Analog Optical Links*, Cambridge University Press, Cambridge, 2004

Chapter 2
Communication Networks

2.1 Introduction

In the previous section we reviewed the basic blocks that make up a fiber optic system. We divided a link vertically into two domains: the processing layers and the physical layer. The function of the physical layer is to convert an electrical signal into an optical signal, transmit the optical signal through the fiber, and convert it back to an electrical signal at the receiver. The performance of a physical layer system can be measured by the signal fidelity at the receiver end, i.e., by comparing the regenerated electrical signal at the receiver to the original electrical signal at the transmitter.

The processing layers, on the other hand, include the hardware and the software that handle tasks such as switching, routing, error detection, multiplexing, demultiplexing, as well as a wide range of other data processing tasks. Oftentimes these higher layers are generically called networking layers. In this book we are mainly concerned with fiber optics at the physical layer. However, before getting into the details of the blocks that make up the physical layer, we need to gain a better understanding about the networking layers.

The networking layers make up the backbone of the infrastructure that connects the users around the world together. These layers fill the space between the end users and the physical layer systems that handle the optical or electrical signals in cables and fibers. In a way, the networking layers use the services of the physical layer in order to move information contents around in communication networks.

Within the last few decades, networking has been a dynamic, fast-evolving area, going through continuous and at times fundamental changes in terms of technology, data rates, nature of traffic, scale, governing standards, and economy [1–2]. In this chapter we will go over some of the fundamental concepts in networking, with an eye on the role of fiber optics. In this way the contents of the data processing layers in Fig. 1.3 would become more meaningful, and we will get a better context for the topics of the remaining chapters.

2.2 Network topologies

In the first chapter we discussed the concept of a fiber optic link, which basically consists of an optical source, a fiber, and an optical receiver. We can think of an optical link as the basic building block for constructing more complex networks. One way to study networks from a more abstract perspective is by modeling them as a graph. A graph can be thought of as a set of connected points. The points, or *nodes*, are connected to each other via *links*. Thus, a graph is a set of nodes and

M. Azadeh, *Fiber Optics Engineering*, Optical Networks,
DOI 10.1007/978-1-4419-0304-4_2, © Springer Science+Business Media, LLC 2009

links. Moreover, by assigning a failure probability to the links in a graph, the performance and reliability of real networks can be modeled [3–6].

Let us consider the example of language again. When two individuals talk, we have a case of a simple network: each individual is a node, and the communication channel between them is the link (Fig. 2.1a). However, we can easily expand this example: we can think of three people sitting around a table, all talking the same language. Here we have a case of a network with three nodes and three links, because each individual can talk to the other two (Fig. 2.1b). On the other hand, consider the case where one individual, A, speaks only English, and the second, B, only French, while only the third, C, speaks both English and French. Here we have a case of a network with three nodes and two links. The individual who speaks both languages is in the middle, separately linked to the other two nodes (Fig. 2.1c). If C leaves the room, the link between A and B is broken, i.e., they cannot talk to each other because they can only communicate through this third node (Fig. 2.1d).

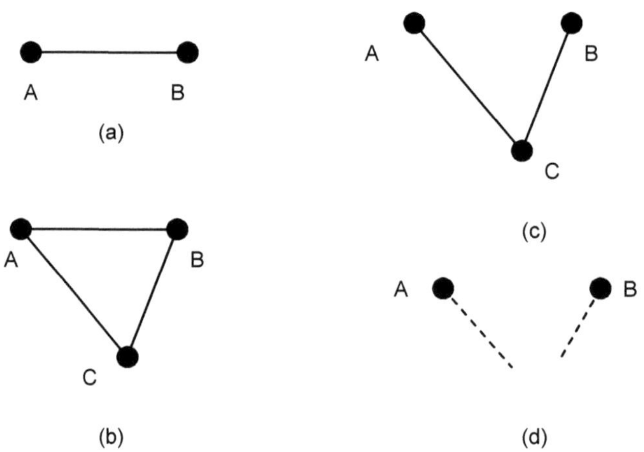

Fig. 2.1. Logical structure of a simple network

This simple example serves well to demonstrate the logical structure, or *topology,* of a network, as represented by a graph. The graph is a schematic representation of the flow of information between various nodes and how nodes are positioned with respect to each other. Each node can only talk directly to the nodes to which it is connected. The number of links is a measure of the interconnectedness of the network. Moreover, a more interconnected network is (naturally) more complex, but more resistant against disruption. For example, Fig. 2.1b represents a more robust network compared to Fig. 2.1c, because if any of the individuals in Fig. 2.1b leaves the room the other two can continue to talk, whereas in Fig. 2.1c, C is a *critical node*, because A and B are linked through C, and if C leaves the room, the communication between C and B is broken.

Communication networks work very much under the same principles. Each node is a point in the network that is capable of sending and receiving information, and the links are channels that they use to communicate to each other. The performance of a network is strongly affected by its topology [7]. Naturally, two nodes can communicate as long as they "talk the same language." Networking languages are defined in networking standards such as SONET and Ethernet. We will cover some of the main standards in fiber optics in Chapter 12.

We can now examine some of the common network topologies and their properties. Some of these common structures are shown in Fig. 2.2. One of the most straightforward ways to connect a certain number of nodes together is the linear or bus topology. Although connection details may vary, in general in a bus topology all the nodes have access to the same cable or channel.

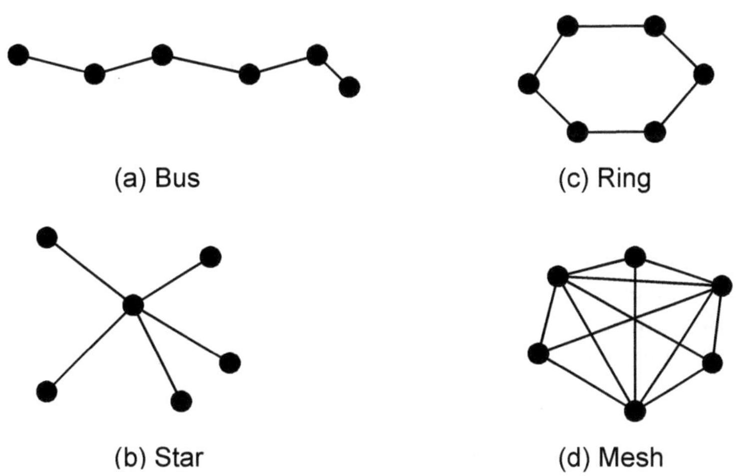

Fig. 2.2. Logical network topologies

If a node wants to talk, it broadcasts a message on the bus, but only the destination node responds to the message. The bus topology has the advantage of simplicity, and in terms of physical cabling requirement it needs the minimum amount of cables. It is also very scalable: adding one node requires adding one link. However, there are several shortcomings. For one thing, the *diameter* of the network is large. We can think of the diameter as the maximum length a signal must travel between any two nodes. If the two nodes at the edges of the bus want to talk, the signals must travel the entire network back and forth. This also exposes another weakness in the bus architecture: a single failure anywhere in the link brings down the whole network. Thus, bus topologies are less reliable compared to other topologies.

The star topology shown in Fig. 2.2b resolves some of these limitations. It consists of a central node, called a *hub*. All the other nodes have to talk through the hub. Thus, the hub's load is disproportionably heavier than the rest of the nodes, because it must handle the traffic from all nodes. Compared to the bus architecture, the star topology is more complex. In practice, it also requires more cabling. Consider the case where two of the nodes are close to each other but physically far from the hub. We need to have two separate links running side by side to the hub. However, the star topology has several advantages. In a star network, if any of the links fails, only the node connected to that link will fail, and the rest of the network can function normally (the exception is the central node or the hub, which if failed, brings down the entire network). The diameter of the network is also much smaller. It takes a maximum of two links for any two nodes to talk to each other. The star topology is also easily scalable. Adding one node requires adding one link from that node to the hub.

Another common topology is the ring topology shown in Fig. 2.2c. Here the load is more evenly distributed, resulting in a more organized network. Bi-directional rings also are advantageous in terms of fault tolerance. A single failure in the network still leaves the other path for the nodes to communicate. Effectively, a single failure reduces a ring to a bus-type topology. The diameter of the network is somewhere between those of the bus and star structures. If the two nodes happen to be at the opposite ends of the ring, the number of links required for them to talk to each other is approximately half the number of nodes. A disadvantage of ring topology is that any change in the network may cause disruption in the entire network. Also, a ring topology is not easily scalable. If a new node is to be added, it has to go in between two other nodes, and generally two additional cables are needed.

The mesh topology, shown in Fig. 2.2d is the most interconnected network topology. In a fully connected mesh, all nodes are connected to all other nodes. The obvious advantage is that the network becomes very *fault tolerant*. Each node is connected to the rest of the network via multiple links; and therefore single or even multiple failures tend to remain contained. A fully connected mesh also provides the minimum diameter possible: each node is only one link away from any other node. But these advantages come at a price. A mesh network requires the maximum amount of physical cabling. Moreover, unlike the other topologies, the signals have more than one path to go from one node to another. This means that switching or routing decisions are needed. Obviously, then, a mesh network can become very complex if the number of nets gets large.

In practice, the choice of network topology depends on the specific situation and is dictated by many parameters. For example, for small networks within a house or a small office, a bus or star network may be an optimal solution. However, for larger networks, it becomes increasingly difficult to justify a single topology, and therefore a mixture of topologies must be used. Still larger networks may not even belong to a single company and could consist of many patches developed separately for various applications and purposes. This is especially true with the telecommunication infrastructure, which expands whole cities or continents. For example, individual users may be connected to a local central hub in a star configuration, the local hubs connected to each other through a ring topology that circles a city, and the ring topologies in several cities connected to each other in the form of a mesh, either partially or fully connected.

In fiber optic networks, the nodes in a particular topology consist of optical transmitters and receivers. We shall discuss these from a hardware perspective in more detail in Chapters 4 and 6. The links themselves are made of optical fibers that are connected to the nodes (and to each other) through optical couplers and other passive components. These components are the subjects of Chapters 5 and 7.

2.3 Telecommunication networks

Fiber optics is an integral part of the telecommunication infrastructure that has come to be such a dominant feature of modern life. Thus, in order to gain a better understanding of fiber optic networks, it is necessary to gain a general perspective on the telecommunication infrastructure [8–10]. This perspective clarifies the position of fiber optics within the broader context of telecommunication industry and provides insight into the forces behind the evolution of the fiber optic industry.

The oldest and perhaps the most familiar telecommunication network is the telephone system. Currently, most major points on earth are interconnected via the telephone networks. Each user has direct access to this network via a dedicated line, which is typically a real pair of copper wires that connect the user to the rest of the system. Voice signals do not require much of a bandwidth, typically only a few kilohertzs of bandwidth is sufficient for a telephone conversation. This bandwidth is provided to each user separately. Telephone networks are usually *circuit switching* networks, which means the system, at least on a conceptual level, provides a direct path between the two users at the two sides of a phone conversation. This path is provided by the switching systems that use the information in the phone number to connect the source with the intended destination, and it is established as long as the phone connection is continuing, regardless of the actual signal contents. In other words, a channel with the required bandwidth is established between the two sides, regardless of the two sides using it or leaving it idle. The technical details are hidden from the users: they can be in two rooms next to each other or at opposite sides of the world. As far as they are concerned, it is as if a pair of wires is connecting the two sides together.

In reality, the signals from individual users go to local telephone centers, where they are converted to digital signals and aggregated through time division multiplexing with signals from other users. Thus, a hierarchy is created: lower data rates are combined together to form higher rates. In case of long-distance calls, these higher data rate signals are typically carried on optical fibers to their destination, where they are de-multiplexed into separate signals, and ultimately switched to their individual destination.

The other major telecommunication system is the Internet. Compared to the telephone network, the roots of Internet are much more recent. What is currently called Internet originated in military applications in the 1960s and 1970s, with various academic institutions joining later. The idea was to connect various computers together through a network based on *packet switching*, as opposed to circuit switching, which was the basis of the telephone system. By mid 1990s the exponential growth of Internet had already started, and the rest is history.

Unlike the telephone system which is designed for the transfer of real-time audio signals, the Internet is designed with the purpose of transfer of digital data. Individual users are connected to the Internet through a variety of ways. These typically include traditional phone lines, digital subscriber line (DSL), and services provided by cable companies. Larger users such as companies or universities have direct access to the Internet. The principle of signal aggregation is valid for Internet too. For long-distance transmission, several low data rate signals are combined into fewer high data rate signals through time division multiplexing. Once data is aggregated into these high data rate streams, it is often mixed with other data formats and carried on the same backbone infrastructure that carries high-speed switching-based signals. At the destination, the signals are unpacked and then routed toward their individual destinations.

Another major component of the telecommunication infrastructure is the radio and TV broadcasting systems. Unlike the telephone and Internet networks, the radio and TV systems were primarily designed for one-way transmission: the same data stream flows from a central TV radio station to all the users within the reach of the signal. Cable TV companies are more or less based on the same model: they transmit the same signal to all the users that are connected to their services, although they may scramble certain prime channels and only provide the descrambler equipment to those subscribers who are paying for those prime channels.

In general, TV and radio broadcasting have their roots in analog transmission formats, based on frequency multiplexing of different channels with different carriers. This means broadcast signals generally cannot be baseband, otherwise all the sources will have to use the same frequency band for their signals, which causes significant interference and renders all the signals useless. Frequency multiplexing thus allows each user to tune in to a specific signal. A clear example is radio: all the stations broadcast their signals on individual carriers at various frequencies that span from approximately 500 to 1600 kHz for AM bands and from 88 to 108 MHz for FM bands. Thus, frequency multiplexing provides a solution for broadcast communications. However, as far as the analog nature of broadcasting is concerned, things are changing. This is especially evident in TV broadcasting. Many TV stations and cable TV companies are already broadcasting in high-definition digital formats (HDTV), and the other stations are following.

Telephone, data, and broadcasting applications are, of course, by no means separate or isolated. Many individual subscribers are connecting to Internet via telephone or cable TV lines, just as many broadcast stations receive or transmit their contents between themselves and other major content sources through backbone Internet structures. In general, the farther we get from the individual users, the more convergence we encounter in terms of data formats, speeds, and standards. This is a requirement based as much on technological reasons as on economic grounds. If a specialized content is to be transmitted across the globe, and if it is going to use the existing long-haul telecommunication structures such as the submarine fiber optic cables, it has to conform to the standards and formats of digital optical signals. As a result, sooner or later it has to be transformed into a digital format and multiplexed within the rest of the traffic that the fiber optic link will ultimately carry. Fortunately, the hierarchical structure of the networks hides these complications from users. The content is converted into the right format by the intermediate processing layers and converted back to its original form on the other side. The end users do not have to worry about these conversions. On the other hand, as far as the fiber optic link (physical layer) is concerned, everything is 1s and 0s: it receives a digital stream at the source side from the higher processing layers, and it must deliver the same signal at the other side of the link to the higher processing layers.

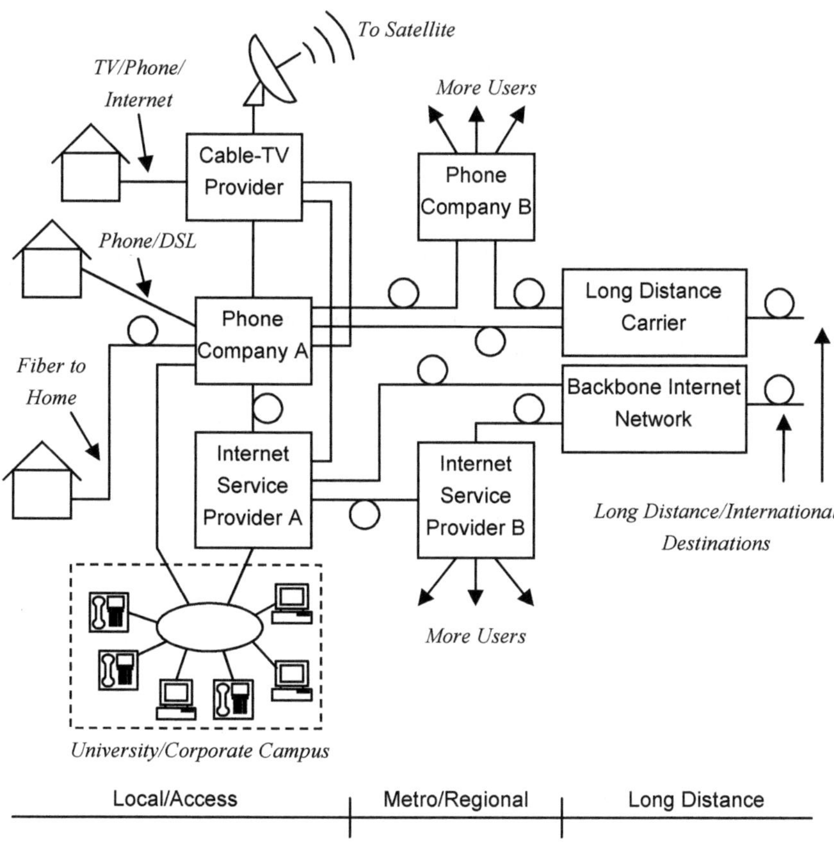

Fig. 2.3. A simple illustration of telecommunication infrastructure

Figure 2.3 is a simple illustration of the telecommunication infrastructure. It should be noted that what is depicted in this figure is merely an illustration since each particular network differs from others in many details. Thus, this figure is like a representative map that only serves to clarify and illustrate the concepts. In general, this figure is arranged in terms of physical distance from left to right. Local or *access* networks that are limited to smaller geographical areas are in the left side. These could be the networks within a corporate or university campus or the infrastructure that connects individual homes to the main networks. The reach of access networks is typically up to a few kilometers. Regional or *metro* networks are the next level. These are networks that connect local telephone companies or Internet providers within a larger geographical area such as a city. The reach of these networks can be up to a few hundred kilometers. The next level comprises the long-distance carriers. These carriers transfer the aggregated data from regional phone companies and Internet providers and transfer them across long distances, for instance across a country or from one continent to another. An example of long-distance networks is the submarine fiber optic cables that transfer the traffic across the Atlantic or Pacific oceans.

We can see that individual homes are connected to the network via regular phone lines, DSL lines, or cable connections to the cable companies. It used to be the case that various companies offered distinct services, for example cable companies offered TV programming, while phone companies offered standard phone services. But we see an increasing merger of these services. With the expansion of Internet, phone lines were also used to carry data signals, although in slow rates, while digital subscriber line (DSL) modems offered faster data rates. However, increasingly, cable companies offer both phone and internet services over their copper cables. A new development, as far as fiber optics is concerned, is the emergence of fiber to the home (FTTH) technology, which satisfies all user data requirements with a single strand of fiber while providing very high data rates by taking advantage of the high bandwidth of the optical fibers.

There are also campus-wide networks (usually called LANs or local area networks) that connect the computers within a university or company campus. These networks then usually have a direct access to the main data networks through Internet service providers. Typically, they also have an internal phone network, and thus they require dedicated connections to the local phone companies as well.

The next level up in the network is the regional or metro networks, which typically include local phone companies and various Internet providers. If a call originates from a subscriber, it typically goes to the local switching board where it is handled by a local phone service company. If the destination is another local subscriber within the same area, the signal stays within the same area. If the destination is a subscriber in the same region, the call may be switched from one regional office to another, or even it may be handled by more than one phone company. If the destination is farther away, for example in an international call, the signal is passed on to a long-distance carrier, where it is transferred on specialized high bandwidth links to other geographical areas.

Long-distance carriers typically function on a different business model. They do not interact with individual users directly. Instead, they are specialized in receiving relatively high-speed signals from regional phone companies and Internet providers and multiplexing them in yet higher data rates, in formats suitable for long-distance transmission usually in fiber optic links. Before the growth of fiber optics, these long-distance communications had to be handled by microwave links or specialized copper-based systems. However, with the advantages offered by fiber optics, long-distance communication is now almost exclusively the realm of fiber optics.

Figure 2.3 also illustrates the role of fiber optics in the telecommunication infrastructure. In this figure the links that are likely to be fiber optic based are marked with a circle on the connections between the nodes, while copper connections are just simple lines. It can be seen that in general, the farther the distances, and the higher the data rates, the more likely it is for a link to be fiber based. This is of course a direct consequence of the characteristics of fiber optics that makes it the main technology of choice for long-distance communications. For example, submarine fiber optic cables transfer data streams at speeds of many gigabits per second under the oceans for thousands of kilometers. The only practical alternative to long rang fiber-based links are satellite links. For example, as shown in Fig. 2.3, a cable TV company may receive some of its contents through a satellite link from a main feed. Satellite links are also useful for areas where the fiber-based infrastructure does not exist.

However, as we get closer to the individual users, where the distances are shorter and the data rates are lower, more competing technologies exist. Depending on the special circumstances and the existence of legacy infrastructure, local phone companies or switching offices may be interconnected to each other through either copper cables or fiber links. Likewise, depending on their sizes and complexities, local area networks can be copper or fiber based, although in many cases new installations tend to be fiber.

Individual homes, on the other hand, are still connected mostly through copper wires and cables. On one side, this is a result of legacy: copper cables and telephone wires have a long history, and telephone and cable companies are reluctant to incur cost and modify existing (and working) systems. On the other hand, expanding fiber toward individual subscribers must result in an economically viable system, and for that to happen, the cost of the fiber optic box that goes in each subscriber's house along with the cables that connect each house to a central office should be low enough to make the resulting service a consumer product. In spite of these challenges, the fiber is finding its way to individual homes. This is an area at its beginning phases, but is expanding fast. In practice, this means in many cases the fiber does not stop at a local cable or phone company office, but comes closer to individual users, perhaps to an apartment complex, or in many cases even to individual homes. The fiber to the premise (FTTP) or fiber to the curb (FTTC) or fiber to the home (FTTH) infrastructures are all variations on the same idea and are sometimes categorized together under the more generic name of FTTX. We will examine this subject again later in this chapter.

2.4 Networking spans

Topology is just one attribute that can be used to describe and categorize networks. Another useful way to categorize networks is based on their span or range. From this perspective, networks are usually divided into local area networks (LAN), metropolitan area networks (MAN), and wide area networks (WAN). Although these categories are already evident in Fig. 2.3, we should discuss them in some more detail for future reference.

2.4.1 Local area networks (LANs)

A LAN is typically a small network that connects computers and other resources in a small geographical distribution. Examples include university campuses, hospitals, and corporation or factory buildings. As a result, usually LANs are maintained by and belong to the same organization. The maximum distance a LAN covers is determined by the mode of communication. Fiber optic links can extend to tens of kilometers, whereas copper cables are more limited. The reach of a WiFi LAN is limited by the power of the transceivers, and it could extend to tens of meters.

The important feature of a LAN is that it allows the connected resources to share information with each other. For example, in a company, all computers, several printers, and several servers may be connected together through a LAN. A very popular protocol for LANs is the Ethernet, specified in various flavors of the IEEE 802.3 standard, which in its original form defined a 10 Mbps protocol on a coaxial bus topology [11–12].

However, the bus topology is not very successful for fiber optic links. The reason is that without using optical to electrical converters, tapping a link off an optical bus requires a coupler, and using couplers causes loss of power, which in turn limits the number of nodes that can be connected to a LAN.

In a LAN, each node can access the network at any time, and this may cause access collisions. Thus, it is the responsibility of protocols to resolve any potential conflicts arising from the use of the same resources by many users. Usually LANs are not isolated, rather, they are connected to larger and more extended networks.

2.4.2 Metropolitan area networks (MANs)

The next level of span in a network is when the network range extends beyond that of a LAN, perhaps to several blocks, multiple neighborhoods or districts, or an entire city. Such a network is usually known as a metropolitan area network or a MAN.

Unlike LANs, MANs are usually not managed or used only by a single individual or organization. For instance, a group of service providers may share the ownership of a MAN in a city, although it is possible for a company to lease bandwidth on a MAN from another company. Generally speaking, individual users do not have access to a MAN directly. Instead, MANs are used to interconnect other networks such as individual LANs. Therefore, individual users can connect to a MAN either through a LAN or through an access network, such as community access TV (CATV) network or a DSL line.[1]

Because MANs typically cover longer distances and higher data rates compared to LANs, they are more frequently based on fiber optic links. A MAN can have a physical reach of up to 100 km and may carry data rates as high as 40 Gbps. Moreover, because of the pressure toward larger bandwidths and because the costs can be shared between a larger number of end users, MANs afford to use more sophisticated technologies and devices. Routers and switches, and bandwidth-enhancing techniques such as WDM, can be used in a MAN. A typical protocol for MANs is asynchronous transfer mode (ATM), although migration toward Ethernet standards such as 10 Gbps Ethernet is taking place as well [13].

2.4.3 Wide area networks (WANs)

Networks that extend to many cities, countries, or even continents are designated as wide area networks or WANs. Because WANs have the longest physical range among the networks, they are usually not owned by a single entity. Instead various entities may own and operate various segments of WANs. Like the case of metropolitan area networks, individual users do not have direct access to WANs. Instead, WANs are used to connect smaller MANs and LANs.

WANs handle large amounts of data at high data rates and over large distances. As a result, fiber optics plays a dominant role in WANs. Moreover, advanced multiplexing techniques, especially WDM techniques, are more likely to be used in WANs, because they increase the effective bandwidth of existing fiber. A clear example is the submarine cables that carry the traffic between the continents. It is much more expensive to deploy a submarine cable than it is to add sophisticated equipment at the two sides of an existing link to improve the bandwidth efficiency.

Because of the large reach of WANs and the fact that they typically connect many other types of networks with a vast array of devices and technologies, WANs tend to use a mixture of different topologies. WANs, like other kinds of networks, have to follow well-defined specific protocols. Common protocols include ATM and SONET (synchronous optical network).

---

[1] That is why sometimes access networks are also considered a form of LAN.

2.5 Hierarchical structure of networks

So far we have described the networks from a "horizontal" perspective. Describing a network from a topological viewpoint, for instance, is essentially a description of how various nodes are connected to each other and communicate with each other. Likewise, categorizing networks as LANs, MANs, or WANs is a horizontal categorization, because it divides the networks based on their geographical reach. However, a complete description of networks requires some level of hierarchical or vertical description as well. A vertical description refers to the fact that information must go through several layers of processing before it can be converted to a physical signal and transferred in a physical medium. In the example of human language in Chapter 1, this hierarchy refers to the processes that take place in the brain in the form of structuring an idea in terms of language, organizing and refining the language into a set of letters, words, and sentences, changing these linguistic units into neural signals, and finally modulating the air through the vocal system driven by those signals. Once the resulting sound waves travel through the medium, the opposite processes must take place to convert the physical signal back into ideas.

While this vertical or hieratical process can be immensely complex and "fuzzy" in case of human conversation, for communication networks it must be well defined and logically clear. All the required processes must be well thought of and be divided into manageable, clear steps. These steps or *layers* must be defined in a way that each has a clear and separate function, so that the whole task of communication between two separate points can be divided into small, manageable, logically separate chunks. Moreover, each layer should ideally interact only with the layers above and below it.

Such a division of labor is clearly advantageous for many reasons. It simplifies the task of each layer and makes improvements easier, because the changes in each layer do not have to affect other layers. Thus, each layer can specialize and excel in a particular task, resulting in networks with improved design, better efficiency, and more reliability.

2.5.1 Open System Interconnect (OSI) model

One of the most common models for this hierarchical approach is the Open System Interconnect (OSI) model developed in 1984 by International Organization for Standardization (ISO) [14–16].[2] The OSI model itself is not a protocol or standard, and it is not even followed in all practical systems. However, it defines a conceptual framework that can be used as a reference for other hieratical structures and, as such, is a widely used and referenced model. Figure 2.4 shows an overview of the OSI model.

---

[2] For more information on ISO and its activities see Chapter 12.

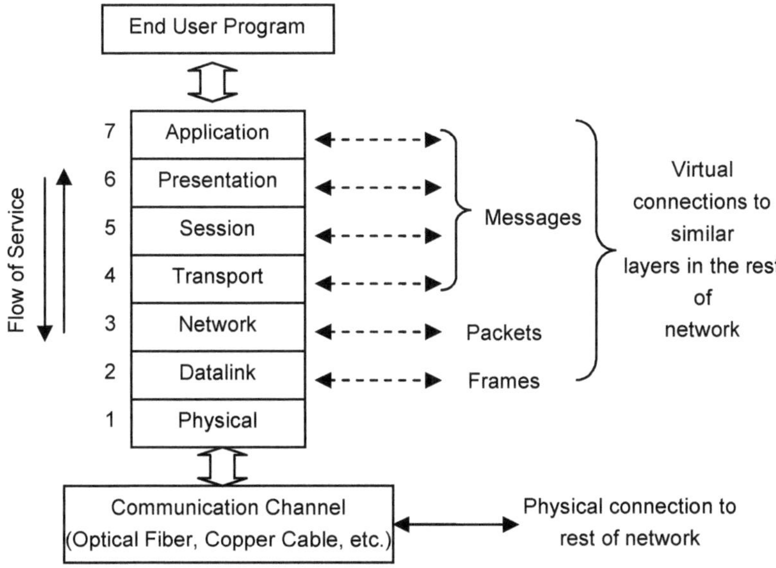

Fig. 2.4. The OSI hierarchical network model

The higher layers in this model are closer to the end users, while the lower layers deal with more hardware-oriented tasks such as addressing, switching, and multiplexing functions. The physical layer, which is the lowest layer, deals with physical signals such as electrical voltages and optical signals. It receives the information from the datalink layer and sends it over the physical channel, which can be fiber, cable, free space, etc. The physical layer on the other side of the link receives the signal and sends it up to the datalink layer. The highest layer, or the application layer, is the closest layer to the end users. Although in the OSI model it does not interact directly with the end users, both the end user and the application layer interact with various programs that have some form of communication function, for example web browsers, and email programs.

Figure 2.4 also illustrates another feature of the OSI model. In general, each layer communicates both vertically and horizontally. Vertically, it communicates with the layers below or above itself. A higher layer requests some kind of service from the layer immediately below itself, and the lower layer provides that service to the higher layer. The interface between the two layers is called a service access point (SAP). Therefore, the relationship between the layers is a *client–server* relationship.

Horizontally, each layer interacts with its *peer* layer on the other side of the link. For example, the network layer is receiving data from the transport layer and passing it down to the datalink layer. At the same time, it is interacting with its peer network layer on the other side of the link, although it does not care about the details of how that communication has actually been achieved through the lower layers. That is why in Fig. 2.4 all the horizontal connections are designated as virtual: only the physical layer provides physical connection between the nodes.

As noted before, the major focus of this book is fiber optics communication from the physical layer perspective. The physical layer interacts directly with the physical medium. In case of fiber optics, this includes the modulation of light in the form of optical signals and sending and receiving it through the optical fiber. We will cover these topics in detail throughout this book. However, we should have some insight into the higher layers, especially those that are closer to the physical layer. Therefore, we will have a brief review of these higher layers, with an emphasis on those closer to the physical layer.

2.5.2 Datalink layer

The datalink layer (layer 2) is the closet layer to the physical layer. In the OSI model, the datalink layer is responsible for reliable transfer of data between adjacent nodes. In other words, the datalink layer does not concern itself with sending data to its ultimate destination; it is just concerned with sending data out to the next available node(s).

The unit of data in the network layer is usually called a frame. A frame at the datalink layer consists of data from layers above encapsulated in headers and trailers added by the datalink layer. These headers and trailers are intended for use by the datalink layer in the destination and are transparent to higher layers on both sides. Functions of the datalink layer generally include such tasks as encapsulating data into frames, ensuring the frames are transmitted and received without errors, and controlling the flow of frames.

Following the IEEE 802 standard, sometimes the datalink layer itself is divided into two sublayers: the logical link control (LLC) and media access control (MAC) sublayers, where the LLC layer is located above the MAC layer. In 802.2, the LLC is responsible for communication between devices located on a single link. This is done by dedication of various overhead fields in the frame such that higher layers can use and share the same physical link through these fields. The MAC layer is responsible for organizing access to the physical layer. For example, in many cases the same medium is shared between several nodes, such as in a bus topology where all the nodes use the same link. Therefore, a set of rules must be established to avoid conflict. IEEE 802 also defines physical MAC addresses that uniquely identify each device on the network (not to be confused with the IP address, which is a higher layer concept). Note that the datalink layer is the lowest level in which data frames and various fields within a frame are meaningful, because the lower physical layer does not distinguish between these various fields and simply treats data as a serial digital stream.

Some of the most popular protocols that define the datalink layer include IEEE 802.3 Ethernet (carrier-sense multiple access with collision detection LAN) and IEEE 802.5 (token ring LAN).

2.5.3 Network layer

The network layer (layer 3) is the next layer up in the OSI model. The main task of network layer is the transfer of information from source to the final destination, which in general involves routing functions. The unit of information in the network layer is sometimes called a packet. The network layer makes packets by receiving data from higher layers and encapsulating them with its own headers and trailers. These headers and trailers are intended for network layer operation and are transparent to layers below and above. Thus, the datalink layer below treats a packet as data and adds its own headers and trailers, as explained before.

2.5.4 Higher layers

The first three layers in the OSI model (physical, datalink, and network layers) are sometimes grouped together as interface layers. These layers are directly involved in the telecommunication functions. On the other hand, the higher layers (transport, session, presentation, and application) are involved with end-to-end (also called peer-to-peer) tasks. The functions of these layers include establishment and termination of sessions, data compression and decompression, interactions with operating systems, translating data into standard formats, user interface services, etc. Moreover, these higher layers are very "fluid" and are highly dependent on evolving technologies and protocols [17]. We do not go into the details of these layers, because for the most part, these layers are not directly involved in the telecommunication aspects of networking.

As noted before, the OSI model is a useful tool for understanding network concepts. In reality, not all networks conform to this model. In practical networks some of these layers may be omitted, expanded, or mixed. It is also possible that some technologies define their functions in a way that does not clearly correspond to the standard OSI layers. For instance, SONET has provisions to deal with end to end connections defined in the network layer. However, SONET has other functions too that do not correspond to the network layer. Despite these conformance details, the OSI model provides a useful conceptual tool for mapping other protocols, standards, and technologies.

2.6 Circuit switching and packet switching networks

Communication services generally fall in two categories: connection oriented and connectionless. These types of services are closely related to the ways networks (and related protocols that operate over the networks) function.

Generally, there are three different approaches to digital communication. In a *broadcast* network, all the messages that are sent from any of the nodes are received by all other nodes. For example, consider an Ethernet LAN, where all computers are connected to a single bus. When one computer sends a message, all other computers on the network receive that message. It is only the content of certain fields within the message that marks the destination. Obviously things can get more complicated, for instance, when two computers send their messages simultaneously and a collision happens. The network protocol must therefore have provisions to resolve the contention. A broadcast network is thus like a gathering at the dinner table, where everybody can hear what everybody else says. It can be seen that a broadcast approach is not suitable for large networks with multiple nodes. The alternative to a broadcast strategy is to implement some form of switching or routing, so that data from one node can be directed only to its destination node. This gives rise to two alternative approaches: circuit switching and packet switching [18–19].

2.6.1 Circuit switching

The circuit switching approach has its origins in the telephone network. In a phone conversation, the two sides are connected together through a real circuit. In other words, there must be a channel that connects the two sides for the duration of conversation. As long as the two sides keep the connection (even if they do not talk) they have the channel only to themselves. This is because voice is a *time-sensitive* application. If I talk to my friend, I like my friend to hear my words in the same order that I speak them. Indeed, all real-time audio and video applications are time sensitive. The receiver must receive the signals in the same order and at the same rate that the transmitter is sending them. This could automatically be obtained in a single point-to-point link: after all, the signal sent through a fiber link reaches the other end of the link at the same rate and in the same order.

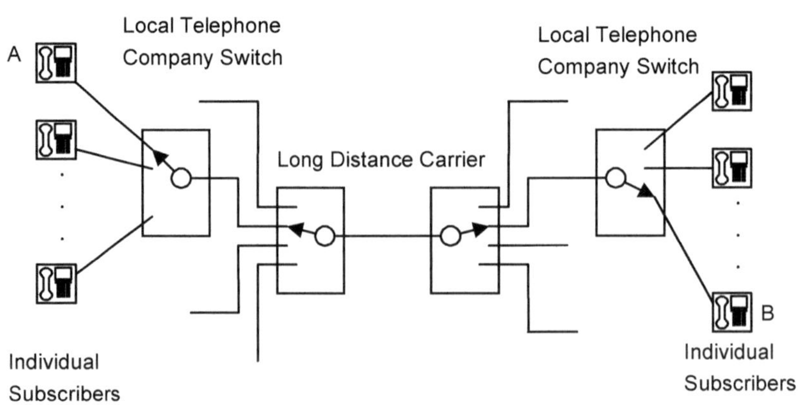

Fig. 2.5. Schematic of a circuit switching network

However, in a network where signals must be multiplexed and routed, the issues can be more complex and require special attention. This is why telephone networks are based on the concept of circuit switching. As shown in Fig. 2.5, when two individuals talk over the phone, various switches in the network provide a direct path between them, so that during their conversation they are connected in real time and have the entire bandwidth of a standard telephone line to themselves.

Of course in reality the signal will be multiplexed into successively higher rates and then demultiplexed back as it approaches its destination. Nevertheless, these details are transparent to the two subscribers at the two sides. A prominent example of circuit switching architectures is SONET. We will talk more about SONET later in this chapter.

2.6.2 Packet switching

Unlike voice and video applications, data applications are not so sensitive to time delays. Moreover, data traffic tends to come in bursts. At one point in time the application may produce a large amount of data that needs to be transmitted, while in the next moment it may not need to transmit any data. Obviously dedicating an entire channel to this kind of traffic is not efficient, because the channel may have to be idle most of the time. The solution is to break down the data into packages and send each packet independently over the network. Here a technique called *statistical multiplexing* can be used: packets from various nodes will be transmitted through the same channel as they arrive. If the number of packets exceeds the available bandwidth of the channel, they must be buffered or stacked and sent later once the bandwidth is freed up. In this way, digital content can be broken down into several pieces and transmitted separately, even over different paths. Once all the pieces arrive at the destination, the original content can be reconstructed.

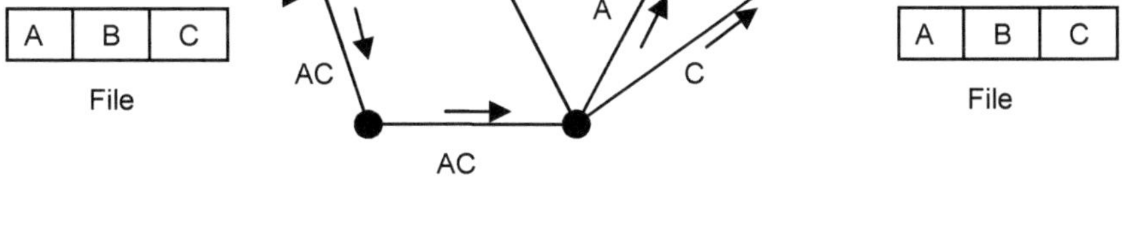

Fig. 2.6. Packet switching network

This process is schematically shown in Fig. 2.6. Here some digital content needs to be sent across the network. Instead of having a direct dedicated channel, the file is divided into several pieces or packets. Next, the packet is embedded with additional data called *headers*. The header includes information such as the address of the destination to which the file must be sent. The packets are then sent across the network. Various routers and switches in the way decide on the path that each packet will take by looking at the destination address in the header, as well as by considering the load on each available link. If no link is available at the time, the packet may be buffered and sent in a later time where a link becomes available. As a result, depending on the particular protocol in use, not all pieces may travel along the same path, and they may not even arrive at the destination in the same order that they were sent (more on this later). Once all the packets arrive, the original content will be reassembled based on the additional information in each packet.

As noted above, the advantage of packet switching is that available bandwidth is used much more efficiently. However, the disadvantage is that time-sensitive applications may suffer, because unless explicit provisions are made, no real-time relationship between the source and the destination exists. Examples of packet switching technology include IP, ATM, and MPLS.

Naturally, one may ask if there is any way to accommodate time-sensitive applications with packet switching. Conceivably, a direct connection may be "emulated" by a packet switching network through provisions that ensure packets arrive at their destination in the same order and time frame that they were sent. As a result of such considerations, packet switching networks can be divided into two further categories: connectionless and connection oriented.

In a *connection-oriented* network, a virtual path between the source and the destination must be established first. In the next phase, all the packets are sent through this same path. Finally, once all the packets are transferred, the connection will be terminated. A connection-oriented network is different from a circuit switching network in that several virtual paths may share the same bandwidth, i.e., the channel is not solely dedicated to any single virtual path. Primary examples of connection-oriented networks include ATM and MPLS.

In a *connectionless* packet switching network, on the other hand, no virtual path between the source and the destination is established and each packet is treated individually. The fact that no prior connection has been established means that the information may be sent without verifying if the receiver has a valid address or is actually capable of receiving the information. A primary example of connectionless networks is the Internet protocol (IP).

2.7 SONET/SDH

As noted above, a prominent example of circuit switching is SONET, which is an acronym for synchronous optical network. Closely associated with SONET is its international counterpart, the synchronous digital hierarchy or SDH. SONET and SDH are well-established robust technologies in wide use, especially in backbone networks. Since their introduction in the late 1980s and early 1990s, SONET/SDH have had a fundamental role in optical networks by defining a clear set of optical signals, hierarchical multiplexing schemes, bit rates, and signal management rules [20–23].

The origins of SONET and SDH go back to the telephone network and bandwidth requirements for voice. Although humans can be sensitive to audio frequencies as high as 20 kHz, for a normal conversation only a fraction of this bandwidth is sufficient. Therefore, a standard telephone channel assigns 4 kHz of bandwidth for a normal conversation. The audio signal is then sampled at twice this rate or 8000 times per second. This assures that later on it can be faithfully reproduced when it is converted back to analog format. Moreover, each audio sample is digitized with a resolution of a byte, which represents 128 possible values. This brings the number of bits per second for a digitized signal to 8000×8=64,000. Thus, we can think of the rate of 64 Kbps as a "unit" of bandwidth.

To utilize the available bandwidth of communication channels more effectively, several voice channels can be multiplexed. The DS standard, adopted in North America, defines such a multiplexing scheme for higher data rates. The next level of multiplexing defined by the DS standard is when 24 voice channels are combined, which results in a bandwidth of 24×64 Kbps=1.544 Mbps. Such a signal is known as a digital signal 1 (DS-1), or T1. Higher data rates can be achieved by multiplexing more voice channels. Of particular interest to our discussion is the DS3 (also known as T3) signal which consists of 672 voice channels and corresponds to a data rate of 44.736 Mbps.

Outside North America, the same hierarchy was defined slightly differently through the ITU-T standards. ITU-T defines the first level of multiplexing with 30 voice channels. The result is called an E1 signal and corresponds to a data rate of 30×64 Kbps=2.048 Mbps. Multiplexing more voice channels results in successively higher data rate signals, for instance, the E3 signal consists of 480 voice channels and corresponds to a data rate of 34.368 Mbps.

The SONET standard originally proposed by Telcordia (formerly known as Bellcore) developed out of the need to multiplex several DS signals such as the T1 signal and transmit them optically. Because SONET was not exactly conforming to ITU's digital hierarchy signal structure, ITU defined a similar set of standards to SONET called synchronous digital hierarchy (SDH), which was suitable for multiplexing ITU's digital signals. We should note, however, that the SDH standard was defined in a way to accommodate SONET signals too, and therefore all SONET standards comply with SDH.

Both SONET and SDH handle data in the form of frames. Figure 2.7 shows the basic structure of a SONET frame. The SONET frame can be represented as a two-dimensional array of bytes with 90 columns and 9 rows.

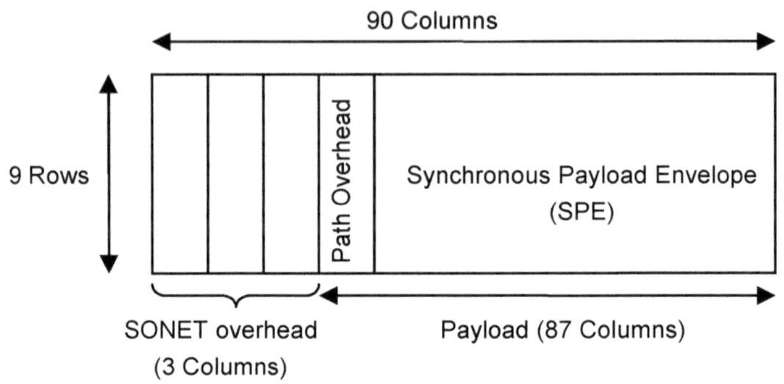

Fig. 2.7. Structure of the basic SONET frame

Data is transmitted serially starting from left to right and from top to bottom, with the most significant bit in each byte transmitted first. The first three columns are set aside as transport overhead and are used for SONET's internal functions. The remaining 87 columns are called synchronous payload envelope. Also, another column is set aside as path overhead, which leaves 86 columns for users' data.

Each SONET frame is defined to be 125 μs long, corresponding to 8000 frames per second. Since each frame has $9\times90=810$ bytes, and each byte is 8 bits, the base bit rate in a SONET signal would be: $8000\times810\times8=51.84$ Mbps. This base rate is defined as a STS-1 signal. Note that not all this bandwidth is user data. From Fig. 2.7 it can be seen that the payload is 86 columns out of the total 90 columns, or 774 bytes out of the 810 bytes. This means a SONET signal can support user payloads of up to $8000\times774\times8=49.536$ Mbps. Looking back at the DS standard, we see that this bandwidth is sufficient for a DS-3 signal.

The STS-1 signal described above is a *logical* signal. By itself, this signal may not be very easy to transmit over the fiber, because it may include long sequences of identical bits, a situation that can cause problems for an optical transmitter. To avoid these problems, and also to make clock recovery easier on the receiver side, the STS-1 signal is scrambled before being converted to an optical signal. The resulting optical signal, after this scrambling, is called an OC-1 signal, where OC stands for optical carrier.

SONET allows multiplexing several STS-1 signals into a higher level STS-N signal. This is done by dividing each column in the SONET frame into N columns. Obviously, then, an STS-N frame contains N times as many bytes as an STS-1 signal and involves a physical layer data rate N times higher. The corresponding optical signal would then be called an OC-N signal. According to existing SONET standards, values of N equal to 3, 12, 24, 48, 192, and 768 are allowed. These values correspond to approximate data rates of 155 Mbps, 622 Mbps, 1.25 Gbps, 2.5 Gbps, 10 Gbps, and 40 Gbps, respectively. In practice, the electrical STS-N and optical OC-N signals are not usually differentiated, and both signals are referred to under the generic OC-N designation.

Table 2.1 SONET and SDH signal levels and physical layer data rates

SONET		SDH	
Logical signal	Optical signal	SDH signal	Data rate (Mbps)
STS-1	OC-1	Not defined	51.84
STS-3	OC-3	STM-1	155.52
STS-12	OC-12	STM-4	622.08
STS-24	OC-24	STM-8	1,244.16
STS-48	OC-48	STM-16	2,488.32
STS-192	OC-192	STM-64	9,953.28
STS-768	OC-768	STM-256	39,813.12

The SDH signal structures are very similar to SONET, with the difference that SDH does not have an equivalent for STS-1. Instead, it starts from the equivalent of STS-3, or 155.52 Mbps. In SDH, such a signal would be called a synchronous transport module level 1, or STM-1 signal. The designation for higher data rates is STM-N, where N=1, 4, 16, 64, and 256. Thus, in order to obtain the equivalent SONET signal we should multiply N by 3. For example, a STM-16 signal in SDH is equivalent to an OC-48 signal in SONET. Moreover, in SDH no distinction is made between the logical and optical signals, and they are both represented under the same STM level signal designation. Table 2.1 includes a summary of the SONET and STM signals.

The above discussion gives an overview of SONET and SDH. For the most part, from a physical layer perspective the most important parameter of a SONET/SDH signal is its data rate. There are many details about the internal structure of a SONET/SDH frame and methods of multiplexing lower signal levels into higher levels. But such details fall outside the scope of this book [22,24].

2.8 WDM networks

Throughout the evolution of telecommunications, one of the few constants in the otherwise evolving scene has been the need for higher data rates. This has been a direct consequence of the replacement of text-oriented traffic with bandwidth-intensive graphics and video-rich applications. As a result, the need to squeeze more bandwidth out of a channel has always been present and is likely to continue for the foreseeable future. One solution to this problem has come from the time division multiplexing schemes like SONET which allow for packaging data in higher and higher data rates.

Time division multiplexing is a very effective tool, but it has its own limitations. In particular, it becomes increasingly difficult to handle higher speed signals at the physical layer. It is more difficult to modulate light at higher data rates, and higher data rate optical signals suffer more from effects such as dispersion. As a result, the physical reach of an optical link decreases as the data rate increases.

A different approach to use the wide bandwidth potential of optical fibers is the wavelength division multiplexing (WDM) technology [25–28]. In this technology, multiple wavelengths of light are modulated separately and sent into the fiber simultaneously. As long as the power within each signal is not too high, the fiber acts as a linear medium, the interaction of different wavelengths on each other will be negligible, and each wavelength propagates in the fiber independent of the others. [29]

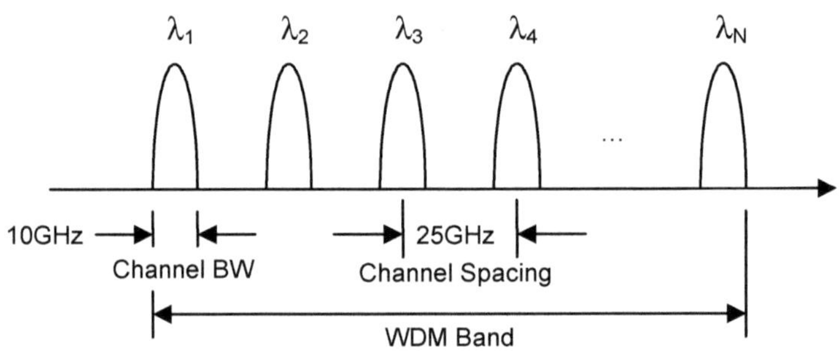

Fig. 2.8. WDM multiplexing

As a result, the effective usable bandwidth of the channel will be multiplied by the number of wavelengths in use. Figure 2.8 illustrates an example of WDM multiplexing. Here *N* wavelengths have been used, each modulated with 10 GHz of bandwidth. The wavelengths are spaced 0.2 nm apart in wavelength, which corresponds to a frequency spacing of 25 GHz. All these signals can travel together in a fiber, as long as they all fall within a low-loss transmission window. For instance, the C band is defined between 1530 and 1565 nm, which covers approximately 4.3 THz of bandwidth.[3] Given a 25 GHz channel spacing, we can fit approximately 170 WDM channels in the C band, while the electronic circuits that modulate and demodulate light operate at the reasonable speed of no more than 10 GHz. In practice, the wavelengths used in WDM systems are set by standards [30]. The recommended value for channel spacing is 50 GHz (0.4 nm) and 100 GHz (0.8 nm), referenced to a frequency of 193.1 THz (1552.524 nm). In a practical WDM link, as many as 80 wavelengths can be used, which would result in a tremendous increase in bandwidth otherwise unachievable.

---

[3] For a discussion of optical spectral band designations and transmission windows, see Chapter 5.

WDM technology has also opened new possibilities in the area of optical networking. The different types of networks we have discussed so far use a mixture of optical and electrical technologies. As we noted, the common trend is that as the range of a network and/or the amount of traffic a network carries increase, optical fibers become more attractive and are more widely used. However, the role of optics in these networks is limited mainly to data transmission. Higher level functions are realized in the domain of electronics.

What makes it hard for optics to play a more active role at the higher layers is the difficulty of processing signals in the optical domain. Thus, we have to use optical-electrical-optical (OEO) converters, where signals must be converted from the optical domain to electrical domain, processed in the electrical domain, and then converted back into optical domain and sent into fibers toward their destination [31]. Such a network is sometimes called a *non-transparent network*, meaning there is no direct optical path between the source and the destination.

A disadvantage of an OEO link is that at higher data rates the burden on the electronics in the routers and switches increases, causing higher cost and complexity. Thus, it would be useful to find ways to keep optical signals within the optical domain, and only convert them to the electrical domain at the destination. WDM technology provides one of the few practical ways of achieving this goal. In principle, this is doable because WDM allows access to components of an aggregate signal in the optical domain. Without WDM, all the various streams multiplexed within a signal are bundled into one optical signal. Therefore, any data processing on any of the individual signals requires a conversion to electronic format. In other words, the optical signal must be terminated, and after processing, regenerated.

This is what happens, for instance, in a SONET Add/Drop Multiplexer (ADM), where a higher level SONET signal can be decomposed into its constituent lower level SONET signals. Some of those signals whose destination is different can be extracted (or dropped), and other SONET streams whose destination is the same can be added into the stream from. The regenerated SONET signal can then be sent back into the fiber to continue its path toward its destination. In this way the ADM is very much like a bus station, and the individual lower rate signals like the passengers in a bus. The station provides an opportunity for individual passengers to get off the bus, while allowing other passengers to get on. The bus can then continue its trip toward its destination.

WDM allows a similar process to take place only in optical domain. The reason is that the individual data streams that have made up the multiplexed signal can be distinguished optically. This optical "marker" is the wavelengths of individual streams, which provides granularity at optical level. This allows for certain routing functions to be carried out completely in optical domain. In this way, we get closer to the realizations of all optical networks.

WDM technology depends on advancements in several fronts. From the standpoint of networks, the appearance of wavelength switching and routing devices has been a key enabling development. Two examples of these devices are *Optical Add/Drop Multiplexers* (OADMs), and *Optical Crossconnects* (OXCs).

An OADM is a device that operates on a WDM signal and allows extracting or adding some wavelengths at a node while allowing other wavelengths to pass through without interruption. For example, the signals whose destinations are the local node can be terminated and extracted from the WDM signal, while the signals that the local node needs to add to the stream can be added. Other wavelengths whose destinations are elsewhere continue through the OADM.

An OADM can be implemented in a variety of ways, but Fig. 2.9 shows a typical block diagram. Let us consider one direction of traffic flow, say, from left to right. A WDM signal comprising N wavelengths λ_1 to λ_N enters a WDM demultiplexer which separates the signal into individual wavelengths. Each wavelength goes through a 2×2 switch, which can be in one of the two states: it can either directly connect its input and output ports or it can cross connect the inputs and outputs to the local ports. In the former case, the wavelength passes through the switch undisturbed. In the latter case, the wavelength is dropped from the traffic to be used locally. If needed, a different signal can be added to the traffic on the same wavelength. The outputs of the switches are then multiplexed again and sent out on the other nodes.

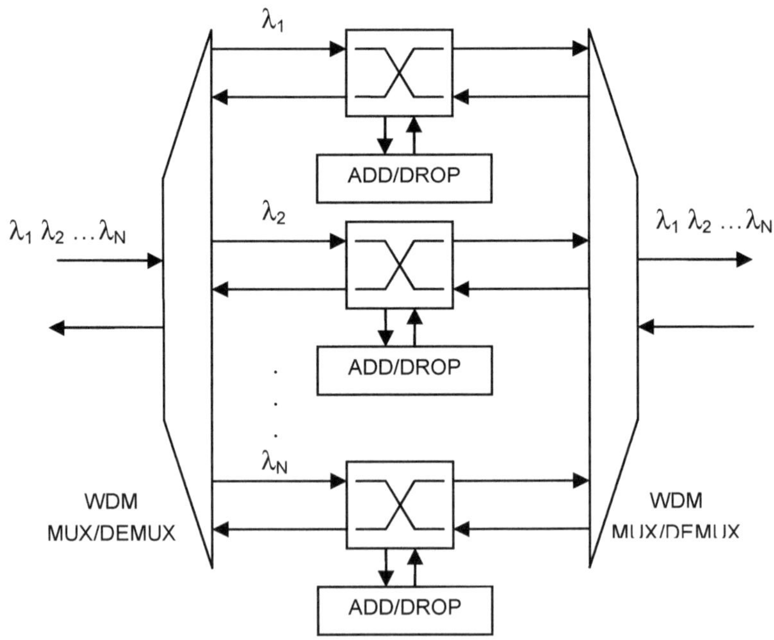

Fig. 2.9. An optical add/drop multiplexer (OADM)

An OADM operates on a single fiber and is a useful device for networks with simple topologies. However, for more complex topologies, a multi-port device is needed which would be capable of arbitrary routing of wavelengths between several fibers. This is a function an OXC can fulfill. Figure 2.10 shows the concept of an OXC. Each optical port is connected to a fiber which carries *m* wavelengths. Each fiber goes to a demultiplexer which decomposes the signal into various wavelengths. There are also *m* $N \times N$ optical switches. Each switch is dedicated to a particular wavelength and can connect one of its inputs to one of its outputs. The outputs from all the switches are then multiplexed, and are coupled back into the *N* output ports. It can be seen that an OXC is a very flexible device, as it is capable of selectively switching and routing different wavelengths between various fibers.

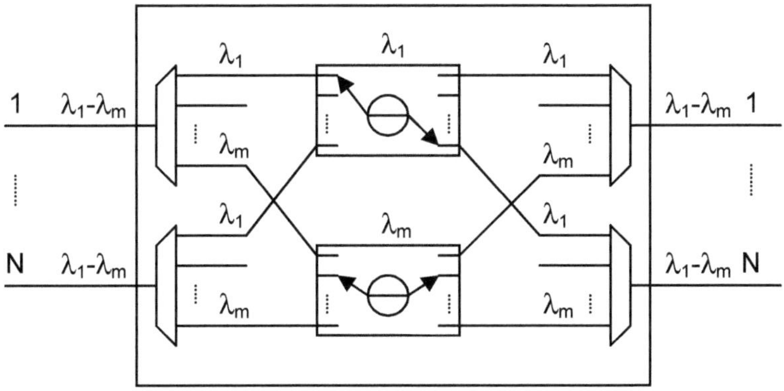

Fig. 2.10. An optical cross-connect (OXC)

To illustrate the application of these devices in an optical WDM network, let us consider Fig. 2.11 which depicts a network with five nodes, A–E, and four wavelengths, λ_1–λ_4. In this network node A transmits signals on three wavelengths λ_1–λ_3. These wavelengths go through an OADM which drops λ_1 to node B. Therefore, node A communicates to node B through λ_1. The other two wavelengths λ_2 and λ_3 pass through the OADM and arrive at the OXC. The OXC has two input and two output ports. One of the input ports receives λ_2 and λ_3 from the OADM. The other input port receives λ_4 from node C. The OXC is configured to separate λ_2 and λ_3 and combine λ_3 with λ_4 and route them to node E, while sending λ_2 to node D. Thus, A is connected to B via λ_1, A is connected to D via λ_2, A is connected to E via λ_3, and C is connected to E via λ_4.

This example can be used to highlight a few important concepts. First, note that connections are established between nodes via *lightpaths*. A lightpath is a continuous uninterrupted optical link, akin to a virtual fiber, along which signals travel without being converted back and forth between electrical and optical domains. An example of a lightpath is the dashed line shown in Fig 2.11 representing λ_2 that originates from A, passes through the OADM, and is routed to D via the OXC. This makes such a WDM network an example of a circuit switching network, because the lightpath acts as a real-time connection that establishes a dedicated channel between two nodes.

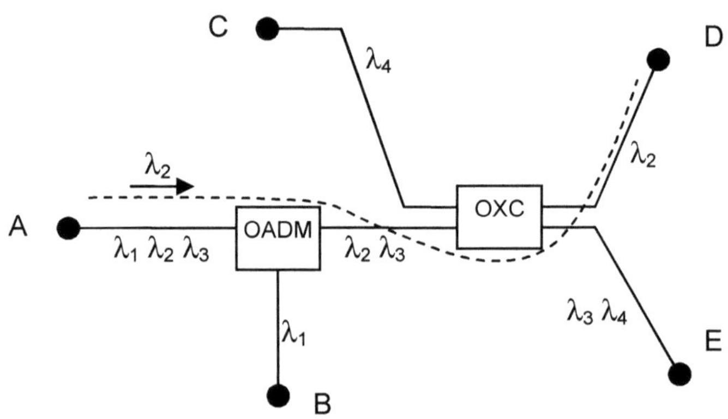

Fig. 2.11. Wavelength routing in a WDM network

Moreover, because each wavelength is routed and switched in optical domain, the lightpaths are *protocol agnostic*. Another way to state this is to say that several protocols can simultaneously use the networks without interfering with each other. This is a very important feature, because it makes the network very flexible. In practice, for instance, a wavelength can be leased to a customer, and the customer chooses the protocol and the data rate of the leased channel without having to worry about compatibility issues or interference with other traffic that use the same fibers.

Note also that the addition of the OXC adds considerable flexibility to the network. Obviously, as a router it can control the flow of traffic. For instance, it can swap λ_2 and λ_4 by routing λ_4 to D and λ_2 to E. In that case C would be connected to D and two channels would be established between A and E. This *dynamic reconfigurability* allows for the topology of the network to be changed based on real-time parameters such as the status of the traffic or load on the nodes.

From this simple example it can be seen that WDM networks offer great potentials for wideband optical networking. This is further illustrated by the fact that the standards bodies have recognized the potentials in this technology and are already moving toward providing encompassing standards and protocols for taking advantage of these capabilities. Emerging technologies and protocols such as optical transport networks (OTN) are establishing the basis upon which next-generation optical networks are expected to be based [32–33].

2.9 Passive optical networks (PONs)

The need for wider bandwidths is not limited to large companies. Increasingly, smaller customers such as residential complexes and individual homes require wider bandwidths. This need is driven by data-intensive applications such as high-definition TV (HDTV), bandwidth-intensive Internet applications, as well as other voice and video services. As a result, network architectures known as fiber to the curb (FTTC), fiber to the home (FTTH), fiber to the premise (FTTP), or more generically fiber to X (FTTX) have been developed [34–36].

In an FTTX network some or all of copper cables that are normally used to establish the communication link between end users and the larger communication infrastructure are replaced by fibers. The basis for most FTTP architectures is a passive optical network (PON), shown in Fig. 2.12.

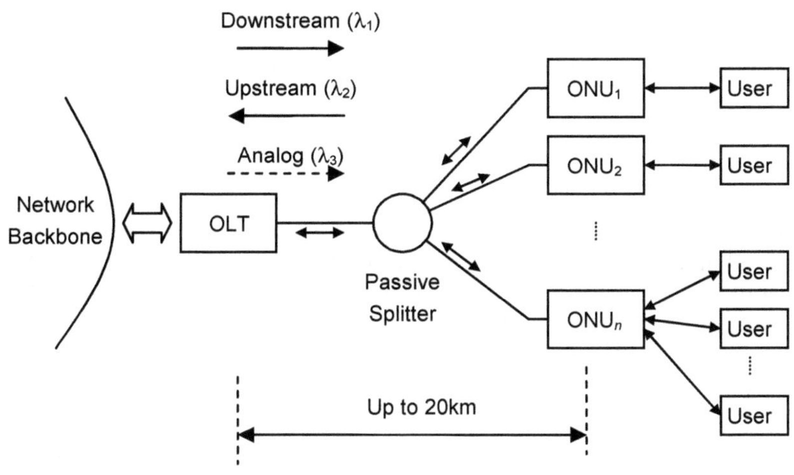

Fig. 2.12. Typical passive optical network (PON) architecture

A PON system consists of an optical line terminal (OLT) and a number of optical network units (ONUs). The ONU is also called optical network terminal or ONT. Thus, the terms ONU and ONT are often used interchangeably. The OLT is located at a central office, and the ONUs are located close to the end users. The connection between the OLT and the ONUs is established by optical fibers and through a passive splitter. The OLT provides the connection of the PON to the network backbone. On the other hand, the ONUs establish the connection to the individual users. The individual user can be a single subscriber, or a number of subscribers, like in the case of a residential complex.

In a PON architecture, the traffic flow from the OLT to the ONUs is usually referred to as *downstream traffic*, while the direction from the ONUs to the OLT is referred to as *upstream traffic*. The downstream and upstream traffics are typically carried over different wavelengths. Some PON deployments also use a third wavelength in the downstream direction for an analog overlay, usually for a video signal. Therefore, a PON deployment is an example of WDM architecture. Moreover, PONs can be symmetric or asymmetric in terms of data rate, as the transmission rates in the downstream and upstream directions are usually not equal. Most PON systems are *asymmetric* in this respect, i.e., the downstream data rate is higher than upstream data rate. This is because the end user's receiving bandwidth needs are typically higher than their transmitting bandwidth needs.

As can be seen from Fig. 2.12, the central piece in a PON is a passive splitter/coupler. We will discuss passive devices such as couplers in Chapter 7 in more detail. The transmitted signal from the OLT is divided equally between the ONUs through this splitter. Thus, if the output power of the OLT is P, the power received by each ONU is roughly P/n, where n is the splitting number, also known as the *splitting ratio*. A 50:50 split represents a 3 dB loss, thus, a splitting ratio of n represents a loss of $3\times\log_2(n)$. For example, a splitting ratio of 32 represents a splitting loss of 15 dB. Obviously, there is a trade-off between the splitting ratio and the physical distance the PON can support. Most PONs are implemented with a splitting ratio of 16 or 32, with a typical distance of 20 km. Another constraining factor is data rate, as higher data rates imply reduced distance.

Note that in this architecture, in the downstream direction, all ONUs receive the signal transmitted by the OLT. As such, a PON is an example of a broadcast network. Consequently, if the OLT intends to address only a specific ONU, it must add additional addressing information to the data to specify which ONU is the target of the message. On the other hand, in the upstream direction, the signals sent by individual ONUs are added together by the coupler and received simultaneously by the OLT. To prevent the mixing of signals, a time division multiplexing scheme is used that ensures only one ONU transmits at any given time. This scheme is shown in Fig. 2.13. Each ONU is assigned a particular time slot during which it can transmit its data, while it should stay off in the remaining time slots. In this way, the OLT can distinguish between the signals from various ONUs.

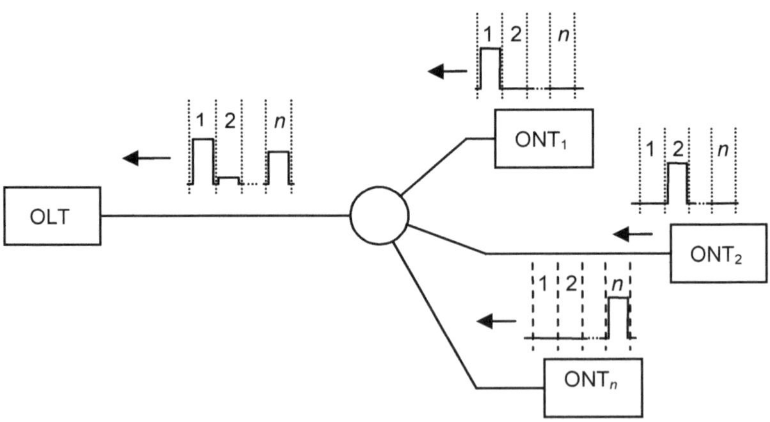

Fig. 2.13. Upstream burst mode traffic in a PON

The fact that the ONUs can be at different physical distances from the OLT complicates the structuring of traffic in the upstream direction. When the PON is being set up, through a process known as *ranging*, the OLT interrogates each ONU separately to determine the round trip delay associated with each branch. Only after knowing each individual delay exactly can the OLT assign time slots to the ONUs; otherwise the packets issued by the ONUs can collide. Because these delays are not completely deterministic and because the ONUs need some time to turn on and off, a *guard time* is inserted between packets to reduce the collision probability.

Another consequence of the difference in physical distances is that the optical powers of signals originated from different ONUs are different. The signals from closer ONUs are stronger than signals from farther ONUs. As a result, once the signals are added, the average power in the resulting stream may vary from one packet to the next. The *burst mode* nature of traffic in the upstream direction creates special challenges both for ONU transmitters and OLT receivers. We will discuss some of these challenges in Chapters 8 and 9.

From the above discussion it becomes clear that a particular PON implementation involves several parameters, among which are the upstream and downstream data rates and wavelengths, the splitting ratio, and the distance. Several standards exist that determine these parameters. Currently the most popular standards are broadband PON (BPON), Ethernet PON (EPON), and Gigabit PON (GPON). Work has also started in defining 10G-based PONs. We will revisit these standards again in Chapter 12.

2.10 Summary

Fiber optics is a crucial part of the communication networks. Therefore, a full understanding of fiber optics is not possible without having a general perspective on some main networking concepts and the role that fiber optics plays within the telecommunication infrastructure. Thus, this chapter presented an overview of networking concepts with an emphasis on the role of fiber optics communications.

We started first by examining networks from a topological perspective and reviewed common topologies such as star, ring, and mesh. The topology of a network does not say anything about its physical properties. Instead, it describes the logical structure of a network.

From the perspective of rang, networks are generally divided into LANs, MANs, and WANs, in the order of physical area they cover. Although the distinction between these categories can sometimes be fuzzy, this classification is still a useful framework to keep in mind, especially because a network's management requirements oftentimes depend on the geographical area it is covering.

It is also useful to look at networks from both a "horizontal" and "vertical" perspective. Aspects such as topology or physical distance can be considered horizontal, as they deal with connectivity between various nodes in a network or how far a network is extended. However, equally important are issues related to the hierarchy within each individual node, i.e., the way data is organized and handed from higher levels down to the physical layer and vice versa. A popular model for this purpose is the seven-layer OSI model. Not all networks can be mapped directly into the OSI model; nevertheless, the OSI model gives a useful framework for understanding and categorizing networking hierarchies. In the OSI model each layer acts as an interface between the layer above and the layer below it and hides the unnecessary details of each from the other.

Another way to categorize networks is according to the way they transfer information. From this point of view, networks can be divided into packet switching and circuit switching. In packet switching networks, data is divided into separate packets, and each packet is individually routed by the network. Thus, in principle, various packets belonging to the same set of data can travel through different routes and arrive at their destinations at different times. The advantage of this approach is bandwidth efficiency. However, time-sensitive applications such as voice or video may suffer. On the other hand, in circuit switching networks a direct connection in real time is established between the source and the destination. This may give rise to inefficient use of bandwidth, because the actual amount of data that needs to be transferred may not need a full dedicated channel. However, time-sensitive applications can run more reliably because a certain level of quality of service is guaranteed by the nature of the link. A well-established example of circuit switching protocols is SONET, which is in wide usage in fiber optic communications.

Traditional networks utilize optical links at the physical layer, i.e., mainly for transfer of information from one node to another. The need to conduct all higher level networking functions in the electronic domain is a big burden and a limiting factor in achieving higher speeds. It is expected that transferring more complex tasks into the optical domain will increase the speed and efficiency of networks. Wavelength division multiplexing (WDM) technology is the primary candidate for realization of this goal. In WDM networks, the ability to dedicate various wavelengths to different and independent data streams has created exciting new opportunities, yielding to concepts such as wavelength switching and wavelength routing. WDM technology has also allowed for much more efficient use of fiber bandwidth, because each wavelength can be modulated separately without much interference from other wavelengths present in the fiber.

Fiber optics is also expanding its reach toward individual users. An exciting new development in this area involves fiber to the home (FTTH) networks. These networks provide very wide bandwidths to the end users through fiber. FTTX networks are mainly based on passive optical network (PON) architecture, where several users are connected to a central office through optical fibers passively coupled together. Traffic is bidirectional, and separate wavelengths carry the upstream and downstream data. These networks provide another example of the potentials of fiber optics and the key roles it can play in modern telecommunication infrastructure.

References

[1] J. Berthold, A. A. M. Saleh, L. Blair, and J. M. Simmons, "Optical networking: past, present, and future," Journal of Lightwave Technology, Vol. 26, pp. 1104–1118, 2008

[2] Recommendation G.Sup42, "Guide on the use of the ITU-T, Recommendations related to optical technology," ITU-T, 2008

[3] H. Masahiro and T. Abe, "Evaluating reliability of telecommunications networks using traffic path information," IEEE Transactions on Reliability, Vol. 57, pp. 283–294, 2008

[4] S. Y. Kuo, F. M. Yeh, and H. Y. Lin, "Efficient and exact reliability evaluation for networks with imperfect vertices," IEEE Transactions on Reliability," Vol. 56, pp. 288–300, 2007

[5] S. Soh and S. Rai, "An efficient cutset approach for evaluating communication network reliability with heterogeneous link-capacities," IEEE Transactions on Reliability, Vol. 54, pp. 133–144, 2005

[6] Y. Chan, E. Yim, and A. Marsh, "Exact & approximate improvement to the throughput of a stochastic network," IEEE Transactions on Reliability, Vol. 46, pp. 473–486, 1997

[7] N. F. Maxemchuk, I. Ouveysi, and M. Zukerman, "A quantitative measure for telecommunications networks topology design," IEEE/ACM Transactions on Networking, Vol 13, pp. 731– 742, 2005

[8] A. Valdar, *Understanding Telecommunications Networks*, Institution of Engineering and Technology, London, 2006

[9] T. Anttalainen, *Introduction to Telecommunications Network Engineering*, Artech House, London, 2003

[10] M. El-Sayed and J. Jaffe, "A view of telecommunications network evolution," IEEE Communications Magazine, Vol. 40, pp. 74–78, 2002

[11] IEEE802.3, 2005, available from www.ieee.org

[12] M. Huynha and P. Mohapatra, "Metropolitan Ethernet network: a move from LAN to MAN," Computer Networks, Vol. 51. pp. 4867–4894, 2007

[13] Javvin Technologies, *Network Protocols Handbook*, Javvin Technologies, 2005

[14] Recommendation X.200 [ISO/IEC 7498-1.1994] "Information technology-open systems interconnection-basic reference model: the basic model," ITU-T, 1994

[15] T Tuma, et al., "A hands-on approach to teaching the basic OSI reference model," International Journal of Electrical Engineering Education, Vol. 37, pp. 157–166, 2000

[16] L. Raman, "OSI systems and network management," IEEE Communications Magazine, Vol. 36, pp. 45–53, 1998

[17] D. Wetteroth, *OSI Reference Model for Telecommunications*, McGraw-Hill, New York, 2002

[18] W. Stallings, *Data and Computer Communications*, Prentice Hall, Englewood Cliffs, NJ, 2007

[19] R. A. Thompson, "Operational domains for circuit- and packet-switching," IEEE Journal on Selected Areas in Communications, Vol. 14, pp. 293–297, 1996

[20] GR-253, "SONET Transport Systems: common Criteria," Telecordia, 2005

[21] G.957, "Optical interfaces for equipments and systems relating to the synchronous digital hierarchy," ITU-T, 2008

[22] H. G. Perros, *Connection-Oriented Networks, SONET/SDH,ATM, MPLS, and Optical Networks*, John Wiley & Sons, Hoboken, NJ, 2005

[23] V. Alwayn, *Optical Network Design and Implementation*, Cisco Press, Indianapolis, IN, 2004

[24] U. Black, *Optical Networks, Third Generation Transport Systems*, Prentice Hall, Englewood Cliffs, NJ, 2002

[25] B. Mukherjee, *Optical WDM Networks*, Springer, New York, 2006

[26] B. Mukherjee, "WDM optical communication networks: progress and challenges," IEEE Journal on Selected Areas in Communications, Vol. 18, pp. 1810–1824, 2000

[27] J. Zhang and B. Mukherjee, "A review of fault management in WDM mesh networks: basic concepts and research challenges," IEEE Networks, Vol. 18, pp. 41–48, 2004

[28] G. Keiser, "A review of WDM technology and applications", Optical Fiber Technology, Vol. 5, pp. 3–39, 1999

[29] A. R. Chraplyvy, ''Limitations on lightwave communications imposed by optical-fiber nonlinearities,'' Journal of Lightwave Technology, Vol. 8, pp. 1548–1557, 1990

[30] G.692, "Characteristics of optical components and sub-systems," ITU-T, 1998

[31] J. M. Simmons, *Optical Network Design and Planning*, Springer, New York 2008

[32] ITU-T, "Optical Transport Networks & Technologies Standardization Work Plan," 2007. Available from http://www.itu.int/itudoc/itu-t/com15/otn

[33] G.709, "Interfaces for the Optical Transport Network (OTN)" ITU-T, 2003. Available from www.itu.int.

[34] L. Hutcheson "FTTx: current status and the future," IEEE Communications Magazine, Vol. 46, pp. 90–95, 2008

[35] M. Nakamura et al., "Proposal of networking by PON technologies for full and Ethernet services in FTTx," Journal of Lightwave Technology, Vol. 22, pp. 2631–2640, 2004

[36] G. Keiser, *FTTX Concepts and Applications*, John Wiley & Sons, Hoboken, NJ, 2006

Chapter 3
Signal Characterization and Representation

3.1 Introduction

Fiber optic systems utilize the wide bandwidth of the fiber to transfer very high data rates over long distances. As noted in Chapter 1, in order to represent information, a physical quantity must be modulated. A "signal" is the representation of information, and a physical signal is realized when a voltage, a current, or an electromagnetic wave is modulated. In fiber optics, we deal with electrical and optical signals. At the level of physical layer, a fiber optic link converts an electrical signal to an optical signal, transmits it over the fiber, and converts it back to an electrical signal at the other side.

From a design perspective, it is crucial to understand the nature of these electrical and optical representations of information. In the electrical domain, working with high-speed signals requires special attention both in design and in implementation. In the optical domain, several parameters affect signal integrity both during propagation through the fiber and at detection. Thus, because optical and electrical signals are converted back and forth, any limitation or degradation in one directly affects the other.

In this chapter, we focus on the characterization of high-speed optical and electrical signals, both in time domain and in frequency domain. We first start by reviewing some mathematical foundations, including Fourier transform and its properties. Fourier transform is the main tool for frequency domain analysis and provides valuable insight into issues such as modulation and multiplexing. After that we proceed to characterization of physical waveforms, that is, electrical and optical signals that transfer the information across physical media.

3.2 Signal analysis

A signal can be represented as a waveform which varies with time. Mathematically, a signal is represented by a function, $x(t)$. However, certain important features of a signal such as its bandwidth and spectral shape are not immediately obvious from a time domain representation. Frequency domain techniques provide powerful tools which make these hidden features much easier to examine and analyze. From a mathematical perspective, frequency domain analysis is based on Fourier transforms. Therefore, we start this section by a review of the Fourier transform and some of its important properties.

M. Azadeh, *Fiber Optics Engineering*, Optical Networks,
DOI 10.1007/978-1-4419-0304-4_3,

3.2.1 Fourier transform

The Fourier transform of a time signal, $x(t)$, is defined as [1]

$$X(f) = \int_{-\infty}^{\infty} x(t)\exp(-j2\pi tf)dt \tag{3.1}$$

where $X(f)$ is the *frequency spectrum* of $x(t)$. The inverse Fourier transform is defined as

$$x(t) = \int_{-\infty}^{\infty} X(f)\exp(j2\pi tf)df \tag{3.2}$$

Therefore, $x(t)$ and $X(f)$ form a Fourier pair, representing the same signal in the time and frequency domains. We use the ⇔ notation to show this, thus, we can say $x(t) \Leftrightarrow X(f)$. In general, $X(f)$ is a complex quantity, which means it has both amplitude and phase. However, Fourier transforms of real functions have spectral symmetry:

$$X(-f) = X^*(f) \tag{3.3}$$

where the * represents complex conjugate. Moreover, often we are interested not in the phase information in $X(f)$ but only in the *energy spectral density* of the signal, $S(f)$, which is defined as

$$S(f) = |X(f)|^2 \tag{3.4}$$

A direct consequence of Eqs. (3.3) and (3.4) is that the energy spectral density of real signals is an even function, i.e., $S(-f) = S(f)$. A number of results with important physical interpretations follow from these definitions.

• **Rayleigh/Parseval energy theorem:** In time domain, the total energy of a signal can be obtained by integrating the signal over time. Likewise, in frequency domain, the energy can be obtained by integrating the spectrum over all frequencies. The energy theorem states that these two quantities are equal:

$$\int_{-\infty}^{\infty} |X(f)|^2 df = \int_{-\infty}^{\infty} |x(t)|^2 dt \tag{3.5}$$

• **Multiplication and convolution:** One of the results of Fourier transform is that multiplication of two signals in time domain is equivalent to convolution in frequency domain (and vice versa). In other words, xy and $X*Y$, or $x*y$ and XY, form Fourier pairs:

$$\begin{aligned} x(t)y(t) &\Leftrightarrow X(f)*Y(f) \\ x(t)*y(t) &\Leftrightarrow X(f)Y(f) \end{aligned} \tag{3.6}$$

Here the operator "*" (not to be confused with complex conjugate) defines the convolution of two signals:

$$X(f)*Y(f) = \int_{-\infty}^{\infty} X(\tau)Y(f-\tau)d\tau \tag{3.7}$$

• **Signal scaling:** Changing the time scale of a signal is equivalent to a reciprocal change of scale in frequency domain (and vica versa):

$$x(at) \Leftrightarrow \frac{1}{|a|}X\left(\frac{f}{a}\right) \tag{3.8}$$

In the next section we will discuss the physical implications of these results with regards to physical signals.

3.2.2 Fourier analysis and signal representation

As noted before, Fourier transform is a representation of a signal in frequency domain. The frequency contents of a signal are important in how linear systems interact with the signals. For example, if $y(t)$ is the response of a linear system to an input signal $x(t)$, we have

$$Y(f) = H(f)X(f) \tag{3.9}$$

where $H(f)$ is the *transfer function* of the system obtained by taking the Fourier transform of the impulse response of the system. This is a direct result of Eqs. (3.6), because the response of a system in time domain can be obtained by convolution of the input signal and the system's impulse response.

Another interesting insight can be gained from Eq. (3.8). Essentially, this equation states that localizing a signal in time inevitably results in increasing the bandwidth of the signal. On the other hand, localizing a signal in frequency domain (i.e., band-limiting it) increases its time span. This is known as *Fourier uncertainty principle* [2]. The extreme example is a pure harmonic function whose time profile extends to infinity, but whose frequency spectrum is a delta function with an infinitely small span at the harmonic frequency. The other extreme case is a delta function in time, which is completely localized temporally, but whose frequency components extend to infinity. As an example of the application of this principle, ultrashort pulse shaping techniques are usually based on manipulating the spectrum of signals in frequency domain [3,4].

Fourier analysis is also very useful in providing insights about standard communication techniques such as modulation and frequency multiplexing. As an example, let us assume we need to transmit several *baseband* signals over a single channel. Let us assume these signals are given by $a_1(t)$ to $a_n(t)$ in time domain and $A_1(f)$ to $A_n(f)$ in frequency domain. We face two problems in transmission of these signals. First, many useful communication channels (including optical fibers) are not efficient at low frequencies, and therefore they cannot be used. Second, because the frequency spectrum of these signals overlap, we cannot transmit them over any channel simultaneously.

An efficient solution to both these problems is *modulation* and *frequency multiplexing*. If $a_1(t)$ is multiplied by a harmonic function such as $\cos(2\pi f_1 t)$, the result is the shifting of the frequency profile $A_1(f)$ to the *carrier frequency* f_1. This is a consequence of Eqs. (3.6) and (3.7). As mentioned before, the Fourier transform of a harmonic wave is a delta function. Thus, modulation of $a_1(t)$ by $\cos(2\pi f_1 t)$ results in the convolution of $A_1(f)$ with a delta function at f_1. This, in turn, converts $A_1(f)$ to a *bandpass* signal centered at f_1. Therefore, efficient bandpass channels like free space, waveguides, or optical fibers can now be used to transmit $a_1(t)$.

Of course modulation is widely used in radio and microwave applications. By the same token, by using different carrier frequencies, we can stack $A_1(f)$ to $A_n(f)$ next to each other in the frequency domain (Fig. 3.1), and as long as the carrier frequencies are far enough from each other, no overlap occurs.[1]

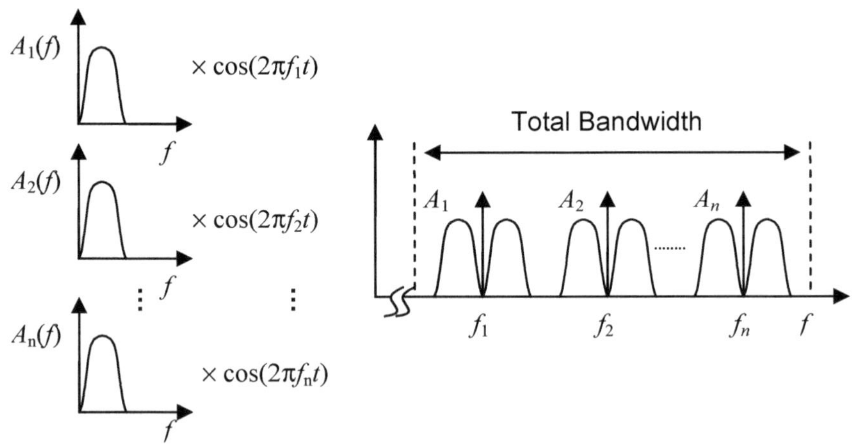

Fig. 3.1. Spectrum shifting and signal aggregation through frequency multiplexing

[1] Notice that the overall bandwidth requirement is twice that of the sum of each signal's bandwidth, as each baseband signal yields two sidebands after frequency shifting. Single side band modulation can improve bandwidth efficiency by transmitting only one sideband for each signal.

The same principle is at work in fiber optic communication. Light is used as an *optical carrier*, so the frequency contents of the signals are shifted to optical frequencies where fibers can be used to transfer them. Moreover, in analog fiber optic links, it is not uncommon to frequency multiplex several analog signals at different frequencies and then use the resulting composite signal to modulate an optical carrier. On the other hand, in wavelength division multiplexing (WDM), several wavelengths are used as optical carriers, each being modulated by a separate signal. So a WDM scheme is also a frequency multiplexing scheme, in which f_1 to f_n in Fig. 3.1 will be optical frequencies.

3.2.3 Digital signals, time and frequency domain representation

So far we have not made any particular assumptions about $x(t)$. However, digital signals have additional time and frequency structuring. A simple square wave can provide a good starting point for analyzing a digital signal. The Fourier transform of this function is shown in Fig. 3.2 [5]. We see that the square wave has diminishing components at harmonics of the fundamental frequency. Theoretically, the spectrum extends to infinity. This should not be surprising, because the square wave has infinitely fast rise and fall times, corresponding to infinite bandwidth in frequency domain.

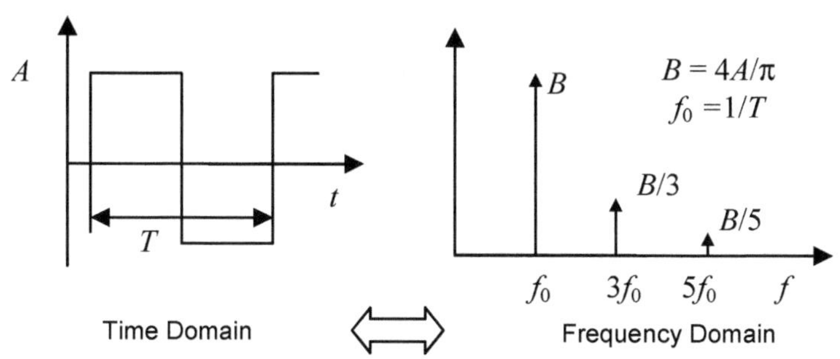

Fig. 3.2. A square wave in time and frequency domains

A simple square wave, like a simple sinusoidal wave, cannot carry information. To transfer information, the square wave must be modulated, and for this we need to somehow change one of its features. That brings us to the topic of coding.

3.2.4 Non-return-to-zero (NRZ) and pseudorandom (PRBS) codes

In fiber optic applications, the most common coding scheme for digital signals is the non-return-to-zero (NRZ) code, where a logical 1 is represented by a constant high value, and a logical 0 is represented by a constant low value [6–8]. Once converted to a physical signal, an NRZ signal is typically a voltage or an optical waveform that toggles between two physical levels, as shown in Fig. 3.3.

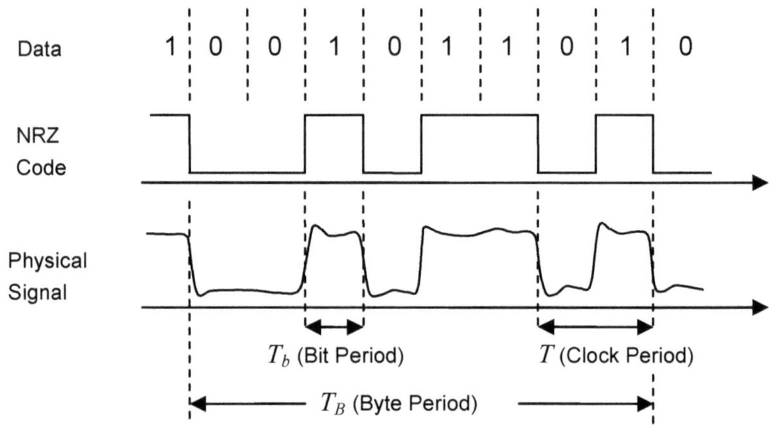

Fig. 3.3. Binary data, NRZ representation, and physical signal

Note that the physical signal deviates from an ideal NRZ code in several aspects, i.e., the physical signal has both *amplitude noise* and *phase noise* or *jitter*. A very convenient way of representing a digital signal in time domain is through an *eye diagram*. Because of their importance to optical signals, we will discuss eye diagrams separately later in this chapter. In an NRZ signal the bit period determines the bit rate. In Fig. 3.3, a sequence of 10 bits is shown: 1001011010. For instance, if the length of time shown in the figure is 10 ns, this data stream represents a data rate of 1 bits per ns, or 1 Gbps.

Another point to emphasize is the random nature of an information carrying NRZ stream. Assuming a balanced signal, each bit has a 50% probability of being 1 or 0, regardless of the past bits. In other words, a true random signal is *memoryless*. However, for test purposes, a true random signal has to be approximated by a pseudo-random bit sequence (PRBS) signal. PRBS signals are widely used for testing the performance of digital systems, including in fiber optic applications. Typically a PRBS generator consists of a series of shift registers with an XOR feedback tap. Applying a clock to the shift registers results in the generation of the pseudo-random patterns [9–11].

A PRBS pattern is characterized by its length. A PRBS pattern of a particular length has the property of including all combinations of bits in a word with that length, except for an all-zero combination. For example, a PRBS-7 pattern generates all the possible 7-bit long combinations of 1s and 0s, expect 0000000. Thus, such a pattern is 2^7-1 or 127 bits long and is therefore called 2^7-1. In other words, such a code repeats itself after 127 bits. On the other hand, a PRBS-31 pattern, which is typically the longest pattern used in practical systems, is 2, 147, 483, 647 bits long. Common PRBS pattern lengths include 7, 11, 15, 23, and 31.

Naturally, longer PRBS patterns are better approximations for true random NRZ data. A longer PRBS pattern has more *consecutive identical digits* (CIDs), and therefore it has more low-frequency contents. This raises the question of bandwidth requirements and frequency contents of random and pseudorandom NRZ signals, a topic which we turn to next.

3.2.5 Random and pseudo-random signals in frequency domain

In order to obtain the spectrum of random NRZ signals, a mathematical result can be used, which states that the energy spectral density of a function is equal to the Fourier transform of its autocorrelation [12]. Using simple arguments based on the random nature of an NRZ data stream, it can be shown that the autocorrelation of a random NRZ stream is a triangular function, as shown in Fig. 3.4a [13]. This result can also be obtained by more rigorous mathematical approaches [14]. Thus, the Fourier transform of this triangular function, which is a Sinc function, gives us the energy spectrum of the NRZ signal, *S*(*f*):

$$S(f) = A\left(\frac{\sin(\pi T_b f)}{\pi T_b f}\right)^2 \tag{3.10}$$

where T_b is bit period, and all the proportionalities are lumped into a constant *A*. This function is shown in Fig. 3.4b. Note that the spectrum has very low frequency contents, in fact theoretically all the way down to dc. Also note that the spectrum has zeros at the fundamental frequency and its harmonics. For example, a random NRZ stream at 1 Gbps has nulls at 1 GHz, 2 GHz, etc.

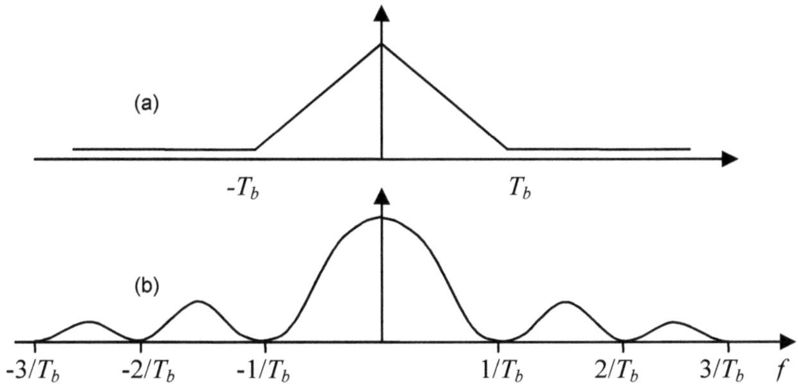

Fig. 3.4. (**a**) The autocorrelation and (**b**) the energy spectrum of a random NRZ signal

Moreover, the spectrum diminishes with the square of frequency, and in fact most of the power in the signal is concentrated in the first lobe, below the first zero. This fact can be used to set the bandwidth of circuits that handle random NRZ signals.

The energy spectrum of a PRBS signal is similar to that of a random NRZ signal, except that the Sinc becomes an envelope for a series of discrete frequency components that are spaced at $1/T_{PRBS}$ intervals, where T_{PRBS} is the time period at which the PRBS pattern repeats itself. This spectrum is shown in Fig. 3.5. As the code-length of the PRBS signal increases (i.e., from PRBS-7 to PRBS-15, to PRBS-31, etc.), the frequency spacing between the components decreases, and at the limit of infinite code-length (true random NRZ signal) the spectrum reduces to the waveform shown in Fig. 3.4b.

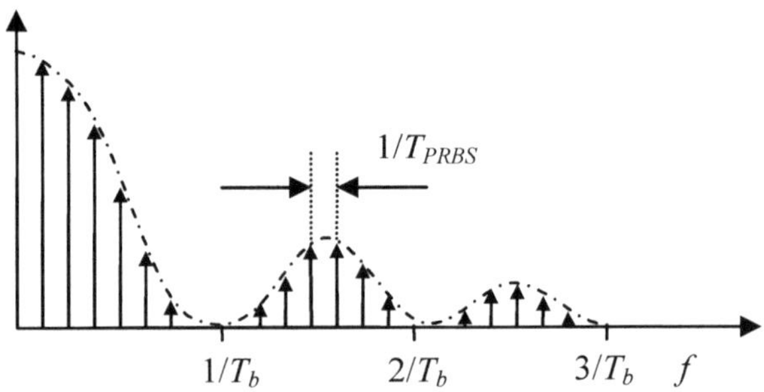

Fig. 3.5. Spectrum of a PRBS signal

With this brief mathematical introduction, we will now turn our attention to physical waveforms, starting with electrical signals.

3.3 High-speed electrical signals

Working with high-speed electrical signals is a crucial part of design process in modern fiber optic systems. In a fiber optic transceiver, usually the transmitter and the receiver have to operate at very high speeds, and therefore management of high-speed signals becomes a crucial factor in the design. Here we will review several important concepts related to characterization and handling of high-speed electrical signals.

3.3.1 Lumped and distributed circuit models

One point that needs to be clarified in advance is the definition of a "high-speed signal" [15–17]. In terms of operating frequency, circuits can be divided into two categories: *lumped* and *distributed*. Lumped circuits consist of resistors, capacitors, inductors, and electrical interconnects and obey Kirchhoff's laws of current and voltage. The critical (but often unstated) assumption is that of infinite speed for voltage and current signals. In other words, once a signal is applied to a node, it appears instantaneously on all the points connected to that node. The only source of delay in a lumped circuit is time constants created by circuit components, such as charging of capacitors, etc.

We know in reality that nothing travels instantaneously. As noted in Chapter 1, Maxwell's equations result in a wave equation for electromagnetic fields with the traveling speed of $c=1/(\varepsilon\mu)^{1/2}$. In fact, Kirchhoff's laws are approximations of Maxwell's equations. Kirchhoff's voltage law can be obtained from Eqs. (1.1) and (1.6) by setting $\mu=0$. Likewise, Kirchhoff's current law can be obtained from Eqs. (1.2) and (1.5) by setting $\varepsilon=0$ [15]. Setting permittivity and permeability to zero is equivalent to setting the speed of light to infinity. Therefore, Kirchhoff's laws hold as long as the signals travel instantaneously.

This simplifying assumption is valid when the physical size of the circuit is small compared to the wavelength of the signals involved, and therefore delays can be neglected. For instance, for a dc voltage, the wavelength is infinite, and therefore all circuits regardless of their size can be modeled as lumped elements. On the other hand, if the size of the circuit is comparable to the wavelength of the signal, not all points in the circuit will be in phase at a given point of time, and therefore delays become noticeable. In this case, lumped models are no longer valid and the circuit must be considered a distributed circuit. We can summarize the requirement for lumped circuits as

$$D \ll \frac{\lambda}{10} = \frac{c}{10nf} \tag{3.11}$$

where D is the physical dimension of the circuit, λ is the wavelength of signal, c is the speed of light in vacuum, f is the highest frequency in the signals, and n is the index of refraction of the medium which for typical circuit materials is in the range of 2–3. Although the factor 10 in this expression is somewhat arbitrary, it serves well as a rule of thumb for most cases. For example, if $f=5$ GHz and $n=2$, the right-hand side of Eq. (3.11) is 0.3 cm. This means that any circuit (or circuit element) larger than 0.3 cm behaves like a distributed element at this frequency, and therefore lumped models cannot be used for it anymore.

But what exactly is the highest frequency involved in a circuit? The situation is clear for narrowband or passband signals that are concentrated around a carrier frequency. This is usually the case for analog signals. In these cases, the carrier frequency can be used to calculate the relevant wavelength. The situation is somewhat more complicated for broadband digital signals, such as the random or PRBS NRZ streams that we discussed earlier. As we saw, the frequency contents of such signals extend beyond the nominal bit rate. However, the energy content of the higher lobes is indeed small compared to the first lobe. In fact, the higher lobes correspond to the edges of the signal, where the rate of change of the signal is fastest. So the question of highest frequency contents of the signal is indeed related to the purpose of the circuit in question. If the intention is signal fidelity, especially when it comes to maintaining rise and fall times, then perhaps higher lobes must be maintained. However, for most applications, it is sufficient to consider only the first lobe. Thus, the highest frequency in a PRBS or random NRZ signal can be approximated by the bit rate.

Once we determine that Eq. (3.11) is not satisfied for a given circuit, we need to treat it as a distributed circuit. The first elements that are affected as frequency increases are electrical interconnects, because typically they have the largest dimensions in a circuit. Therefore, we need to treat them not as simple interconnects, but as *transmission lines*. However, we should note that at still higher frequencies lumped models are not adequate for other common elements such as resistors and inductors, and distributed models must be used for them also. However, in this chapter we only cover transmission lines, because in most circuits they are the main elements that must be specifically designed and managed for signal integrity purposes.

3.3.2 Transmission lines

A transmission line is composed of two parallel, uniform conductors that are separated by a uniform dielectric [18–20]. Two conductors are needed to eliminate low-frequency cutoff, in other words, a transmission line is a baseband channel that can conduct electric energy at very low frequencies, including dc.

The important property of transmission lines is that as long as frequencies are not too high, they can support only a transverse electromagnetic (TEM) wave. TEM waves are electromagnetic waves in which the electric and magnetic fields are perpendicular to the direction of wave propagation. Figure 3.6 shows the electric and magnetic fields around a two-wire transmission line.

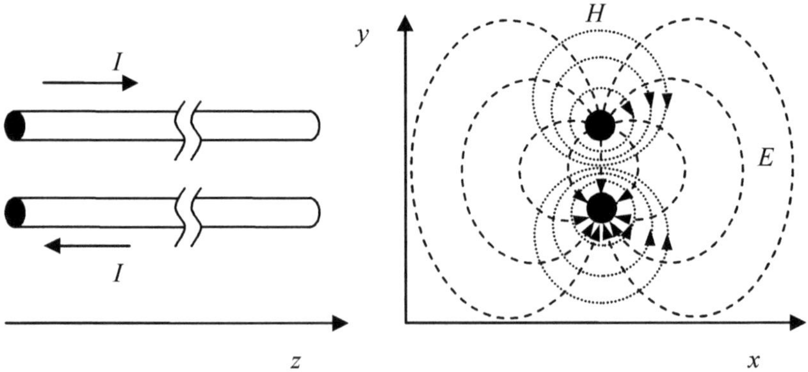

Fig. 3.6. TEM wave propagation in a transmission line

As noted in the previous section, a transmission line is inherently a distributed element and is best described through Maxwell's equations. However, it is possible to approximate a transmission line with a series of lumped circuit elements, i.e., series inductances and shunt capacitances. A small length of transmission line, dz, is then analyzed using Kirchhoff's laws which are assumed to be valid within that small length. The result is the following differential equation for voltage:

$$\frac{\partial^2 v}{\partial t^2} - \frac{1}{LC}\frac{\partial^2 v}{\partial z^2} = 0 \tag{3.12}$$

The same equation can be obtained for current. In this equation L is inductance per unit length and C is capacitance per unit length. Equation (3.12) can readily be recognized as the wave equation. The solutions are right or left traveling waveforms of arbitrary shape, with their velocity given as

$$V = \frac{1}{\sqrt{LC}} \tag{3.13}$$

If we take the approach of Maxwell's equation we reach the same conclusion, provided we neglect non-TEM solutions. The phase velocity for the TEM waves is found to be $V = 1/(\varepsilon\mu)^{1/2}$ where ε and μ are the permittivity and permeability of the dielectric that surrounds the conductors. Comparing this with Eq. (3.13), we conclude that for a transmission line, $LC=\varepsilon\mu$. Moreover, the velocity of the TEM voltage waves is the same as the speed of light in the medium.

3.3.3 Characteristic impedance

Perhaps the most important characteristic of a transmission line from a design point of view is its characteristic impedance. Following circuit theory, the impedance of a transmission line is defined as the ratio of voltage to current phasors for the voltage and current waves that propagate along a transmission line. For low-loss transmission lines it is found that [18]

$$Z = \sqrt{\frac{L}{C}} \tag{3.14}$$

where Z is impedance in Ohms. This relationship shows that impedance of a transmission line is a function of its geometry and the dielectric separating the two conductors. Moreover, for a lossless transmission line of infinite length, the impedance is a real number, which means voltage and current signals are in phase. Equation (3.14) confirms the intuition that as the shunt capacitance of a transmission line increases, its impedance drops. On the other hand, as the series inductance of the transmission line increases, its impedance also increases. Typical values for Z are 50–200 Ω.

The importance of characteristic impedance for signal integrity is the requirements of *impedance matching* [21–23]. Consider the case of a transmission line with characteristic impedance of Z_0 that is terminated by a load with impedance Z_L (Fig. 3.7). To understand what happens at the load, we can apply Kirchhoff's voltage and current laws at the termination point. If $Z_0 = Z_L$, that means the ratio of voltage to current within the transmission line is equal to the ratio of voltage to current in the load. This ensures that the load does not introduce any discontinuity as far as the traveling waves are concerned. In other words, to a wave traveling to the right, it appears as if the transmission line is continuing to infinity. Now consider the case of $Z_0 \neq Z_L$. Here the ratio of voltage to current at the two sides of the termination point is not equal, which means the load is introducing a discontinuity in the transmission line. To satisfy Kirchhoff's law, part of a right traveling wave reaching this point must reflect back toward left. The ratio of this left traveling reflected wave to the original right traveling wave is given by

$$R = \frac{Z_L - Z_0}{Z_L + Z_0} \tag{3.15}$$

R is called the *reflection coefficient* of the load. We can see that the special case of $Z_0 = Z_L$ results in a reflection coefficient of zero, which means all the energy in a right traveling wave is absorbed by the load. This situation is often desirable in high-speed design and is a requirement for signal integrity. Impedance mismatch causes part of the energy to reflect back from the load, causing degradation in the shape of the waveforms. If there is impedance mismatch at both ends of a transmission line, multiple reflections can take place and the wave may bounce back and forth between the two ends.

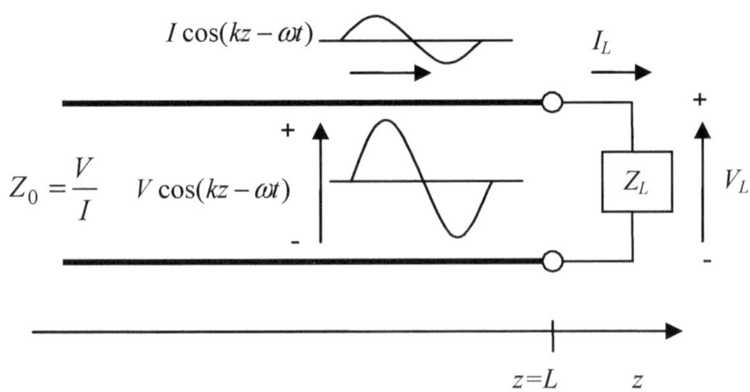

Fig. 3.7. Voltage and current waves in a transmission line and its load

An important part of handling high-speed electrical signals is impedance matching between signal sources, transmission lines, and loads. This could mean controlling the impedance of the transmission line by tweaking its geometrical parameters or adding additional elements to achieve impedance matching.

3.3.4 Microstrip and striplines

Transmission lines between electronic systems are typically coax cables. These cables provide acceptable performance up to tens of gigahertzs for short distances. However, using coax cables is not practical in printed circuit boards. As a result, microstrip and stripline structures are commonly used for implementing transmission lines [24–26]. Figure 3.8 shows the cross section of these structures.

A microstrip is a rectangular conductor that runs in parallel with a reference plane with a dielectric filling the space between the two. A stripline is a conductor that is buried inside the dielectric between two reference planes. In standard printed circuits the conductors are copper and the dielectric is usually FR-4, a material made of woven glass and epoxy-resin [27–29].

A microstrip is technically not a transmission line, because the dielectric surrounding it is not uniform. The waves above the conductor travel in air, and the waves under the conductor travel in the dielectric with a slower speed. As a result, true TEM modes cannot be supported. However, if W and h are small compared to wavelength, the waves are close to TEM modes. In practical circuits W and h are usually less than a millimeter, which makes a microstrip a good approximation of a transmission line for frequencies up to many gigahertzs.

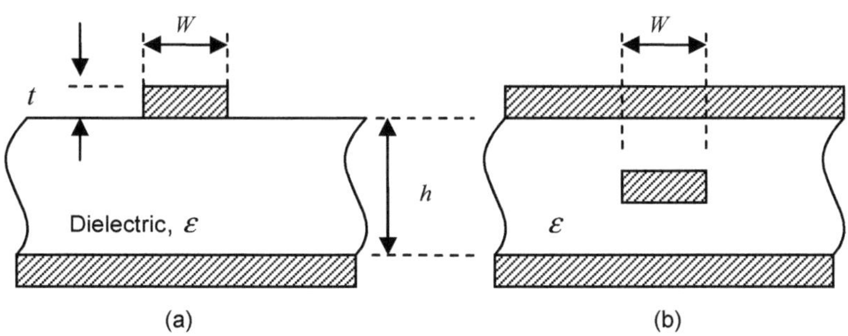

Fig. 3.8. (**a**) Microstrip and (**b**) stripline transmission lines

We can use Eq. (3.14) to estimate the characteristic impedance of the microstrip. To do this, we need to estimate the capacitance and inductance per unit length. For capacitance, we can use the standard formula for a parallel plate capacitor. However, in this case the width of the two plates are not equal, one is W (the trace) and the other is infinite (the reference plane). Therefore, we can expect the capacitance to be larger than if both plates had a width of W. As a first-order approximation, we assume, somewhat arbitrarily, that the "effective width" is $2W$. Based on this assumption we get

$$C = \varepsilon \frac{2W}{h} \tag{3.16}$$

To estimate the inductance per unit length, we can assume that the wave velocity in the medium is given by both $V = 1/(\varepsilon\mu)^{1/2}$ and Eq. (3.13). Equating these two expressions for wave velocity yields

$$\frac{1}{\sqrt{\mu\varepsilon}} = \frac{1}{\sqrt{LC}} \tag{3.17}$$

If we substitute C from Eq. (3.16) in this equation and solve for L and insert the results for L and C back into Eq. (3.14), we arrive at

$$Z = \frac{h}{2W}\sqrt{\frac{\mu}{\varepsilon}} \tag{3.18}$$

This is a simple (but approximate) formula for impedance of a microstrip line. For example, in a standard FR-4 printed circuit board, $\mu = \mu_0 = 1.25\times10^{-6}$ H/m, $\varepsilon=\varepsilon_r\varepsilon_0$, and $\varepsilon_r \cong 4.5$ (typical), and $\varepsilon_0=8.85\times10^{-12}$ F/m. If we assume $W=2h$, Eq. (3.18) yields an impedance of 44 Ω.

More precise formulas are available based on more accurate assumptions and careful approximations to numerical results. For example, a common formula is [30]

$$Z = \frac{87}{\sqrt{\varepsilon_r + 1.41}} \ln\left(\frac{5.98h}{0.8W + t}\right) \tag{3.19}$$

Note that here the thickness of the conducting trace t has also entered the equation. If we use the same numbers that we used to evaluate Eq. (3.18) with this equation, and set $t=0$, we get the value of 47 Ω, about 6% higher than the estimate of Eq. (3.18). More accurate estimates can be obtained by more complicated formulas or impedance calculating softwares.

The stripline structure shown in Fig. 3.9b can also be analyzed using the same techniques. We can repeat our simple analysis if we note that the separation between the track and reference plane is half, which makes the capacitance twice as large. Moreover, there are two reference planes, one on each side, and this also doubles the effective capacitance. Based on these observations, we expect an approximate impedance of

$$Z = \frac{h}{8W}\sqrt{\frac{\mu}{\varepsilon}} \tag{3.20}$$

for the stripline. Here too more precise estimates are available. As a reference, a more accurate formula for the impedance of stripline is [31]

$$Z = \sqrt{\frac{60}{\varepsilon_r}}\ln\left(\frac{3.8H + 1.9t}{0.8W + t}\right) \tag{3.21}$$

The microstrip and stripline line structures use a single conductor along with a reference plane. A different implementation of these structures that is very popular in high-speed connections is a *differential transmission line*. Differential transmission lines are typically used for differential signaling, a topic we will discuss later in this chapter. A differential microstrip consists of two parallel traces with a common reference plane. Similarly, a differential stripline consists of two parallel traces sandwiched between two reference planes. To utilize the advantage of differential signaling, usually the two tracks run close to each other, and this causes mutual inductive and capacitive coupling between the traces.

For these structures, just like their single-ended counterparts, it is important to characterize impedance and propagation speed. However, here the situation is slightly more complex. In principle, two modes can exist in a coupled differential structure: an odd mode and an even mode [32]. Odd modes consist of signals of opposite polarity traveling next to each other. This is indeed the mode under which differential signaling is to operate. Even modes, on the other hand, consist of signals of the same polarity. In differential signaling schemes, (ideally) the main signal is carried purely by differential modes, whereas any noise signal is in the form of even modes. Figure 3.9 is an illustration of the odd and even modes.

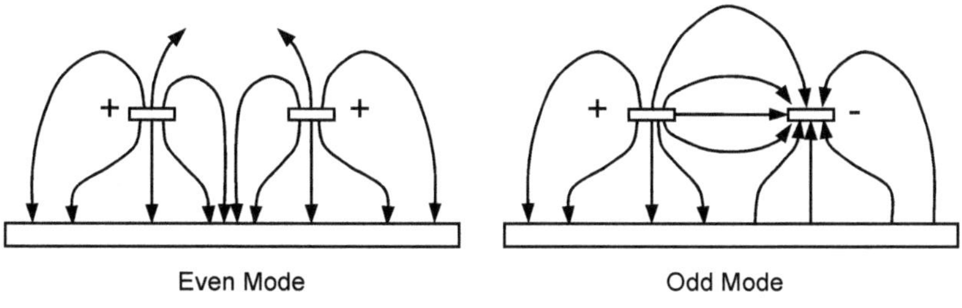

Fig. 3.9. Odd and even modes in a differential transmission line

To characterize the odd and even modes, in addition to the inductance and capacitance per unit length, we need to take into account the mutual inductance per unit length, L_m, and mutual capacitance per unit length, C_m. The odd and even mode impedance of the line can be calculated as follows [32]:

$$Z_{\text{even}} = \sqrt{\frac{L+L_m}{C-C_m}} \tag{3.22}$$

$$Z_{\text{odd}} = \sqrt{\frac{L-L_m}{C+C_m}} \tag{3.23}$$

And the velocity for each mode is

$$V_{\text{even}} = \left[(L+L_m)(C-C_m)\right]^{-1/2} \tag{3.24}$$

$$V_{\text{odd}} = \left[(L-L_m)(C+C_m)\right]^{-1/2} \tag{3.25}$$

From these equations it can be seen that when the mutual coupling terms L_m and C_m are zero, the odd and even modes become identical and these expressions reduce to Eqs. (3.13) and (3.14). But as the coupling increases, the impedance of the odd mode reduces while that of the even mode increases. Moreover, as the mutual coupling increases, the velocities of the two modes start to deviate from the single-ended transmission line. As a result, for a differential coupled transmission line we have four impedance terms:

- odd mode impedance, which is lower than a similar single-ended trace;
- differential impedance, which is twice the odd mode impedance;
- even mode impedance, which is higher than a similar signal-ended trace;
- common mode impedance, which is double the even mode impedance.

3.3.5 Differential signaling

As noted above, differential transmission lines are commonly used with differential signaling schemes. In differential signaling, two out-of-phase copies of the signal are produced and sent simultaneously. If a differential transmission line is used, each trace carries one of the signals. Differential signaling obviously has some disadvantages in terms of complexity. It requires two conductors instead of one, and it requires differential sources and loads. However, there are some important advantages that make differential signaling unavoidable in many high-speed applications.

A key advantage of differential signaling is immunity to common mode noise. In any real system there are many noise sources that could cause problems for signals. However, in most cases these noise sources affect both signals in a differential scheme in the same way, i.e., the noise source adds the same amount of noise to both signals, in terms of both amplitude and phase. On the other hand, differential receivers are intentionally designed to respond to differential modes and to reject common modes, a property known as common mode rejection ratio (CMRR). Therefore, differential receivers are generally less sensitive to noise.

Moreover, differential signaling reduces the amount of noise generated by the source. This is a consequence of the fact that a differential source generates both polarities of the signal, and thus the total current draw remains constant. On the other hand, the current draw in a single-ended source varies greatly during the generation of the signal.

Differential signaling is also beneficial in terms of radiative noise. A single-ended line that carries a high-frequency signal can act as an antenna and radiate part of that energy, causing interference problems. However, in a differential scheme usually the two transmission lines run side by side. As a result, the net current passing through the small cross section of the differential structure is zero.

Because of these advantages, differential signaling is used extensively especially in high-speed digital signals. Here we will review several widely used differential signal families.

• **CML:** One of the most common families is current mode logic (CML). Figure 3.10 illustrates a CML transmitter. The source consists of a differential pair of transistors biased with a constant current which is switched between the two outputs by the transistors. The output transistors are set up in a common emitter configuration with collectors terminated to the supply voltage through load resistors. These load resistors, which are usually 50 Ω, provide a bias path for the collectors of the transistors. As a result, CML signals are referenced to supply voltage. The resistors also provide back termination for any possible back reflection. The bias current for the differential pair is typically in the range of 10–20 mA, which results in a single-ended voltage swing of few hundred millivolts. Therefore, CML is considered a low swing family. A lower voltage swing corresponds to less supply current, less power supply switching noise, and less electromagnetic radiation.

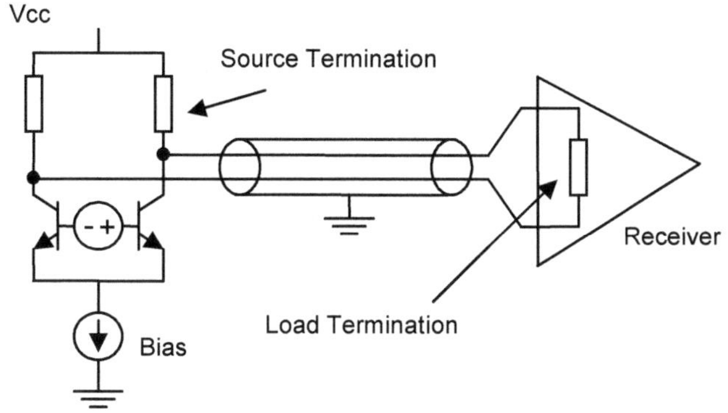

Fig. 3.10. CML signal structure

This, added to the fact that a CML receiver also has a nominal differential impedance of 100 Ω, makes the CML signal both source and load terminated. As a result, CML signals are easy to use and generally have very good signal integrity. Another advantage of a CML structure is that both the source and the receiver have their own biasing circuits. This allows for ac-coupling the two sides, which is convenient in cases where the dc operating points of the CML source and CML receiver are not the same.

• **PECL:** Another family of differential signals is the positive emitter coupled logic (PECL) and its low-voltage counterpart, LVPECL. The supply voltage for PECL is 5 V and for LVPECL is 3.3 V. A PECL structure uses a pair of emitter follower transistors as its outputs (Fig. 3.11). As a result, a PECL source has a very low output impedance (in the range of few ohms) and is capable of driving high currents. The source usually does not have a dc path for biasing the output transistors. Thus, a dc path must be provided through external termination. The termination is also important because a typical PECL receiver has relatively high-input impedance in the range of kilo-Ohms. Without proper termination, this high impedance causes reflections back toward the source.

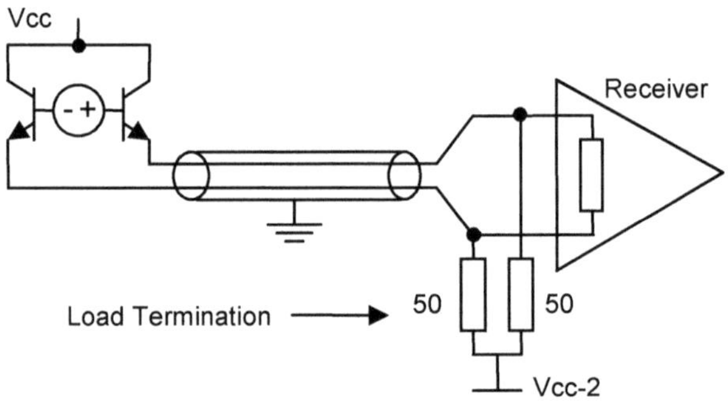

Fig. 3.11. PECL signal structure

A PECL signal must be terminated through a 50 Ω resistor to 2 V below the supply voltage. In practice, this is achieved by a variety of methods. A popular way is through the so-called Y termination, where the two 50 Ω resistors are connected to ground via a third resistor. For PECL signals (5 V supply) the values of the third resistor is around 110 Ω, for LVPECL signals (3.3 V supply) it is 50 Ω. It can be seen from Fig 3.11 that unlike CML, a PECL signal is only load terminated. This usually causes signal integrity degradation. Moreover, PECL signals have higher swings compared to CML. Typical values range from 400 to 800 mV for each signal. The common mode voltage is VCC-1.3 V. The relatively high swing of PECL makes it suitable for high-speed signal transmission over longer ranges in a printed circuit board or back plane.

• **LVDS**: Another common family is low-voltage differential signaling (LVDS). LVDS has a lower voltage swing compared to PECL and CML, and thus it has the least power consumption among the three. Typically the voltage swing is around 400 mV. The low-voltage swing allows operation under low-voltage supplies, commonly 2.5 V. LVDS is generally considered a lower speed signaling scheme compared to PECL or CML. The output stage in an LVDS source is a push–pull transistor pair. LVDS requires a load termination at the far end of the transmission line, with a typical value of 100 Ω differential.

Figure 3.12 gives a summary of logic levels for these common differential families. Also for reference, logic levels for low-voltage TTL (LVTTL) and low-voltage CMOS (LVCMOS) families are included in this figure.

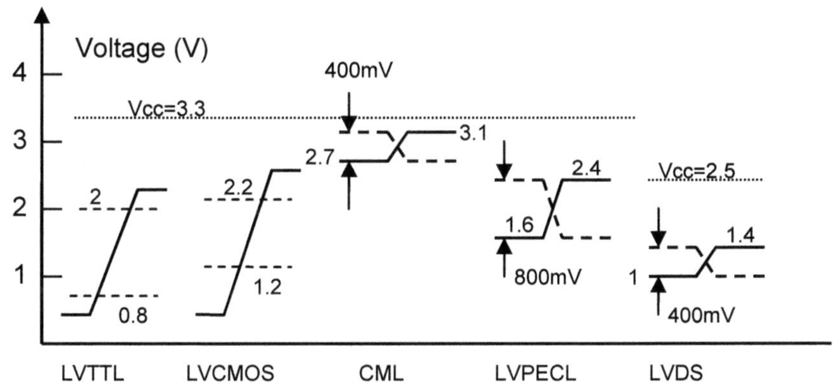

Fig. 3.12. Typical voltage levels for common logic families

Note that for logic families like TTL and CMOS, absolute logic levels are critical. However, for differential families, it is the difference between the two signals that defines the logic state. As a result, the absolute levels may vary from case to case.

3.4 Optical signals

Optical signals are the physical implementation of signals in the optical domain and obviously play the central role in fiber optics. It is important to start our discussion by defining exactly what we mean by an "optical signal." Physically, an optical signal refers to the optical power of light, just like voltage (usually) is the physical representation of an electrical signal.[2] This point, simple as it may seem, requires some clarification.

[2] This assertion holds for almost all practical fiber optic systems. Although it is possible to use the phase or frequency of light to encode information, almost all practical fiber optic systems modulate the amplitude of light and therefore can be considered AM modulation schemes.

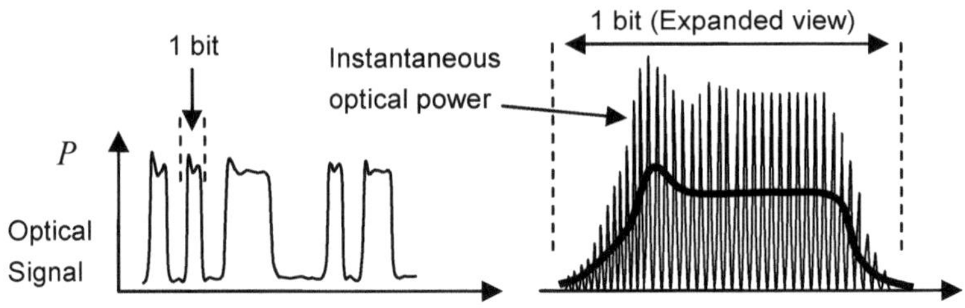

Fig. 3.13. Optical signal as the average of instantaneous optical power

The instantaneous power in an optical wave is the sum of powers in its electric and magnetic fields and is a quantity that oscillates at twice the optical frequency. The instantaneous optical power is almost never a quantity of interest in fiber optic applications, because it is oscillating at a frequency that is far beyond the bandwidth of electronic circuits. However, if we average this instantaneous field power over thousands of optical cycles, we get to a time constant of around 10^{-12} s, which starts to become meaningful for electronics. As a result, this average power (from an optical frequency perspective) becomes the instantaneous power from the viewpoint of electronic circuits. This is what we normally call the optical signal, a quantity that can be modulated in a variety of digital or analog formats (Fig. 3.13). An obvious result of dealing with power is that there is no "negative" light. Unlike electronic signals that can be bipolar, optical signals are only positive.

3.4.1 Average power

One of the most important characteristics of an optical signal is its average power. The output signal from a transmitter, the input signal to a receiver, as well as signal transmission through many active or passive parts are all characterized either directly by average power or by average power loss (or gain). Moreover, as discussed in Chapter 1, average power is a quantity that is used in link budgeting. Later in Chapter 11 when we discuss measurement methods we will see that in many cases the negative effects of a phenomenon such as dispersion are measured in terms of power penalty. The average power of a signal is usually measured in the logarithmic scale of decibel-milliwatt, defined in Eq. (1.18), and repeated here for convenience:

$$P_{\mathrm{dBm}} = 10 \log_{10}(P_{\mathrm{mW}}) \tag{3.26}$$

where P_{mW} is the optical power in milliwatt.

3.4.2 Eye diagram representation

The most common way of representing digital signals is through the eye diagram. The eye diagram is widely used in all digital signals; however, in the case of an optical signal, we may have to pay particular attention to certain characteristics that are not necessarily so important in other systems. In an eye diagram, many cycles of the signal are superimposed on top of each other, which highlights certain features otherwise not readily visible. We will discuss the process behind acquiring an optical eye diagram in Chapter 11. Here, we just focus on signal parameters regardless of how the eye diagram is generated. The main signal parameters characterized by an eye diagram are shown in Fig. 3.14.

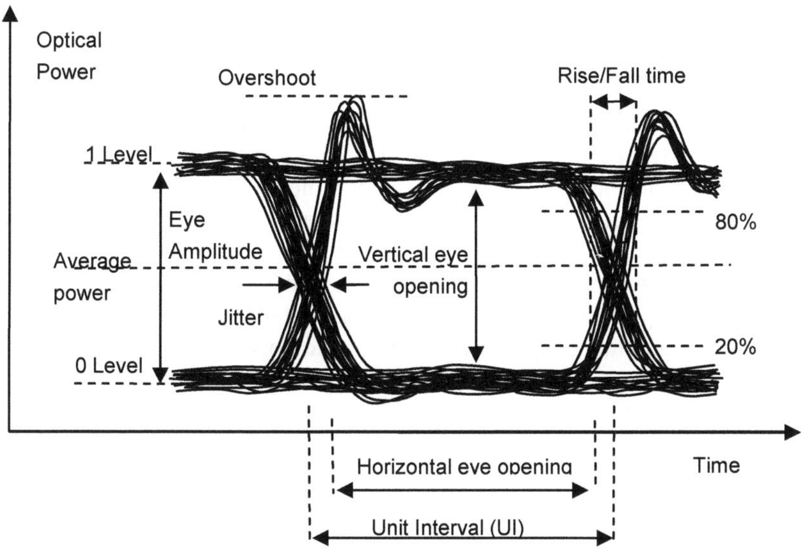

Fig. 3.14. Major parameters in an optical eye diagram

Because an eye pattern represents numerous signal cycles, it is an excellent tool for showing long pattern dependencies in a signal. As a result, it characterizes the "quality" of the signal and to what extent a particular signal deviates from an "ideal" digital signal. Note that the eye diagram for an ideal digital signal is a rectangle. Thus, an eye diagram is not only a tool from which quantitative parameters about the signal can be extracted; it provides a quick snapshot on the quality of a signal as well.

The signal parameters that can be extracted from an eye diagram can be divided into two categories: amplitude-related parameters and time-related parameters. We will discuss each of these separately.

3.4.3 Amplitude parameters

• **Average power:** The eye pattern can provide an easy way of both visualizing and estimating the average optical power. Typically, the average power corresponds to the middle of the eye in terms of the amplitude, where the rising and falling edges cross. The average power is related to the one and zero state power levels as follows:

$$P_{\mathrm{avg}} = \frac{P_0 + P_1}{2} \tag{3.27}$$

The eye diagram is a useful tool to visualize the average power with respect to other features of signal. However, for accurate measurements of average power an eye pattern cannot be used, and optical power meters with accurate low bandwidth broad area detectors should be used.

• **Logical one and zero levels (P_1 and P_0):** An eye diagram can be used to estimate the power level for the zero or one state in an optical signal. Because these power levels cannot be measured with a power meter, the eye diagram is a very useful tool for this purpose. Moreover, the thickness of the one or zero state is an indication of amplitude noise. The noise may originate in the electrical signals that drive the optical source, or they can be generated by the optical source itself. However, it is important to note that the instrument used to measure the eye pattern (usually a sampling scope) always adds some noise to the signal, and the resulting eye pattern includes noise from all the sources, including the equipment. We will come back to this issue in Chapter 11.

• **Eye amplitude:** The eye amplitude, also known as optical modulation amplitude (OMA), is the difference between the one and zero states. Since an eye pattern represents a random digital signal, it includes consecutive 1s and consecutive 0 bits too. Physically, this allows the signal to reach its steady state levels either at 1 level or at 0 level. Eye amplitude is then the difference between the steady-state one and steady-state zero.

• **Vertical eye opening:** This is a parameter that is directly related to the eye amplitude. However, it takes into account the effects of inter-symbol interference and noise. Inter-symbol interference is the inability of a signal to reach its final steady state during a bit. For example, in a 010 sequence, the signal may not have enough time to settle to the 1 state during a single bit period. The result is then vertical eye closure. Vertical eye opening is a critical parameter when it comes to detection of the signal, because it is a measure of the signal to noise ratio (SNR). With decreasing eye opening, the detection of the logical state of the signal becomes more difficult and eventually impossible. This situation corresponds to loss of signal due to insufficient SNR.

• **Overshoot:** Anytime a transition from a low to high (or high to low) takes place, overshoot (or undershoot) may occur. In optical signals, however, overshoot is much more prevalent. One reason behind this is the simple fact that optical signals are always positive. Therefore, even if the electrical signal that drives the optical signal has both overshoot and undershoot, the optical signal reflects the overshoot much more clearly. Another reason which is particularly relevant to optical signals generated by direct modulated lasers has to do with the nonlinear nature of semiconductor lasers which reveals itself in the form of large relaxation oscillations. In general, the overshoot in an optical eye can be a result of a combination of relaxation oscillation and imperfect driving electrical signals. We will discuss this phenomenon in more detail in Chapter 4.

• **Extinction ratio (ER):** This is a parameter that represents the depth of modulation in an optical signal. By definition, it is the ratio of the power in one state to the power in zero state, and like other power parameters, it is commonly expressed in decibels:

$$ER_{\text{dB}} = 10\log_{10}\left(\frac{P_1}{P_0}\right) \tag{3.28}$$

where P_1/P_0 is the linear extinction ratio. In principle, the optical power in the zero stated is "wasted," because only the difference between the one and zero states carries information. Therefore, ideally, we would like the zero state to be completely dark. This requires the optical source to be turned off completely during the zero bits. This may be done at low bit rates. However, at higher bit rates turning the source completely off causes unwanted effects such as turn-on delay and frequency chirping. As a result a very high extinction ratio cannot be achieved, unless an external modulator is utilized. Common values of extinction ratio may be in the range of 8–12 dB.

Combining Eqs. (3.27) and (3.28) for the zero and one power levels we obtain

$$P_1 = \frac{2P_{\text{avg}}ER}{1+ER} \tag{3.29}$$

$$P_0 = \frac{2P_{\text{avg}}}{1+ER} \tag{3.30}$$

where ER is the linear extinction ratio defined as P_1/P_0. Subtracting Eq. (3.30) from Eq. (3.29) yields an expression for eye amplitude in terms of average power and extinction ratio:

$$P_m = \frac{2P_{\text{avg}}(ER-1)}{1+ER} \tag{3.31}$$

3.4.4 Time parameters

• **Unit interval (UI):** The width of an optical eye corresponds to a single bit within the data stream. For example, in a 1.25 Gbp signal, the width of the eye is 800 ps. Sometimes this width is expressed explicitly in units of time (ps). But in many cases it is also convenient to simply refer to it as unit interval (UI). Other time parameters can then be expressed in this normalized unit, which is useful because it provides information about the percentage of the eye width that the particular parameter in question occupies.

• **Rise and fall time (t_r, t_f):** Rise and fall times define the time it takes for the signal to transition from a low to high state or vice versa. Because it is hard to measure (and even define) the exact moment the signal reaches its steady-state power level, it is common to define t_r and t_f in terms of 20 and 80% levels. These numbers are somewhat arbitrary, for instance, one can also use 10 and 90%. However, if these thresholds get too close to the zero and one states the probability of inaccurate measurements increase, because close to the zero and one states measurements may be hindered by the zero and one noise levels.

Notice that rise and fall times are the times the signal is changing the fastest. Consequently, they correspond to the highest frequency components present in the signal. Therefore, it would be very useful to estimate the bandwidth of the signal based on the rise and fall times. The exact bandwidth depends on how we model the rise and fall times. For example, a triangular shaped rise time has different frequency contents compared to an exponential shape [32]. For triangular waveforms a useful rule of thumb is[3]

$$BW = \frac{1}{\tau} \tag{3.32}$$

Here BW is in Hz and τ is the rise or fall time in seconds. For instance, a rise time of 200 ps corresponds to a bandwidth of 5 GHz.

The fact that rise and fall times include the highest frequency components of a signal implies that any time the signal goes through the equivalent of a low-pass filter (either in electrical or optical domain) the rise and fall times will increase, and usually this results in the closure of the optical eye. However, sometimes these edge rates are intentionally slowed down. For example, it is sometimes possible to trade off fast rise and fall times with the amount of noise in the zero and one states, especially in the electrical domain. Faster edge rates cause more overshoot and more ringing, which means it takes longer for the signal to reach steady state. This in turn causes vertical eye closure. In frequency domain, making the

---

[3] A relationship commonly given in literature is $BW=0.35/\tau$. This equation is derived based on the 3 dB bandwidth of a single pole RC-type network. Also, τ is defined based on 10–90% levels. However, in optical signal 20 and 80% levels are more commonly used, and the waveforms are more triangular like than exponential.

rise and fall times faster increases the signal bandwidth, and therefore the parasitic effects become more critical. In such cases slowing down the rise and fall times can in fact help the quality of the eye pattern.

• **Jitter:** One of the most critical parameters in an eye pattern is phase noise, or jitter. Ideally, all transitions should take place in exactly the same phase relationship with the signal. In reality, the transition time varies from one transition to the next, and this causes a distribution of transitions edges in time. Jitter is the deviations of the transition times from the ideal (or expected) point.

Jitter can be divided into two kinds: random and deterministic. *Random jitter*, as the name suggests, is due to random noise sources that are always present in all systems. From a probability point of view, random jitter follows a Gaussian distribution, and consequently it is unbounded. As a result, it is common to use statistical parameters to characterize it. A Gaussian distribution is completely defined by its mean and standard deviation. Therefore, random jitter can be characterized by its mean and standard deviation. The mean value is simply the relative point where the center of distribution happens to be. In eye pattern measurements, we can set this point to t=0 or the starting point of unit interval (Fig. 3.15).

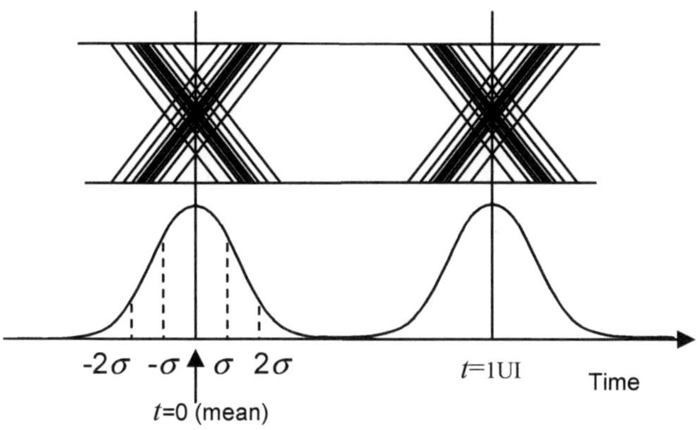

Fig. 3.15. Characterization of jitter in an eye diagram by Gaussian statistics

The standard deviation of the distribution, oftentimes called the rms value of jitter, is the amount of time that extends one sigma from the center on each side. Based on standard Gaussian statistics, this time interval includes 68% of the total number of samples in the distribution. Likewise, moving two sigma away from the center, around 95% of the samples will be included.

It is oftentimes helpful to have an estimate of *total jitter*. However, because a Gaussian distribution is inherently unbounded, to establish total jitter we must specify the percentage of samples that we wish to include within the total jitter time interval. Note that occurrence of a transition represents a possibility that the corresponding bit is sampled incorrectly. Naturally, then, we can connect the total number of samples we wish to include in the interval with a target bit error rate (BER). Thus, to obtain a bit error rate of 10^{-12} (which is a common target rate in many fiber optic applications), we would like to capture $(1-10^{-12})\times 100\%$ of samples. In a Gaussian distribution, this means seven sigma away from the center on each side, or a total of 14 sigma. Therefore, the total jitter for a BER of 10^{-12} is 14 times the rms value of jitter. In the next section we will further elaborate on the relationship between an eye diagram and BER.

Deterministic jitter is caused by factors such as pattern dependence, inadequate bandwidth, crosstalk, and pulse width distortion. Unlike random jitter, deterministic jitter is bounded, because it is caused by a non-random source. In cases where deterministic jitter exists, the total jitter will be the sum of deterministic and random jitters. Therefore

$$J_{\text{total}} = J_{\text{det}} + KJ_{\text{rms}} \tag{3.33}$$

where the constant K is a function of the desired BER. For instance, for a BER of 10^{-12}, K=14. Because of the various factors that may cause jitter, separation of random and deterministic jitter may not always be easy. In practice, a model consisting of both random and deterministic jitter with several free parameters is assumed. The parameters can then be tweaked until an acceptable fit with the actual distribution of the jitter is obtained.

- **Horizontal eye opening (W_o):** Subtracting the jitter from the whole length of the bit yields the horizontal eye opening. This is the effective "jitter-free" width of the eye. In an ideal jitter-free eye pattern, the horizontal eye opening would be equal to the unit interval. However, in reality the horizontal eye opening is always smaller.

3.4.5 Eye pattern and bathtub curves

An optical signal should ultimately be detected by an optical receiver, and the receiver must convert the sequence of optical bits to a sequence of electrical bits. To do this, the receiver must make a decision: it should sample the optical eye at a particular point in time and at a particular threshold. However, not all points in the eye diagram are equally suited for sampling. In order to minimize the probability of error, the detector must sample the eye almost in the middle, in terms of both time and amplitude. From a time perspective, this is a point farthest from transition edges. From an amplitude point of view, this is the point farthest from noise on both zero and one levels.

Obviously, as the sampling point is moved away from this optimal point the probability of error increases, and therefore BER increases. We can therefore think of an eye pattern as a set of equi-BER contours. As we get closer to the edges, the BER increases, until it reaches a maximum of 0.5 at the crossing points and at the one and zero levels. However, as the sampling point is moved away from the edges, the error probability decreases exponentially.

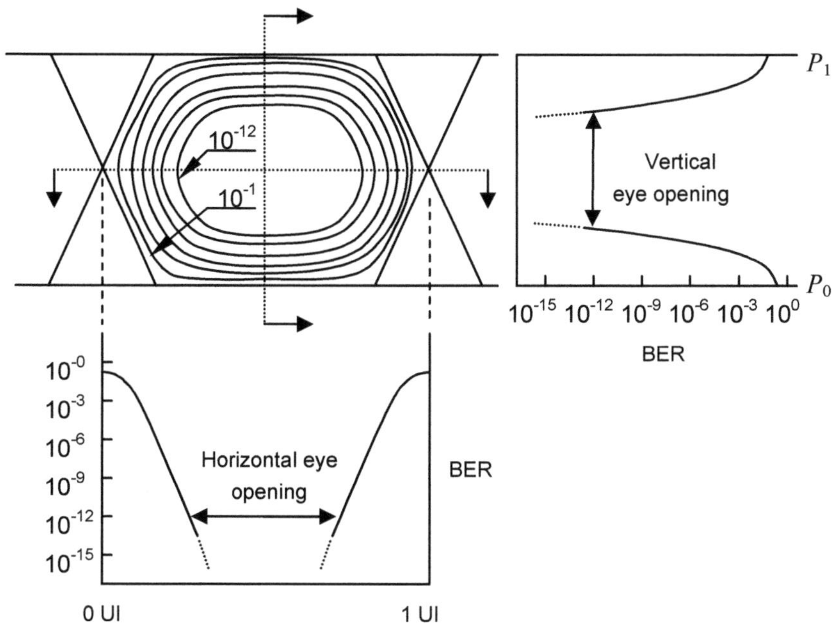

Fig. 3.16. Eye diagram from a bit-error-rate perspective

Figure 3.16 illustrates the equi-BER contours in an eye pattern as well as the BER curves in time and amplitude at the midpoint crossing sections. Usually these BER curves are also called BER bathtub curves. Figure 3.16 also shows the vertical and horizontal eye openings. We see that in fact, speaking of "eye opening" is meaningful only at a specified BER. For example, Fig. 3.16 shows the eye opening at a BER of 10^{-12}. Eye opening at lower BERs can be estimated by extrapolating the bathtub curves, assuming there are no *error floors*. In practice, BERs much lower than 10^{-12} become less meaningful because it is impractical to measure such a BER.

Figure 3.16 provides a useful relationship between the eye opening and jitter at a particular BER:

$$W_O(BER) = UI - J_{\text{total}}(BER) \tag{3.34}$$

Essentially, this equation states that both eye opening W_O and total jitter J_{total} are a function of BER, and at any particular BER, they add up to the unit interval.

Note that Fig. 3.16 depicts a situation where both jitter and amplitude noise are assumed to be purely random. In cases where deterministic noise is present, the bathtub curves and the equi-BER contours within the eye may have more complex shapes. For example, the bathtub curves may be skewed, and thus the optimal sampling point may not be in the middle. This effect is sometimes utilized in digital receivers in order to optimize the performance of the link.

3.5 Spectral characteristics of optical signals

So far we have been discussing the optical signals in time domain. However, it is equally important to pay attention to the spectral characteristics of optical signals. These characteristics are particularly critical in determining the behavior of the optical signal as it propagates in the optical fiber.

As noted before, an optical signal is the result of modulation of an optical carrier. When talking about the spectrum of an optical signal, we can distinguish between two domains: the spectral characteristics of the modulating signal and the spectrum of the carrier itself. For high-speed applications, the spectral characteristics of the modulating signal can extend up to tens of gigahertzs. On the other hand, the *optical spectrum* of the carrier is at much higher frequencies, around 10^{14} Hz. In this section, we are specifically interested in the optical spectrum of the signal, i.e., those that are inherently related to the nature of light.

3.5.1 Single-mode signals

We noted in the beginning of this chapter that the spectrum of a pure harmonic wave is a delta function. This picture holds for optical frequencies too: a "pure" optical signal has only a single frequency (wavelength). In other words, a pure harmonic waveform has *infinite coherence*, i.e., it always remains perfectly in phase with itself over time.

Physically, the output light from a laser is the closest we can get to such an ideal waveform. However, as a result of light generation processes, lasers always produce not a single wavelength, but a range of wavelengths. Moreover, because of inherent noise processes, their light output has a finite non-zero spectral width. Therefore, optical signals always have a *spectral linewidth* and a finite *coherence time*. More coherent sources generate signals with narrower spectral linewidth, and the narrower the linewidth, the closer the signal will be to an ideal harmonic.

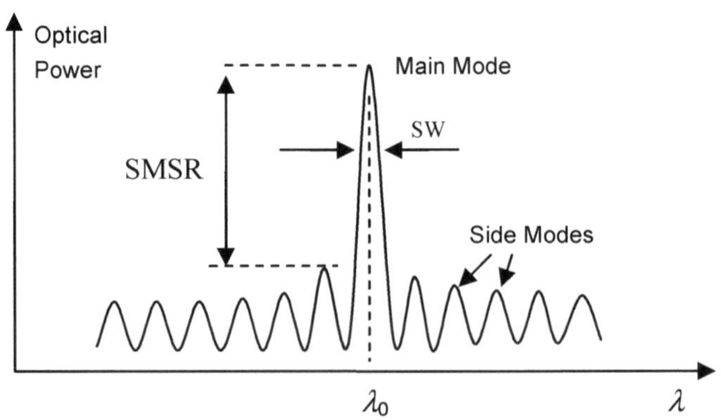

Fig. 3.17. Main parameters of an optical signal in frequency domain

In fiber optics, the most coherent optical signals are generated by single-mode diode lasers. Figure 3.17 shows the optical spectrum of such a signal. Note that the spectrum is plotted against wavelength and not frequency. As we mentioned in Chapter 1, this is because at optical frequencies, it is conceptually easier to work with wavelength instead of frequency. It can be seen that the spectrum consists of several modes that are spaced more or less equally. However, there is only one *dominant mode*. We will discuss the reasons behind this shape in the next chapter. However, here we just concentrate on the main spectral features of the signal.

• **Center wavelength (λ_0):** The main parameter in an optical signal is the center or peak wavelength, where the spectrum is maximum. This is a critical property of the signal because it determines how the signal will travel in an optical fiber. As we will see in Chapter 5, in fiber optic links mainly three wavelength windows are used: 850, 1310, and 1550 nm.

• **Spectral width (SW):** The width of the dominant mode in the spectrum is a measure of coherence for the optical signal and is referred to as spectral width. Because the shape of the peak is not rectangular, one must be more specific about the power level at which the width is measured. To avoid ambiguity, usually the width is defined at 20 dB below the peak power. The spectral width can be specified either in units of length or units of frequency. To convert between the two units, the equation $f=c/\lambda$ can be differentiated with respect to λ, and assuming small bandwidth (which is a valid assumption for coherent sources) the following relationship is obtained:

$$\Delta f = -\frac{c}{\lambda^2}\Delta\lambda \qquad (3.35)$$

For example, a typical single-mode semiconductor laser may have a linewidth of 1 nm. Based on Eq. (3.35), the linewidth of this laser is 117 GHz. This may seem like a huge bandwidth; however, we should remember that compared to the optical frequency, this is indeed a very narrow band. As a side note, it is worth noting that actively stabilized single-mode lasers with sub-hertz linewidths have been demonstrated [33–34].

Earlier we mentioned that the coherence time of a signal is a measure of the coherence of the source. The coherence time t_c may be approximated by the inverse of the bandwidth:

$$t_c \approx \frac{1}{\Delta f} \qquad (3.36)$$

Closely related to coherence time is the concept of *coherence length*, which is the distance the optical signal can travel before it loses its coherence with itself. The coherence length d_c is simply related to coherence time:

$$d_c = ct_c \tag{3.37}$$

Coherence time and coherence length become important when it comes to considering the effects of the interaction of an optical signal with itself. For example, an optical signal may be reflected from a facet and travel back toward the optical source. If the round trip distance from the source to the reflection point is less than the coherence length, the optical signal will interact with itself coherently, i.e., the two waves must be added together as vector fields, and interference patterns may result. On the other hand, if the round trip distance is greater than coherence length, the two signals will interact incoherently with no interference.

- **Side mode suppression ratio (SMSR):** The ratio of peak power in the main mode to the peak power in the next dominant mode is called side mode suppression ratio or SMSR. Typically, SMSR is expressed in decibels. An optical signal from a typical single-mode laser may have an SMSR of more than 40 dB. The higher the SMSR, the closer the optical signal is to an ideal single wavelength waveform. As we shall see in Chapter 5, signal degradation due to dispersion is directly related to spectral characteristics such as SW and SMSR.

3.5.2 Multimode signals

Not all optical signals are single mode. If special care is not taken to suppress other modes, a laser will generally produce an output signal with more than one dominant mode [35]. Therefore, in a multimode signal the optical power is divided between several modes. As a result, the effective spectral width of the signal is much higher, and therefore the signal is much more prone to dispersion.

- **Center wavelength (λ_0):** For a multimode optical signal, the center frequency is defined by a weighted average over the modes:

$$\lambda_0 = \frac{1}{P}\sum_{i=0}^{n} \lambda_i P_i \tag{3.38}$$

where P is the total optical power, and λ_i and P_i are the center wavelength and peak power of the ith mode.

- **rms spectral width (λ_{rms}):** For multimode signals, the root-mean-square value of the central wavelength of each of the modes can be used as a measure of the spectral width of the signal:

$$\lambda_{rms} = \left[\frac{1}{P}\sum_{i=0}^{n} P_i(\lambda_i - \lambda_0)^2\right]^{0.5} \tag{3.39}$$

The spectral width defined by Eq. (3.39) also can be used to determine the coherence time and coherence length of multimode signals. In a typical multimode signal, there may be as many as tens of simultaneous optical modes, and the overall linewidth may be in the range of 10 nm or more. As a result, multimode signals are much more prone to suffering from dispersion effects.

3.6 Summary

In communication systems, signals are the physical representation of information. From a mathematical point of view, signals can be represented either in the time domain or in the frequency domain. Both representations are important as each underlines certain features not immediately obvious in the other. The basis of frequency domain representation is Fourier analysis, which transforms a given function from time to frequency domain. Frequency domain representation is especially useful in characterizing modulation and frequency multiplexing. Modulating a baseband signal will shift its frequency contents to the carrier frequency. In fact, an optical signal is the result of modulating an optical carrier. In frequency multiplexing, many signals can be stacked next to each other in frequency domain. The optical equivalent of frequency multiplexing is WDM, where different wavelengths are used as optical carriers. Digital signals have additional time and frequency structuring. We examined random and pseudo-random NRZ signals, which are widely used in fiber optics.

After this mathematical introduction, we proceeded to physical waveforms, i.e., electrical and optical signals. We reviewed transmission lines, which are the primary means of transferring high-speed electrical signals. Of particular importance to the designer are the impedance of a transmission line and impedance matching. These are issues that, if not handled properly, can greatly deteriorate signal quality. We also considered differential transmission lines as well as differential signaling families, because most high-speed electrical signals encountered in fiber optic systems are indeed differential.

For optical waveforms, we first examined the concept of instantaneous power which is the basis for the concept of an optical signal. We then proceeded to time domain characteristics of optical signals. We divided time domain characteristics into amplitude and time. Amplitude characteristics include extinction ratio, optical amplitude, and overshoot. Time characteristics include rise and fall time and jitter.

For frequency characterizations, we distinguished between the optical and electrical frequency contents. Electrical frequency components of an optical signal are similar to other electrical signals as they cover ranges of frequency that can be handled by electronic circuits. These frequencies typically coincide with the data rate of the signal. On the other hand, the optical frequency contents are far beyond data rate frequencies. The most important optical frequency characteristics include spectral width and center wavelength, which we considered for both single-mode and multimode optical signals. The center wavelength is an important parameter that determines the loss and dispersion that the signal experiences as it propagates through the fiber. Spectral width is a measure of the coherence of the optical signal and determines to what extent a signal is close to an ideal harmonic wave.

References

[1] J. F. James, *A Student's Guide to Fourier Transforms*, 2nd Ed., Cambridge University Press, Cambridge, 2002

[2] S. Shinde and V. M. Gadre, "An uncertainty principle for real signals in the fractional Fourier transform Domain," IEEE Transactions on Signal Processing, Vol. 49, pp. 2545–2548, 2001

[3] Y. T. Dai and J. P. Yao, "Arbitrary pulse shaping based on intensity-only modulation in the frequency domain," Optics Letters, Vol. 33, pp. 390–392, 2008

[4] E. M. Kosik et al., "Complete characterization of attosecond pulses," Journal of Modern Optics, Vol. 52, pp. 361–378, 2005

[5] Y. S. Shmaliy, *Continuous Time Signals*, Springer, Dordrecht, 2006

[6] C. C. Chien and I. Lyubomirsky, "Comparison of RZ versus NRZ pulse shapes for optical duobinary transmission," Journal of Lightwave Technology, Vol. 25, pp. 2953–2958, 2007

[7] P. J. Winzer et al., "Optimum filter bandwidths for optically preamplified (N)RZ receivers," Journal of Lightwave Technology, Vol. 19, pp. 1263–1273, 2001

[8] M. I. Hayee and A. E.Wilner, "NRZ versus RZ in 10–40Gb/s dispersion managed WDM transmission systems," IEEE Photonics Technology Letters, Vol. 11, pp. 991–993, 1999

[9] H. Knapp et al., "100-Gb/s 2^7-1 and 54-Gb/s 2^{11}-1 PRBS generators in SiGe bipolar technology," IEEE Journal of Solid-State Circuits, Vol. 40, pp. 2118–2125, 2005

[10] F. Schumann and J. Bock, "Silicon bipolar IC for PRBS testing generates adjustable bit rates up to 25 Gbit/s," Electronics Letters, Vol. 33, pp. 2022–2023, 1997

[11] O. Kromat, U. Langmann, G. Hanke, and W. J. Hillery, "A 10-Gb/s silicon bipolar IC for PRBS testing," IEEE Journal of Solid-State Circuits, Vol. 33, pp. 76–85, 1998

[12] L. Cohen, "Generalization of the Wiener–Khinchin Theorem," IEEE Signal Processing Letters, Vol. 5, pp. 292–294, 1998

[13] J. Redd and C. Lyon, "Spectral Contents of NRZ test patterns," EDN, pp. 67–72, 2004

[14] L. W. Couch, *Digital and Analog Communication Systems*, 6th Ed., Prentice Hall, Englewood Cliffs, 2001

[15] T. H. Lee, *Planar Microwave Engineering*, Cambridge University Press, Cambridge, NJ, 2004

[16] L. Zhu and K. Wu, "Comparative investigation on numerical de-embedding techniques for equivalent circuit modeling of lumped and distributed microstrip circuits," IEEE Microwave and Wireless Components Letters, Vol. 12, pp. 51–53, 2002

[17] M. Feliziani and F. Maradei, "Modeling of electromagnetic fields and electrical circuits with lumped and distributed elements by the WETD method," IEEE Transactions on Magnetics, Vol. 35, pp. 1666–1669, 1999

[18] S. E. Schwarz, *Electromagnetics for Engineers*, Saunders College Publishing, Berkeley, CA 1990

[19] G. Vandenbosch, F. J. Demuynck, and A. R. Vandecapelle, "The transmission line models-Past, present, and future," International Journal of Microwave and Millimeter-Wave Computer-Aided Engineering, Vol. 3, pp. 319–325, 1993

[20] E. Bogatin, *Signal Integrity Simplified*, Prentice Hall, Englewood Cliffs, NJ, 2004

[21] M. Khalaj-Amirhosseini, "Wideband or multiband complex impedance matching using microstrip nonuniform transmission lines," Progress in Electromagnetics Research-PIER, Vol. 66, pp. 15–25, 2006

[22] G. B. Xiao and K. Yashiro, "Impedance matching for complex loads through nonuniform transmission lines," IEEE Transactions on Microwave Theory and Techniques, Vol. 50, pp. 1520–1525, 2002

[23] S. Kim, H. Jwa, and H. Chang, "Design of impedance-matching circuits with tapered transmission lines," Microwave and Optical Technology Letters, Vol. 20, pp. 403 -407, 1999

[24] T. Chiu and Y. S. Shen, "A broadband transition between microstrip and coplanar stripline," IEEE Microwave and Wireless Components Letters, Vol. 13, pp. 66–68, 2003

[25] A. J. Rainal, "Impedance and crosstalk of stripline and microstrip transmission lines," IEEE Transactions on Components, Packaging, and Manufacturing Technology-Part B, Advanced Packaging, Vol. 20, pp. 217–224, 1997

[26] D. Nghiem, J. T. Williams, and D. R. Jackson, "A general analysis of propagation along multiple-layer superconducting stripline and microstrip transmission lines," IEEE Transactions on Microwave Theory and Techniques, Vol. 39, pp. 1553–1565, 1991

[27] Y. K. Kim, "Viscoelastic Effect of FR-4 Material on Packaging Stress Development," IEEE Transactions on Advanced Packaging, Vol. 30, pp. 411–420, 2007

[28] P. Hutapea and J. L. Grenestedt, "Effect of temperature on elastic properties of woven-glass epoxy composites for printed circuit board applications," Journal of Electronic Materials, Vol. 32, pp. 221–227, 2003

[29] T. M. Wang and I. M. Daniel, "Thermoviscoelastic analysis of residual stresses and warpage in composite laminates," Journal of Composite Materials, Vol. 26, pp. 883–899, 1992

[30] IPC-2141, Controlled Impedance Circuit Boards and High-Speed Logic Design, Association Connecting Electronics Industries (IPC), 1996

[31] IPC-D-317A and IPC-2251, Design Guide for the Packaging of High Speed Electronic Circuits, IPC- Association Connecting Electronics Industries (IPC), 2003

[32] B. Young, *Digital Signal Integrity*, Prentice Hall, Englewood Cliffs, NJ, 2001

[33] J. Alnis et al., "Sub-hertz linewidth diode lasers by stabilization to vibrationally and thermally compensated ultralow-expansion glass Fabry-Perot cavities," Physical Review A, Vol. 77, Article Number 053809, 2008

[34] A. D. Ludlow et al., "Compact, thermal-noise-limited optical cavity for diode laser stabilization at 1×10^{-15}," Optics Letters, Vol. 32, pp. 641–643, 2007

[35] I. Velchev and J. Toulouse, "Fourier analysis of the spectral properties of multi-mode laser radiation," American Journal of Physics, Vol. 71, pp. 269–272, 2003

Chapter 4
Semiconductor Lasers

4.1 Introduction

In this chapter we discuss the basic principles of operation of semiconductor lasers. These devices are by far the most common optical source in fiber optic communication. Properties such as high-speed modulation capability, high efficiency, wavelengths in the infrared communication band, small size, and high reliability make these devices an indispensable part of fiber optic links. This chapter starts with the theory of light amplifiers and oscillators. Next we discuss optical amplification in semiconductors, which is the basis of semiconductor lasers. We will also introduce the rate equations, which are an essential tool in understanding the behavior of semiconductor lasers. Next we will study various properties of these lasers, both in frequency and in time domains. Finally, we will review some of the practical semiconductor devices in use.

4.2 Optical gain and optical oscillation

Any electrical engineer who has worked with a high-gain amplifier is aware of the possibility of oscillation. A high-gain amplifier oscillates if part of the output is fed back to the input with sufficient phase shift to establish a positive feedback loop. This is also the principle behind the operation of a laser. The term laser is an acronym for Light Amplification through Stimulated Emission of Radiation. Basically, a laser consists of a light amplifier combined with optical feedback. The feedback causes the amplifier to oscillate at optical frequencies, generating light as the output of the device.

The concept behind a laser is shown in Fig. 4.1. Here an optical amplifier is placed between two mirrors which provide the needed optical feedback (we will examine the mechanisms behind optical amplification later in this chapter). The light bounces back and forth between the two mirrors, and every time it passes through the amplifier it experiences some gain. An *optical cavity* formed by two reflecting surfaces is called a *Fabry–Perot* (FP) cavity [1–3]. Let us assume that the reflectances of the mirrors are given by R_1 and R_2, and the electric field profiles for the right and left traveling light waves are $A^+(z)$ and $A^-(z)$. Then the electric field itself for either waves can be written as

$$E(z,t) = A(z)e^{j(kz-\omega t)} \tag{4.1}$$

M. Azadeh, *Fiber Optics Engineering*, Optical Networks,
DOI 10.1007/978-1-4419-0304-4_4, © Springer Science+Business Media, LLC 2009

We have dropped the + and – signs from A temporarily to keep the expression general. Equation (4.1) is essentially similar to Eq. (1.8), with vector **r** in Eq. (1.8) having been substituted with z. Because the medium has gain, the z dependence of the amplitudes is exponential.[1] This exponential dependence is usually written as

$$A(z) = A(0)e^{(\Gamma g - \alpha)z} \tag{4.2}$$

where g and α are the gain and loss per unit length of the amplifier, respectively, and Γ is called the *optical confinement factor* and represents the fraction of optical power in the cavity that is subject to optical gain. Now consider a round-trip travel of the optical wave along the cavity, where it propagates to right, experiences a reflection coefficient of R_2, travels back to the left, and experiences a reflection coefficient of R_1. As shown in Fig. 4.1, in steady state, the wave must repeat itself after a round trip, requiring that

$$A(0) = A(2L) = A(0)R_1R_2\ e^{(\Gamma g - \alpha)2L} \tag{4.3}$$

This equation yields

$$\Gamma g = \alpha + \frac{1}{2L}\ln\left(\frac{1}{R_2R_1}\right) \tag{4.4}$$

Equation (4.4) defines the steady-state value of gain needed for laser oscillation. It simply states that in order to have laser oscillation the net gain of the medium must be equal to the loss of optical power due to R_1 and R_2 plus the loss due to the absorption of the amplifying medium.

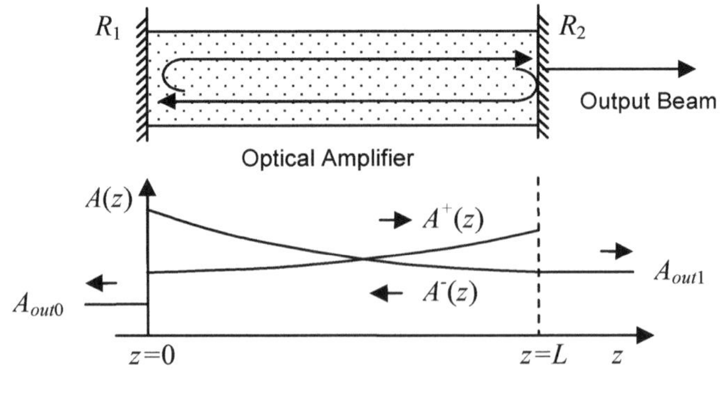

Fig. 4.1. A Fabry–Perot laser and power profiles inside the cavity

---

[1] This assumption is valid as long as the power is low enough for amplifier not to saturate.

Only after the optical gain equals the loss does the lasing action begin. As a result, as shown in Fig. 4.2, the typical light vs. current (LI) curve of a laser shows a *threshold current*, which corresponds to the point where the gain is almost equal to loss. The slope of the LI curve above threshold is called the quantum or slope efficacy of the laser, because it shows the efficiency of electrical to optical conversion. The LI curve shown in Fig. 4.2 is typical of many laser systems, including semiconductor lasers. Above threshold the optical power changes linearly with current and remains linear for reasonably low currents. However, if the current is increased sufficiently, the efficiency starts to decrease and the LI curve flattens.

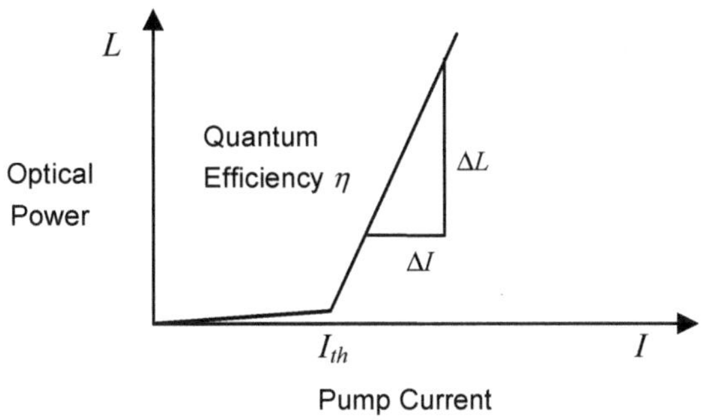

Fig. 4.2. Typical LI curve of a laser

The above description was based on power (or amplitude) of light. We can also analyze the optical wave inside the cavity from the perspective of its phase. The phase condition requires that after a round trip the optical wave should end up with the same phase. From Eq. (4.1), this condition can be expressed as

$$1 = e^{2jkL} \tag{4.5}$$

Using Euler's formula, we can convert this equation into a real cosine part and an imaginary sine part. Equation (4.5) holds if

$$2kL = 2\pi m \tag{4.6}$$

where m is an integer. Using $k=2\pi/\lambda$, $c=f\lambda$, and $c_0=nc$ (n is the index of refraction of medium and c_0 is the vacuum speed of light) we obtain

$$f_m = m\frac{c}{2nL} \tag{4.7}$$

where f_m is the frequency of light. Equation (4.7) defines a set of resonant discrete frequencies f_m that the cavity can support. This is the optical equivalent of a familiar phenomenon. For instance, in musical instruments, an oscillating string can support only certain notes, similar to the discrete frequencies given by Eq. (4.7). These frequencies are known as the *longitudinal modes* of the cavity. It follows that the frequency of the output of the laser must satisfy Eq. (4.7). A given Fabry–Perot cavity could potentially support a very large number of longitudinal modes. The actual lasing frequency is further constrained by the *spectral gain profile* of the amplifier. Any given optical amplifier can provide optical gain only in a certain range of frequencies. This means that the term g in Eq. (4.2) in reality is a function of frequency. Therefore, only those cavity modes that overlap with the spectral gain profile of the amplifier experience enough gain to start lasing. This situation is illustrated in Fig. 4.3.

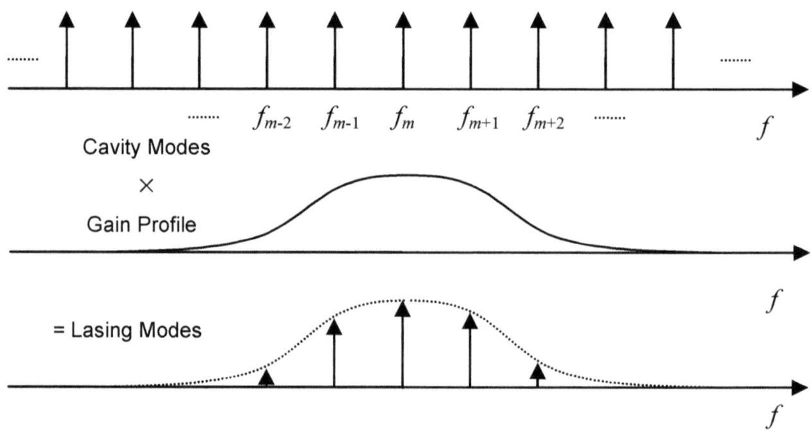

Fig. 4.3. Lasing modes in a Fabry–Perot laser

Because the spectral gain profile is usually wider than the cavity mode spacing, normally many optical modes satisfy the threshold, and the result is a *multi-longitudinal mode* laser. However, for many applications, it is desirable to have a single lasing frequency. To achieve this, further measures must be taken to increase the wavelength selectivity of the optical cavity to increase the loss for all the modes but one. We will discuss some of these methods later when we discuss *single-mode* semiconductor lasers.

4.3 Physical processes for optical amplification

So far we have not discussed how optical gain can be achieved. The phenomenon of optical gain is a result of interaction of light waves with atomic systems. Although there are ways to describe such interactions using classical theories of physics, a full and rigorous description requires quantum physics, where both energy levels in atoms and electromagnetic fields are quantized. However, for almost all practical applications, a semi-classical treatment is sufficient. In a semi-classical theory, the interaction of light and atoms is described in terms of photons of light interacting with electrons that can occupy discrete energy levels. Wave propagation (if taken into account) would be described by Maxwell's equation.

Figure 4.4 illustrates this process. The atom is defined as a system with two discrete energy levels, E_1 and E_2. These two levels define a resonant system, very much like a pendulum or a spring-mass system that can oscillate at its resonant frequency if stimulated by an input with a frequency close to resonance. The resonant frequency is defined by Planck's equation:

$$f = \frac{E_2 - E_1}{h} \tag{4.8}$$

which states that any electromagnetic wave with a frequency close to or equal to that given by Eq. (4.8) is likely to interact with the atom strongly. The electrons normally occupy the lower energy state. However, under external stimulus, they can be pushed to the higher energy level. Because this is not a normal state of affairs for the atom, it is called *population inversion.* There are three main ways of interaction between light and an atomic system, as shown in Fig. 4.4. One way is for a photon with a frequency close to resonance to be absorbed by the atom. The photon's energy is transferred to the electron in the lower energy state and causes it to jump up to the higher energy level.

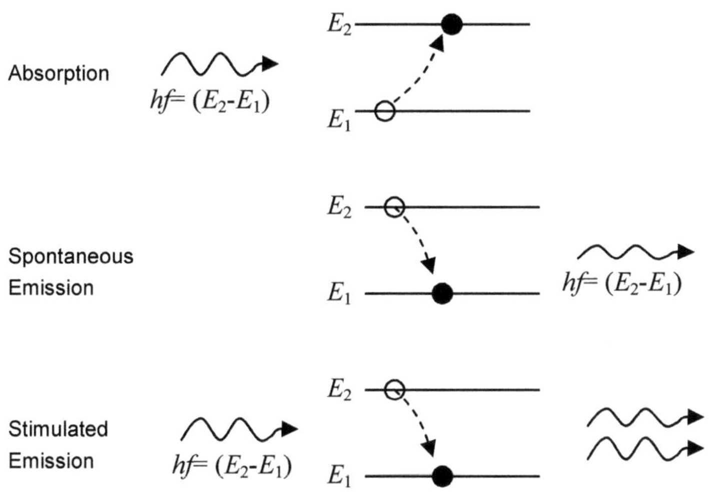

Fig. 4.4. Three main ways of interaction of light with a two-level atomic system

Note that the result in this case is an example of population inversion. Indeed, *optical pumping* is one way of achieving population inversion in many atomic systems. A different case is when an electron that has ended up in a higher energy level jumps down to the lower level spontaneously and releases the energy in the form of a photon with a frequency given by Eq. (4.8). This process is known as *spontaneous emission*. The reason this transition is called spontaneous is that the generated photon does not have a phase correlation with the rest of the electromagnetic fields present in the cavity. As such, spontaneous emission acts like "noise" with respect to the coherent fields.

For laser action the most important case is the third case known as *stimulated emission*. This occurs when a photon close to the resonant frequency causes an electron in the higher state to jump down to the lower state and release its energy in the form of a second photon. The importance of stimulated emission is that the second photon is now in phase with the stimulating photon; as a result, the process starts with one photon and ends with two photons, both at the same frequency. In other words, the energy associated with the optical field has increased, and optical amplification has been achieved.

Stimulated emission is the fundamental physical process behind light amplification. From the discussion above, it can be concluded that here are two requirements for stimulated emission: first, a state of population inversion must be induced, and second, a light wave with resonant frequency must be present to start the light amplification. We will discuss these requirements in the case of semiconductor lasers.

4.4 Optical amplification in semiconductors

Semiconductor lasers are by far the most common source in optical communication. As a result, in the rest of this chapter, we will focus on these devices. From the perspective of energy levels, a semiconductor can essentially be considered a two-level system. These energy levels are derived from Schrödinger's equation which governs the behavior of particles of mass *m* in a potential energy field *V*:

$$\frac{-\hbar^2}{2m}\nabla^2\Psi + V\Psi = i\hbar\frac{\partial\Psi}{\partial t} \tag{4.9}$$

where Ψ is the wave function which describes the probability of finding the particle in a given state and *m* is the mass of electron. The whole semiclassical theory of lasers is derived from combining this equation with Maxwell's equations. In spite of its appearance, Eq. (4.9) states a simple idea: the total energy of a particle can be expressed as the sum of kinetic and potential energies. In an individual atom, the particles are electrons and the potential fields are produced by the nucleus. Steady-state solutions of Eq. (4.9) can be obtained by assuming that Ψ can be expressed as a product of two parts, a space-dependent part and a time-dependant part. Using this assumption, Eq. (4.9) can be reduced to two ordinary differential equations for the two parts.

For individual atoms sufficiently far apart (such as atoms in a gas) Schrödinger's equation results in the familiar sharp energy levels. However, when the atoms get close together and their potential fields overlap, the sharp energy levels of isolated atoms split into *energy bands*, as shown in Fig. 4.5.

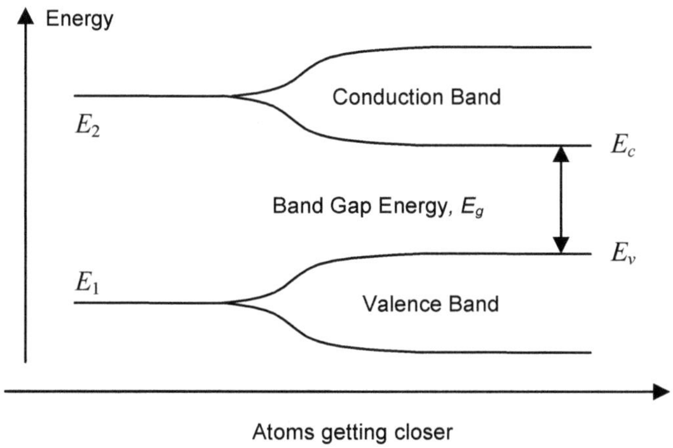

Fig. 4.5. Formation of energy bands in semiconductors

This is the case in semiconductors, where the atoms are covalently bonded in a crystal, and the confining potential of a single nucleus is substituted by the periodic potential function of the lattice. As a result, semiconductors are characterized by a *valence* and a *conduction* band. The top of the valence band E_v and the bottom of the conduction band E_c are separated by an energy gap, called the band gap energy, E_g. In a pure crystal the band gap is a forbidden energy zone, which means no electrons can exist in it.

The valence and conduction bands represent the available energy states that electrons can occupy. However, not all these states are equally likely to be filled by an electron. The probability of actually filling a state with energy E is given by the Fermi–Dirac distribution [4]:

$$f(E) = \frac{1}{1 + \exp[(E - E_f)/kT]} \tag{4.10}$$

Here E_f is the Fermi level, k is Boltzmann's constant (=1.38×10^{-23} J/K), and T is the absolute temperature. At T=0, the Fermi–Dirac distribution reduces to a step function around E_f. That is, the probability of an available state being filled below the Fermi level is 1, and above the Fermi level it is 0. Physically, this means all the electrons occupy the valence band and the conduction band is empty. Such a material behaves like an insulator. At higher temperatures, some electrons are thermally excited and fill some of the available states in the conduction band. When an electron makes a transition from the valence band to the conduction band, it leaves an empty state behind in the valence band, called a hole. Holes behave like positively charged particles and, like electrons, contribute to electric conduction. This process, essentially, yields the two-level system needed for laser action. For an intrinsic material with no impurity and no defect, the Fermi level is located almost in the middle of band gap. Note that $f(E)$ described by Eq. (4.10) approaches 0 or 1 on either side of E_f very quickly and within a few kTs. This means that for an intrinsic semiconductor, relatively few available states in the conduction band are filled.

The quantity that is ultimately important to the conduction characteristics of semiconductors and plays a critical role in the diode laser is the carrier density, which includes the electron and hole densities n and p. In an intrinsic semiconductor, *carrier generation* is the result of transition of thermally excited electrons from valence band to conduction band, and therefore, $n=p$.

The situation can be changed drastically by adding small amounts of impurities. Introduction of such impurities generally creates additional levels within the band gap. In particular, certain impurities result in the creation of *donor levels* close to the conduction band that are filled with electron. As a result, when $T\neq 0$ and even at low temperatures, electrons from this donor level fill in the conduction band in great numbers. These impurities are called *donors*, and the result is an n-type semiconductor. In contrast, some impurities create *acceptor levels* close to the valence band. These levels have plenty of empty states, which are not filled at $T=0$. However, when $T\neq 0$ and even at low temperatures, electrons from the valence band fill these acceptor levels, leaving behind great concentration of holes in the valence band. These impurities are called *acceptors*, and the result is a p-type semiconductor.

The existence of conduction and valence bands in a semiconductor provides a means of realizing a two-level system such as the one we discussed before. However, for light amplification, a situation of population inversion is needed, which can be achieved by forward biasing the pn junction. It is well known that at the interface of the p-type and the n-type materials, electrons from the p-side diffuse into the n-side and holes from the p-side diffuse into the n-side. This diffusion establishes a built-in electric field that ultimately stops the diffusion of more carriers. Applying an external forward voltage reduces the built-in electric field and more carriers start to flow. Figure 4.6 illustrates this process. It can be seen that within the transition region, population inversion is realized: the excess electrons from the n-side and the excess holes from the p-side exist in the same region. This area where population inversion is realized is called the *active region*.

Because a forward biased junction is not in the condition of thermal equilibrium a single Fermi level does not describe the carrier distribution.

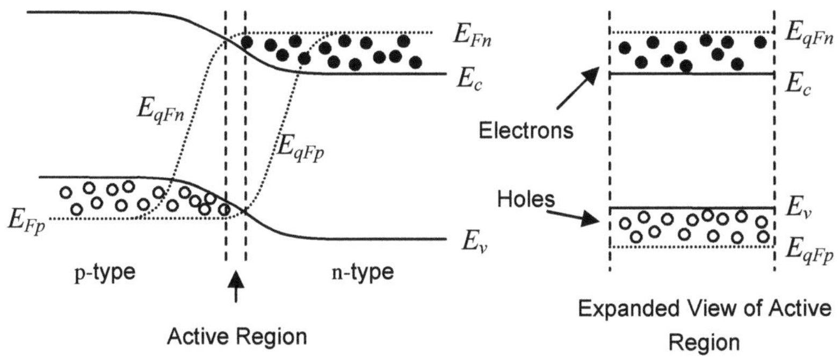

Fig. 4.6. Population inversion in a forward biased semiconductor

However, the Fermi–Dirac distribution is still applicable if a separate *quasi Fermi level* is defined for each of the two populations of electrons and holes. In Fig, 4.6 these quasi Fermi levels are represented by E_{qFn} and E_{qFp} for electrons and holes, respectively. Note that quasi Fermi levels separate in the active region, but combine in the p or n region where they essentially reduce to the ordinary conduction or valence band Fermi levels. Also note that in Fig. 4.6 the Fermi levels are *within* the conduction and valence bands. This is caused by high doping, which indeed is the case for pn junctions used in diode lasers.

Figure 4.6 shows the band diagrams of a *homo-junction.* In a homo-junction, both the p and n semiconductors are based on the same material. Although it is possible to create population inversion in a homo-junction, it is not easy to create high gain, because the carriers diffuse into the opposite region, preventing population inversion to build up. It would be nice to create some sort of barrier to the diffusion of carriers, so that a large carrier density could be built up in the active region. This goal can be achieved by sandwiching a low band gap material in between two higher bandgap materials, effectively creating a potential barrier to the diffusion of carriers. Such a structure is known as a *hetero-junction*, and the resulting lasers are far more efficient. Nevertheless, the principle of operation is the same: by forward biasing a pn junction, a condition of population inversion is created, and stimulated emission yields optical gain.

4.5 Rate equation approximation

As noted before, a rigorous treatment of interaction of light with atoms requires the coupling of Maxwell's and Schrödinger's equations. This approach was first developed by Lamb [5–8] and later on extended and elaborated by numerous others [9–16].

While such an approach can yield rigorous results, generally speaking it is complicated and involves extensive mathematical treatment. An alternative is to use a phenomenological approach usually known as rate equation approximation. In this approach, the behavior of the diode laser is modeled in terms of two independent variables: N, the density of carriers, and P, the density of photons, both in the active region. We start by considering the carrier density rate equation [17–21].

4.5.1 Carrier density rate equation

The carrier rate equation is essentially a book-keeping equation, simply stating that the rate of increase of carrier density equals the rate at which carriers enter the cavity minus the rate at which they leave the cavity:

$$\frac{dN}{dt} = R_{GEN} - R_{REC} \tag{4.11}$$

where R_{GEN} is the carrier generation rate, and R_{REC} is the carrier recombination rate, both in the active region. Note that N can represent either the electron or the hole density, as charge neutrality requires these two to be equal.

The next step is to describe R_{GEN} and R_{REC} more explicitly. In a diode laser R_{GEN} is a result of the junction current I. If the volume of active region is V, the injection current creates $I/(eV)$ electrons per second per unit volume of active region. However, not all the current goes into generating useful carriers, because some of the current will bypass the active region through other conductive paths. As a result, an internal efficiency coefficient η must be added to account for this extra leakage, yielding

$$R_{GEN} = \frac{\eta I}{eV} \tag{4.12}$$

For R_{REC} we must account for the different processes through which carriers recombine. Generally, carrier recombination can be either *radiative* or *non-radiative*. Radiative recombination can be either spontaneous or stimulated. We know from our discussion before that the stimulated recombination is responsible for coherent laser light, whereas spontaneous recombination generates incoherent light which acts as a source of phase noise in the laser light. Non-radiative recombination removes the carriers without any light generation and therefore reduces the efficiency of the laser.

As a result, R_{REC} can be written as

$$R_{REC} = R_{ST} + R_{SP} + R_{NR} \tag{4.13}$$

where R_{ST}, R_{SP}, and R_{NR} are stimulated, spontaneous, and non-radiative recombination rates. Let us examine each of these terms more carefully. The first term, R_{ST}, represents the rate of carrier recombination due to stimulated emission. Therefore, R_{ST} is the underlying mechanism for optical gain. More explicitly, R_{ST} is related to the optical gain and photon density:

$$R_{ST} = v_g gP \tag{4.14}$$

In this equation g is the optical gain, P is the photon density, and v_g is the group velocity of light in the active region. The rationale behind Eq. (4.14) is that in a short length of time dt, the optical signal travels a short distance dz, such that $v_g=dz/dt$. Moreover, R_{ST} can be expressed as $R_{ST}=dP/dt$. If we substitute v_g and R_{ST} in Eq. (4.14), we obtain

$$\frac{dP}{dz} = gP \tag{4.15}$$

which indeed defines the optical gain for a signal traveling in the z direction.

The next task is to express optical gain in terms of carrier density. In general, g is a nonlinear function of N. For instance, a logarithmic fit matches experimental data over a wide range of parameters [17]:

$$g = \frac{g_0}{1+\varepsilon P}\ln\left[\frac{N+N_s}{N_0+N_s}\right] \tag{4.16}$$

Here N_0 is the transparency carrier density, N_s is a fitting parameter, g_0 is a fitting parameter representing the inherent gain of the material, and ε is the *gain compression factor*. Note that ε represents gain saturation at high optical powers. Moreover, note that when $N=N_0$, $g=0$, which is the point that the net gain is equal to the net loss and therefore the material is at the margin of becoming transparent.

Because of the complexity of nonlinear fits, a common approximation is to assume a linear relationship around the transparency point:

$$g = G(N - N_0) \tag{4.17}$$

where G is the *differential gain* at transparency. Therefore, either Eq. (4.17) or Eq. (4.16) can be used in Eq. (4.14) for an explicit expression of R_{ST}.

Next, we need to consider R_{SP} and R_{NR} and express them explicitly in terms of carrier density. One way to proceed is to calculate these rates from first principles of quantum mechanics and semiconductor physics. However, it is common to describe the recombination terms simply in terms of powers of carrier density [17]:

$$R_{SP} + R_{NR} = AN + BN^2 + CN^3 \tag{4.18}$$

Theoretical and experimental considerations show that the term BN^2 corresponds to spontaneous emission rate, where for common semiconductors, $B \cong 10^{-10}$ cm^3/s. The term AN represents *defect recombination*, and the term CN^3 corresponds to *Auger recombination*. Defect recombination refers to additional energy levels in the band gap that act as a trap for carriers and are caused by crystal defects. Auger recombination refers to the collision of two electrons in the conduction band, which knocks one of them to the valence band and the other to a higher energy state in the conduction band.

As a further simplification, we may lump all the non-radiative terms together and approximate them by a linear decay process with a *carrier life time* τ:

$$R_{NR} = AN + CN^3 = \left(A + CN^2\right)N = \frac{N}{\tau} \tag{4.19}$$

It should be noted that in general, the lifetime defined by τ is carrier dependent. However, in cases of low Auger recombination or low carrier density, the dependence of τ on N can be neglected. Equation (4.19) states that without stimulated and spontaneous emission, the carrier density will decay exponentially with a time constant τ.

Now we have all the pieces needed to express the carrier rate equation explicitly in terms of carrier density. By combining Eqs. (4.11), (4.12), (4.13), (4.14), (4.17), and (4.19) we arrive at the most familiar form of the rate equation for carrier density:

$$\frac{dN}{dt} = \frac{\eta I}{eV} - \frac{N}{\tau} - BN^2 - v_g G(N - N_0)P \tag{4.20}$$

The right-hand side terms of Eq. (4.20) represent carrier generation, non-radiative carrier recombination, carrier recombination due to spontaneous emission, and finally carrier recombination due to stimulated emission. Other forms of Eq. (4.20) can also be used, for instance, by using the more accurate Eq. (4.16) instead of the linear approximation of Eq. (4.17). Note that in Eq. (4.20) N is coupled to the photon density, P. Thus, we also need a rate equation for P.

4.5.2 Photon density rate equation

Ideally photons should be generated at the rate of R_{ST}, because R_{ST} represents the depletion rate of carriers due to stimulated emission. However, not all photons generated by stimulated emission contribute to the photon density *in the active region*, because the active region is a fraction of the overall volume of the cavity in which photons exits. As noted in Eq. (4.2), the ratio of photons in the active region to the total number of photons is called the confinement factor and is denoted by Γ. Therefore, the rate at which stimulated photons contribute to the photon density in the active region is ΓR_{ST}, or equivalently, $\Gamma v_g gP$.

Spontaneous emission is another process for photon generation. To account for the contribution of these photons to the laser mode, two factors must be taken into account. First, a fraction of these photons is coupled to the lasing mode. We can represent this fraction by a factor β, called the spontaneous emission factor. The second factor is the confinement factor Γ, which, like its role for stimulated emission, accounts for the fraction of photons that occupy the active region. Combining these two factors, the overall contribution of spontaneously emitted photons to the photon density in the active region would be $\Gamma\beta R_{SP}$.

Finally, to account for photon reduction, we can define a photon life time τ_P. This decay can be due to the loss in the medium or the loss from the cavity mirrors in the form of output light from the laser.

Based on these terms, we can write the photon density rate equation as

$$\frac{dP}{dt} = \Gamma v_g gP + \Gamma\beta R_{SP} - \frac{P}{\tau_P} \tag{4.21}$$

Equation (4.21) clearly shows the contribution of stimulated and spontaneous emissions (the first two terms) as well as absorption and radiation out of cavity (the third term) in the dynamic behavior of photon density. This equation can also be written in a more explicit way, for instance, by expressing R_{SP} as BN^2 and the linear expression for g. The result is

$$\frac{dP}{dt} = \Gamma v_g G(N - N_0)P + \Gamma\beta BN^2 - \frac{P}{\tau_P} \tag{4.22}$$

Therefore, Eqs. (4.20) and (4.22) together constitute a complete set which can describe the dynamic behavior of the photon and carrier densities in a diode laser. In general, these equations can be solved by numerical methods. However, further insight can be gained by considering two special cases of steady-state conditions and small sinusoidal signals.

4.5.3 Steady-state analysis

Steady-state analysis is applicable when the rate of change of variables is much slower than the inherent time constants of the system. For diode lasers, the inherent time constants of the system are τ_p and τ, which are in the picosecond and nanosecond range. Mathematically, the steady-state solutions are obtained by setting the time derivatives in the rate equations to zero. To keep the analysis general, we keep the gain term in its implicit form, g. Setting the time derivatives in Eqs. (4.20) and (4.22) to zero, we obtain

$$I = \frac{eV}{\eta}\left[\frac{N}{\tau} + BN^2 + v_g gP\right] \tag{4.23}$$

$$P = \frac{\Gamma\beta BN^2}{\dfrac{1}{\tau_P} - \Gamma v_g g} \tag{4.24}$$

Moreover, we note that the variable we are interested in is the output power of the laser, L, which is related to photon density as follows[2]:

$$L = \eta_0 hf \frac{VP}{\Gamma\tau_P} \tag{4.25}$$

Here η_0 is the internal optical efficiency of the cavity representing the percentage of photons that reach the mirrors. The physical interpretation of Eq. (4.25) is simple. If P is the photon density, VP is the number of photons in the active region, and VP/Γ is the total number of photons in the optical cavity. Thus, $hf\,VP/\Gamma$ is the total optical energy in the cavity. We also need an efficiency factor, η_0, to model the percentage of these photons that yield useful optical energy in the lasing mode. Finally, dividing the result by the photon life time yields the rate at which this energy is removed from the active region, i.e., optical power of the laser.

Let us first focus on Eq. (4.24). We know that g is a monotonously increasing function of carrier density and therefore current. Thus, as the current increases, g also increases. We can see that for a certain value of g the denominator of Eq. (4.24) will be zero, resulting in an infinite photon density. Therefore, physical requirements should prevent the gain from reach this limiting value. This limiting value of gain is called the threshold gain or g_{TH} and is given by

$$g_{TH} = \frac{1}{\tau_P \Gamma v_g} \tag{4.26}$$

The value of carrier density corresponding to g_{TH} is the threshold carrier density N_{TH}:

---

[2] Normally, we express power quantities by P. However, because in rate equations P is usually reserved for photon density, here we use L for output power. This choice is also consistent with the standard notation of denoting the current–light characteristic of the laser as LI curve.

$$g_{TH} = g(N_{TH}) \tag{4.27}$$

Thus, as N approaches N_{TH}, g approaches g_{TH}, and the optical power increases rapidly, as predicted by Eq. (4.24). This means that N never really reaches N_{TH}, because that yields an infinite value for P, obviously a non-physical situation. Thus, we can conclude that in steady state, as the driving current increases, both g and N must be *clamped* at a value very close to their threshold values. We can use this limiting value and evaluate Eq. (4.23) *above* threshold:

$$I = \frac{eV}{\eta}\left[\frac{N_{TH}}{\tau} + BN_{TH}^2 + v_g g_{TH} P\right] \tag{4.28}$$

Solving for P in Eq. (4.28), using Eq. (4.26) for threshold gain, and using Eq. (4.25) to convert P to L, we get the following relationship for the output power of the laser:

$$L = \eta\eta_0 \frac{hf}{e}(I - I_{TH}) \tag{4.29}$$

where the threshold current of the laser is defined as

$$I_{TH} = \frac{eV}{\eta} N_{TH}\left(\frac{1}{\tau} + BN_{TH}\right) \tag{4.30}$$

Figure 4.7 illustrates the behavior of P, N, and g as a function of the drive current. To summarize, as the current increases, both N and g asymptotically approach their threshold value. This means non-radiative and spontaneous recombination will also clamp at threshold, as suggested by the first two terms in the right-hand side of Eq. (4.28).

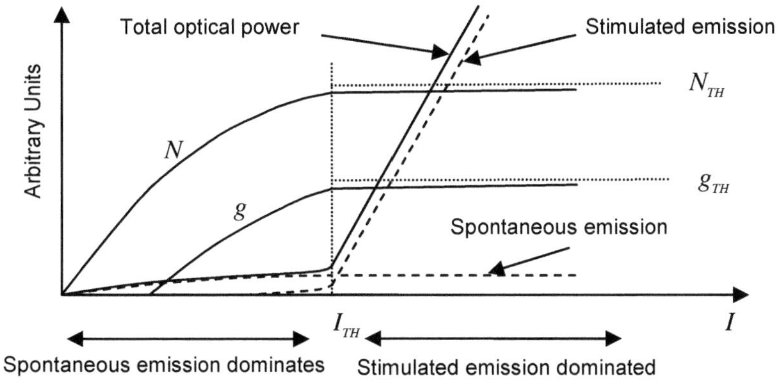

Fig. 4.7. Steady-state behavior of critical laser parameters vs. drive current

Therefore, spontaneous emission will also be clamped at threshold. However, stimulated photon density will increase linearly with I. This can be seen from the last term in the right-hand side of Eq. (4.28) and is expressed explicitly by Eq. (4.29).

4.5.4 Temperature dependence of LI curve

Like any other semiconductor device, diode lasers are sensitive to temperature. In general, as the temperature increases, it takes more current to achieve a given value of gain. This is mainly due to broadening of the Fermi function given by Eq. (4.10). At T=0, the transition from filled states to empty states is abrupt. As T rises, this transition becomes less abrupt and the carrier distribution broadens. So a given carrier density occupies a wider energy band at a higher temperature. Moreover, it can be shown that N_{TH} also increases with temperature. There are also some effects that depend on the material used. For example, in InGaAsP-based lasers Auger recombination seems to have a large impact on the temperature behavior [22]. It is common to model temperature effects by assigning an exponential relationship between temperature and threshold current:

$$I_{TH} = I_0 e^{T/T_0} \tag{4.31}$$

Here T_0 is some characteristic temperature that models the thermal sensitivity of the device. A lower value of T_0 corresponds to higher temperature sensitivity. For GaAs lasers T_0 is in the range of 150–200 K, whereas for InGaAsP lasers T_0 is generally lower and is in the range of 70 K. Strained layer quantum wells achieve somewhat higher values, in the range of 90 K [23].

In addition to threshold current, the slope efficiency of a laser generally deteriorates at higher temperatures. This effect too can be approximated by an exponential temperature dependence similar to Eq. (4.31).

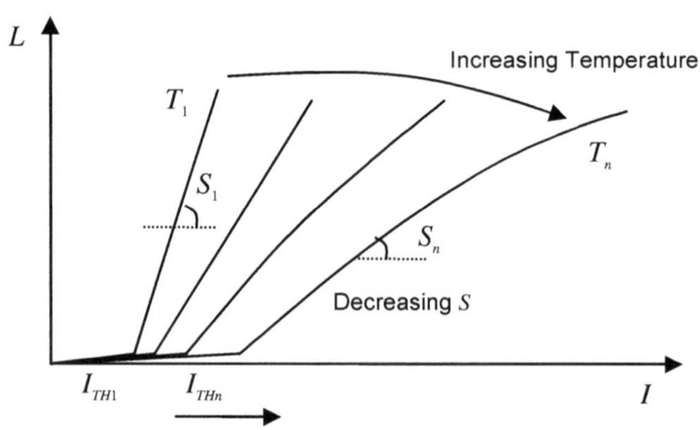

Fig. 4.8. Typical behavior of LI curve with temperature

However, slope efficiency is generally less sensitive to temperature compared to the threshold current. Moreover, long wavelength lasers (1300 and 1500 nm) are more sensitive to temperature degradations [17, 24].

The overall result of temperature degradation on the typical *LI* curve of a laser is shown in Fig. 4.8. As we will see in Chapter 8, temperature effects are very important in designing power and modulation control circuits. We should also note that the behavior shown in Fig 4.8 is typical, but not necessarily universal. One exception can be made in DFB lasers, where the frequency of the dominant mode is selected not by the gain profile, but by the Bragg wavelength of the distributed feedback. Both the Bragg wavelength and the gain profile move with temperature, but usually the gain profile moves much faster. As a result, it is possible to design a laser structure in which the two wavelengths are detuned more at room temperature, but coincide better at higher temperatures. In such a laser, the effective gain can actually increase with temperature, yielding to a lower threshold current at high temperature. However, such anomalous behavior is not typical, and the general rule is that a laser's efficiency deteriorates as the temperature rises.

4.5.5 Small signal frequency response

To study the dynamic behavior of lasers analytically, one method is to assume all signals are small in amplitude and sinusoidal in time. The advantage of this method is that it yields information about the frequency response of the laser.

The small signal frequency response analysis starts with decomposing the quantities of interest into the sum of a (larger) dc value and a (smaller) ac component at a given frequency of ω in the form of $\exp(j\omega t)$. These assumed forms are substituted back into the original rate equations, and are further simplified by noting that the dc parts should satisfy the steady-state equations. Moreover, any harmonic terms at frequencies other than ω are neglected. The remaining terms can be grouped together and a small signal expression for the *modulation transfer function* of the laser above threshold can be obtained:

$$H(\omega) = \frac{L(\omega)}{I(\omega)} = \frac{A\omega_R^2}{\omega_R^2 - \omega^2 + j\gamma\omega} \tag{4.32}$$

where A is the small signal dc gain from current to power. The transfer function defined by Eq. (4.32) is that of a second-order system with a resonance frequency ω_R and a damping factor γ. The resonance frequency can be approximated as [17, 25]

$$\omega_R = \sqrt{\frac{v_g a P_{OP}}{\tau_P}} \tag{4.33}$$

and is known as the *relaxation oscillation frequency* of the laser. In Eq. (4.33) a is the *differential* gain, and P_{OP} is the average photon density, both at the operating point. If we use Eq. (4.16) for optical gain, a can be expressed as

$$a = \frac{\partial g}{\partial N} = \frac{g_0}{(1+\varepsilon P)(N+N_s)} \tag{4.34}$$

The damping factor is defined as

$$\gamma = \omega_R[\omega_R \tau_P + 1/\omega_R \tau] = v_g a P_{OP} + \frac{1}{\tau} \tag{4.35}$$

Although rate equations can be used to model more complicated effects, the simple model we have discussed here is sufficient to highlight many important features [26]. Figure 4.9 illustrates representative plots of the magnitude of the transfer function and the resonance frequencies at several bias points. Below the resonance frequency, the response is essentially flat and the laser optical output can follow the input. The frequency response peaks at a frequency ω_P:

$$\omega_P = \sqrt{\omega_R^2 - \frac{1}{2}\gamma^2} \tag{4.36}$$

After this point the response drops fast and approaches the limit of −40 dB/decade. Note that according to Eq. (4.33), ω_R (and therefore ω_P) is proportional to the square root of optical power.

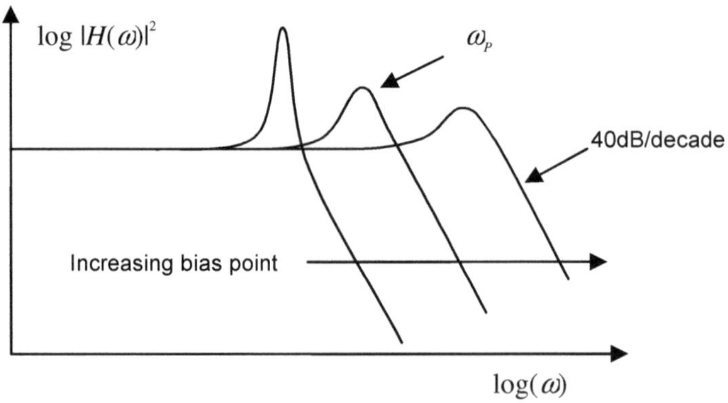

Fig. 4.9. Frequency response of diode lasers

This means that if the laser is biased at a higher operating point, the relaxation oscillation frequency and the bandwidth of the laser will be higher. On the other hand, based on Eq. (4.35), biasing the laser at a higher power will increase the damping and therefore reduces the "peaking" of frequency response near resonance.

4.5.6 Time response

The time response of the laser is especially important in digital modulation where the drive current for each transition is essentially a step function [26, 27]. From linear system theory we know that the response of a second-order system to a step function can be over-damped, marginally damped, or under-damped, and therefore we can expect to observe the same type of behavior in lasers. In particular, if the laser is under-damped, the output goes through a period of ringing before it settles down at the final steady state.

A critical parameter in digital direct modulation transmitters is the maximum bit rate achievable with a particular diode laser. In non-return-to-zero (NRZ) coding, if the length of a bit is shorter than the settling time of the laser, the starting point of the next transition will no longer be the same for all bits. For example, in a pattern of 1110, the output has a long time to settle to the steady-state level during the first three 1s. Thus, the initial condition for the transition to the zero level for the fourth 0 bit is very close to the steady state. However, when a pattern like 0010 is being transmitted, the output does not settle within the 1 bit, and therefore the initial condition for the last zero transition becomes pattern dependent. As shown in Fig. 4.10, this causes pattern-dependence jitter.

Figure 4.10 shows the jitter as a result of only two patterns. It is not hard to imagine that in a random signal where all pattern combinations are present, the jitter starts to accumulate in the form of a jitter distribution, effectively reducing the horizontal eye opening. Thus, the frequency response of the laser ultimately limits the maximum bit rate at which its output can be modulated.

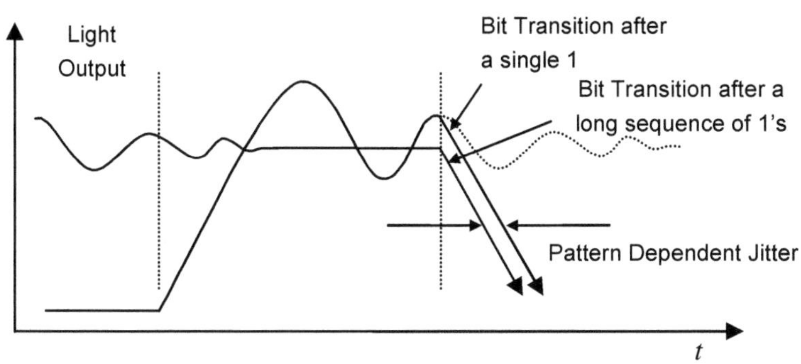

Fig. 4.10. Pattern-dependent jitter as a result of bit rate approaching relaxation oscillation of laser

As a side note, we should add that in reality the frequency or time response of the laser is a result of the dynamic response of the diode itself combined with the response of the electronic circuits that drive it. We will revisit this issue in Chapter 8 where we discuss the design of optical transmitters.

4.5.7 Frequency chirp

The rate equations we have been discussing so far describe the photon and carrier densities. However, any change in the carrier density inevitably will result in a change in the index of refraction of the semiconductor as well. On the other hand, any change in the index of refraction will cause a change in the optical frequency of the lasing mode. Therefore, any time we modulate the intensity of light through direct modulation we are also modulating the optical frequency of the light generated by the laser. This frequency modulation, which is similar to FM modulation in radio frequency applications, is known as *frequency chirping* [28–30]. Frequency chirping implies an increase in the spectral linewidth of the diode laser [31]. As we will see in Chapter 5, an optical signal with a wider spectral width is more prone to degradation due to dispersion. Thus, it is important to characterize the frequency chirping of a laser. Using rate equations, the magnitude of chirp or the change in frequency Δf is found to be [32–33]

$$\Delta f = \frac{\alpha}{4\pi}\left[\frac{1}{P(t)}\frac{dP(t)}{dt} + \kappa P(t)\right] \tag{4.37}$$

where κ is a parameter that characterizes the material, and α is known as the *linewidth enhancement factor*.

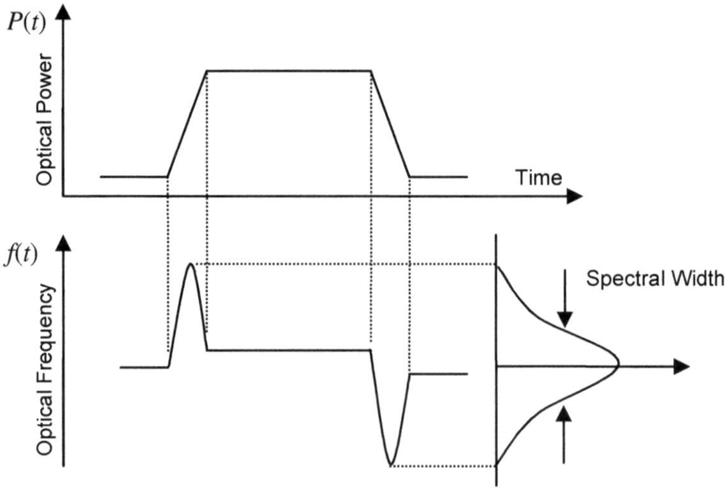

Fig. 4.11. Spectral broadening due to static and dynamic chirp

For common semiconductor lasers, the value of α is in the range of 8–4. Equation (4.37) suggests that frequency chirping can be divided into two components. The first component, known as *dynamic* or *transient chirp*, is proportional to the rate of change in optical power and dominates in large signal modulation or fast transitions. The second term, called *static or adiabatic chirp*, is related to the structure of material and wavelength through the parameter κ [33]. Figure 4.11 illustrates the spectral broadening as a result of the dynamic and static components of chirp in a direct modulation scheme.

The important point to notice is that direct modulation can significantly increase the spectral width of the output light from a diode laser. This is why fiber optic links based on direct modulation schemes are ultimately limited by dispersion. As we will see in Chapter 8, external modulation can significantly reduce chirping and therefore increase the performance of an optical ink.

4.5.8 Large signal behavior

Small signal approximations of rate equations provide great insight into the behavior of diode lasers. However, in digital schemes oftentimes the modulation amplitude can be large. In these cases, although small signal analysis tools can be used with approximation, the laser shows new behaviors without small signal equivalent. Although numerical solutions are the main tool available for analyzing these equations in their general form, it is still possible to describe the large signal behavior of the laser qualitatively.

In digital modulation we are interested in the response of the laser to a pulse of current in the form of a step function. Because for link budget purposes it is better to maximize the modulation depth or extinction ratio, the nonlinear behavior of the laser shows up in the form of asymmetry between rise and fall times, difference in relaxation oscillation in the zero and one states, and turn on delay.

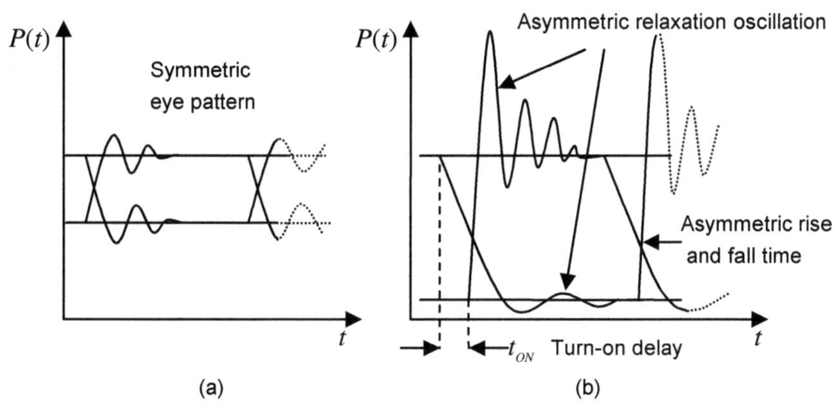

Fig. 4.12. Comparison of small and large signal behavior of a diode laser

Figure 4.12 compares the large signal and small signal behavior of a diode laser as characterized by typical eye patterns in the two cases. Let us consider some of the main features highlighted in Fig. 4.12. When the laser is biased well above its threshold and is driven by a relatively small modulation current, as in Fig. 4.12a, the eye pattern is symmetrical, i.e., the rise and fall times as well as the relaxation oscillation are the same for both a positive and negative transitions.

In Fig. 4.12b the modulating current has been increased by keeping the one level as before, but bringing the zero level closer to threshold. The eye pattern in this case becomes asymmetrical. Some of these features can be understood in terms of the dependence of the small signal parameters on photon density. For instance, according to Eq. (4.33) the relaxation oscillation frequency increases with optical power. As a result, we can expect that the ringing frequency should be higher at the one state compared to the zero state. Moreover, because the photon density cannot become negative, the ringing at the zero state is compressed to ensure the optical output always remains positive.

The asymmetric rise and fall times are another feature of large modulation. Generally, the rise time becomes shorter compared to the fall time as a result of the dynamic interplay between the photon and carrier densities during a positive and a negative pulse of current.

Another phenomenon illustrated in Fig. 4.12 is the turn-on delay. In the case of small signal modulation, the rise and fall time transitions are aligned in time. But for large signal modulation, there is a delay between the time the current pulse is applied to the time optical power starts to rise. As a result, the rise and fall time edges are no longer aligned. This turn-on delay becomes especially noticeable if the laser is pushed below the threshold. It can be shown that in this case the turn-on delay is given approximately by [17]

$$t_{ON} = \tau \ln\left(\frac{I_1 - I_0}{I_1 - I_{TH}}\right) \tag{4.38}$$

In this equation I_1 is the current in the one state, I_0 is the current in zero state, I_{TH} is the threshold current, and τ is the total carrier life time (corresponding to spontaneous and non-radiative carrier recombination).

Turn-on delay can cause significant jitter in directly modulated digital transmitter at high bit rates. The reason is that the delay is a function of I_0 or the current in the 0 level. In practice this current itself is not constant, but goes through ringing after each 1 to 0 transition, before it settles down at its steady state. Therefore, if a 0 to 1 transition occurs before the ringing is damped, pattern dependence jitter follows. This is another limiting factor in achieving signal integrity in high bit rate direct modulated transmitters.

4.6 Semiconductor laser structures

As described in the beginning of this chapter, a semiconductor laser is essentially a forward biased diode where population inversion results from the flow of majority carriers in the active region. Figure 4.13 shows the structure of a simple pn junction and the flow of current and light when the junction is forward biased. The current flows from the p to the n side through the metal contacts. The active region between the p and the n sides is where optical gain is realized, and the light propagates perpendicular to the direction of flow of current. The facets at the two sides of the device act as mirrors and provide the optical feedback necessary for lasing action. In a typical semiconductor, the facets provide a reflectivity of ~30%. In general, light flows out of both facets and forms the optical output of the device.

This structure is called a homo-junction, because the device is made of a single semiconductor material only doped differently on the two sides of the pn junction. However, as noted before, a simple homo-junction is not an efficient structure for laser action, mainly because there are no provisions for light or carrier confinement. This lack of confinement causes a decline in both carrier and photon densities and therefore reduces the optical gain. As a result, this device lacks the performance needed from practical optical sources. We will examine more practical devices shortly.

Modern semiconductor lasers are made through processes based on epitaxial growth of several layers of semiconductor materials with different compositions.

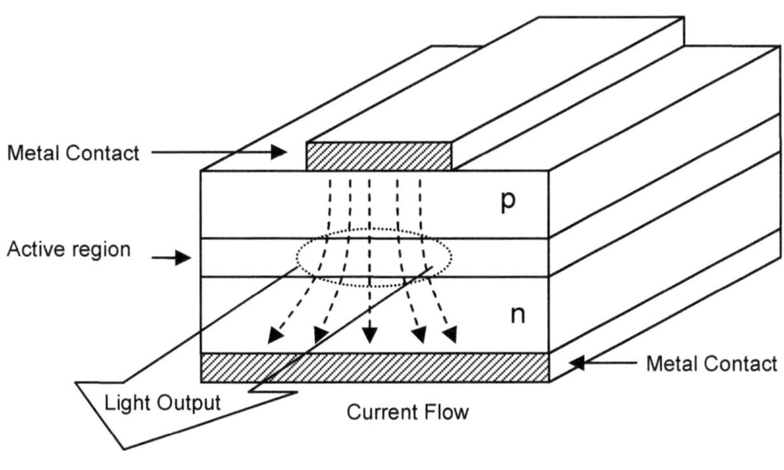

Fig. 4.13. Structure of simple homo-junction diode laser

These structures are called *heterostructures*. The process starts with a base material, usually GaAs or InP. The next layers are grown on top of this substrate. In ternary systems, an additional element is added to the binary compound. Common ternary materials include $Al_xGa_{1-x}As$ and $In_xGa_{1-x}As$. By substituting a fraction x of

Ga with Al or In, the properties of the ternary material can be tailored and tweaked for various applications. The bandgap energy of GaAs corresponds to around 850 nm. Because Al atoms are almost the same size as Ga, $Al_xGa_{1-x}As$ is lattice matched to GaAs and the resulting system is the basis for short-wavelength lasers (700–900 nm) commonly used in multimode fiber links.

It is also possible to add a fraction of a fourth element to a ternary compound and obtain a quaternary compound. One such compound is $In_{1-x}Ga_xAs_yP_{1-y}$. By tuning the ratios of x and y it is possible to shift the bandgap energy to the critical wavelengths of 1300–1600 nm window, where most fiber optic communication systems operate. At the same time, it is possible to obtain lattice-matched or strained epitaxial layers on the substrate, which for long-wavelength lasers is usually InP. As a result, most long-wavelength communication lasers are heterostructures based on the InGaAsP/InP system.

4.6.1 Heterostructure laser

One of the key characteristics of modern lasers is that they achieve higher efficiencies by carrier and light confinement. Modern diode lasers use heterostructures for the purpose of carrier confinement. Figure 4.14 illustrates the cross section of a typical long-wavelength *buried heterostructure* laser [34]. It can be seen that the active InGaAsP layer is sandwiched between InP layers, both vertically and laterally. Because the index of refraction of the InGaAsP is higher, the optical fields will be mostly confined laterally and vertically to the active region. At the same time, because of the insulating SiO_2 structure and the geometry of the p and n junctions, the current is also confined to the active region. The high carrier and photon densities substantially improve the efficiency of the device.

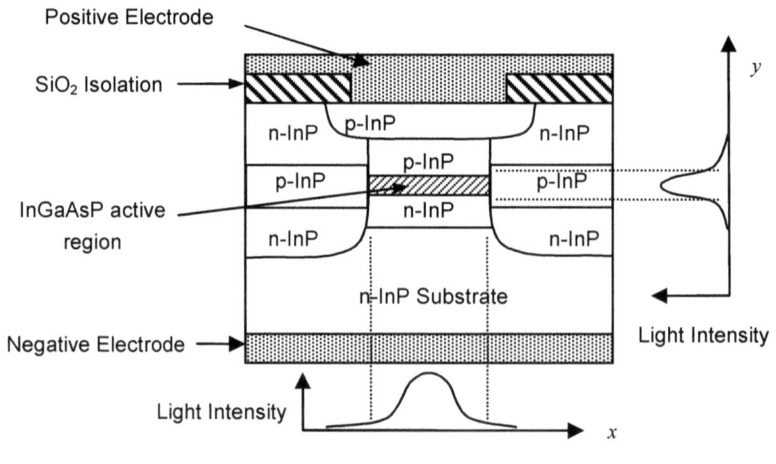

Fig. 4.14. Typical InGaAsP/InP buried heterostructure

A heterostructure such as that shown in Fig. 4.14 is called an *index-guided* laser. This is because optical confinement is achieved by creating an effective waveguide inside the device. It is also possible to achieve optical confinement merely as a result of the higher gain in the active region, because higher gain will by itself cause a change in the index of refraction. Such structures are called *gain-guided* lasers. However, the performance of gain-guided lasers is usually inferior to that of the index-guided lasers. Figure 4.14 also shows the profile of light intensity in the vertical and lateral directions. The light exits the device parallel to the plane of the junction (perpendicular to the page). As a result, this structure is also called an *edge-emitting laser*.

4.6.2 Quantum well lasers

As mentioned above, making the active layer thinner will improve the efficiency of the laser because it yields a higher carrier density. When the thickness of active layer gets in the order of tens of nanometers or smaller, a *quantum well* is formed [35–36]. The waves associated with carriers become quantized very much like the standing waves on a string hold on its two ends. The quantization changes the energy levels of carriers and impacts the gain. Consequently, a quantum well can have a much higher gain compared to a bulk semiconductor. Moreover, the thickness of a quantum well can be used to control the energy band gap between the conduction and valence bands.

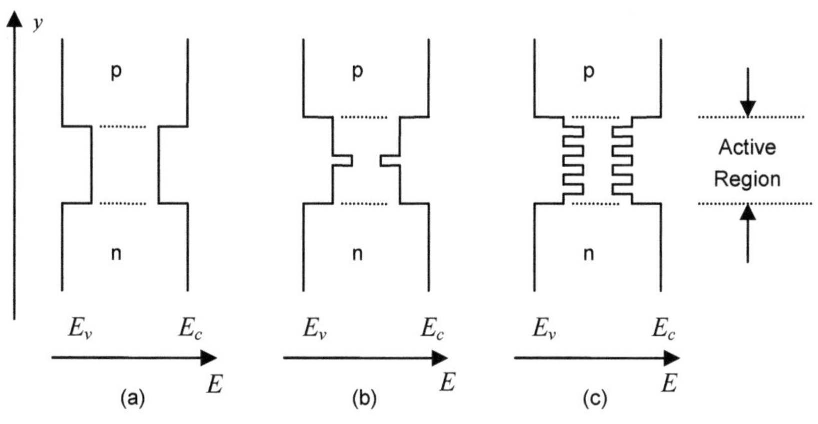

Fig. 4.15. Energy band diagram of (**a**) bulk heterostructure, (**b**) single quantum well, (**c**) multiple quantum well laser

Figure 4.15 shows a comparison of the energy band diagram of a bulk laser with those based on quantum wells. In the left-hand side the band diagram of a heterostructure laser is shown. The vertical or y direction is the direction of crystal growth and the direction of current flow. The electrons flow from the n material on the bottom side into the active region, and the holes flow from the p material on the top side. In a bulk semiconductor laser (Fig. 4.15 a), the confinement is achieved by the large potential well formed by the whole active region. However, in a quantum well structure additional layers of heterostructures are used to realize one or several quantum wells within the active region.

Figure 4.15b shows a single quantum well structure implanted within the active region. The problem with a single quantum well is that it may be so thin that it cannot provide sufficient optical confinement. As a result, the overlap between carriers and the optical field may decrease, causing a decrease in the efficiency of the laser. *Multiple quantum well lasers* (Fig. 4.15 c) address this problem by providing additional quantum wells and thus increasing the overlap between photons and carriers.

4.6.3 Distributed feedback (DFB) lasers

In the laser structures we have discussed so far the optical feedback is provided by the cleaved facets on the two sides of the cavity. As we noted earlier in this chapter, such a laser is called a Fabry–Perot laser. In principle, a Fabry–Perot cavity can support many optical modes. Therefore, Fabry–Perot lasers will generally end up lasing in several *longitudinal modes*, each with a slightly different wavelength. As a result, Fabry–Perot lasers are usually *multimode lasers*.

An ideal optical source must have a narrow spectral width, and therefore all but one of these lasing modes must be suppressed. The most common way of achieving this goal is to provide the optical feedback not through the cleaved edges, but by a periodic diffraction grating along the active region. Such a structure is called a *Bragg reflector*, and these lasers are called *distributed feedback* (DFB) lasers [37]. In a DFB laser, each grating reflects a little bit of light. Reflection is only significant in the wavelength where the reflected waves add constructively, whereas in all other wavelengths they will be out of phase and add destructively.

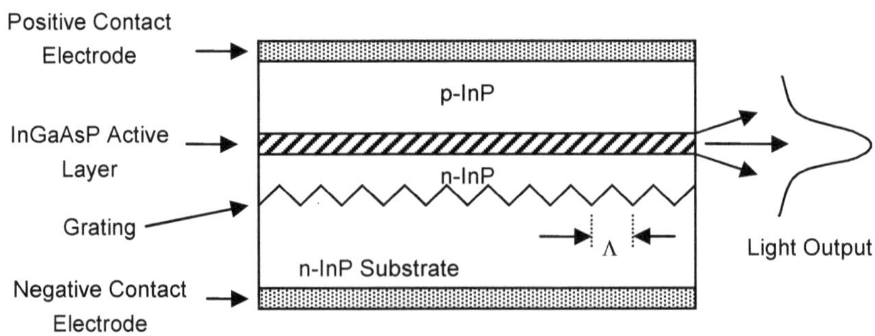

Fig. 4.16. Longitudinal cross section of a DFB laser

Therefore, the cavity provides optical feedback only in a single wavelength. The reflectivity from the two facets are minimized (for instance by the application of anti-reflection coating) to minimize any potential Fabry–Perot mode. Figure 4.16 illustrates the longitudinal cross section of a DFB laser.

The *Bragg wavelength*, that is, the wavelength at which maximum reflection within the active region takes place is given by

$$\lambda_B = \frac{2n_E\Lambda}{m} \tag{4.39}$$

where Λ is the period of grating, n_E is the effective index of refraction for the mode, and m is an integer representing the order of coupling between the incident and reflected waves The strongest reflection takes place for the lowest order (m=1), although in some cases higher orders are also used because they allow for a longer grating period which could be easier to manufacture.

In an actual DFB laser the lasing wavelengths will be distributed symmetrically on both sides of the Bragg wavelength. Thus, in a normal DFB laser without additional provisions, two dominant modes around the Bragg wavelength can exist simultaneously. A common way to remove this degeneracy is to use a *quarter-wavelength shifted grating*, where an additional grating with a (spatial) phase shift of $\Lambda/2$ is introduced in the cavity [38]. This results in a single-mode operation at the Bragg wavelength. In this case, almost all the laser energy is concentrated in one main mode. The ability of a DFB laser to “suppress” all but one single mode is known as *side mode suppression ratio* (SMSR). Modern DFB communication lasers have an SMSR of 45 dB or better, although SMSR can suffer under large direct modulation.

4.6.4 Vertical surface emitting lasers (VCSELs)

The main difference between VCSELs and other types of laser structures lies in the direction of light propagation. In regular edge-emitting lasers the light propagates along the direction of the active region and exits from the two facets on the two sides of the junction. On the other hand, in a VCSEL the light propagates perpendicular to the plane of active region and exits from the top or bottom surface of the device [39–40].

Figure 4.17 shows a typical GaAs VCSEL structure. The active region consists of multiple layers of quantum wells, surrounded on both sides with multiple layers of Bragg reflectors. The light travels in the vertical direction back and forth between the Bragg reflectors. Because of the relatively short length of the active layer along the path of light, the Bragg mirrors must provide very high reflectivity to support the necessary optical feedback needed for lasing. The current is injected from the contact ring on the top and flows down within the structure and through the active region. A second current confinement ring is implanted within the structure to help in focusing the current and achieving higher carrier concentration.

The small volume of active layer in VCSELs makes their threshold current relatively low. Moreover, because the length of the optical cavity is very short, longitudinal modes are spaced very far from each other, and therefore a VCSEL typically has only one longitudinal mode. In edge-emitting lasers because of the aspect ratio of the active region the output light tends to have an elliptical profile, which can reduce the coupling efficiency into a circular aperture like that of an optical fiber. However, because of the symmetrical structure of VCSELs, the output light tends to have a circular profile. Also, because the light exits from the surface of the device coupling it into a fiber can be much easier, as the fiber can be brought in direct contact and mounted to the surface. Another advantage of VCSELs is that they can be tested and operated in arrays, without the need to separate them into individual devices. This makes them very attractive for parallel optical interconnects and smart pixel applications. Also the fact that they can be tested as arrays makes it possible to test them at wafer-level, thus reducing the cost of testing. Because of these advantages, VCSELs are extensively used in low-cost transmission systems, especially in multimode short-distance applications.

However, there are several problems that have prevented the proliferation of VCSELs for higher end data and telecom applications. The most important problem is that long-wavelength VCSELs in the 1300–1600 nm window have been proven difficult to build. Most practical VCSELs operate in the 850 nm wavelength range, which corresponds to the bandgap of GaAs. VCSELs based on strained InGaAs can operate in the 980 nm range. However, construction of InGaAsP based VCSELs for the longer wavelengths has been more challenging because Bragg reflectors based on InGaAsP have a low reflectivity.

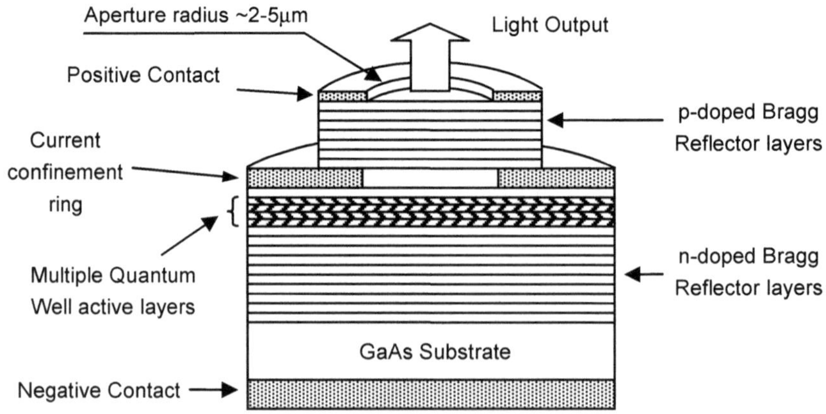

Fig. 4.17. Cross section of a VCSEL

Another challenge in VCSELs is their tendency for multi-transverse-mode operation. Because of their large surface, the optical cavity can support more than one transverse mode, and under modulation, mode hopping between these lateral modes can occur. This becomes more of a problem especially when the bias current gets much higher than the threshold current. These lateral modes have different optical frequencies and thus the overall spectrum can be multimode. Moreover, because of different far-field profile of these transverse modes coupling into a fiber can be greatly affected. As a result of these limitations, VCSELs have not found wide application in long wavelength fiber optic communication yet, although efforts in this direction continue.

4.7 Summary

Semiconductor lasers are the main source of light used in fiber optic communication. Their size, technology, electrical properties, and modulation bandwidth are among the properties that make these devices ideal candidates for telecommunication applications.

We started this chapter by discussing the general concepts behind the operation of a laser. A laser is essentially an optical oscillator made up of an optical amplifier and feedback. In semiconductor lasers, optical gain is realized through population inversion of carriers in a pn junction, and optical feedback is usually achieved by the facets of the semiconductor material.

A simple and effective tool in theoretical studies of semiconductor lasers is the rate equation approach, which is based on a phenomenological description of carrier and photon densities. Using this approach one can gain great insights into the steady-state and dynamic behavior of semiconductor lasers without having to go through complex theories and difficult mathematical formulations. We outlined these equations in some detail and discussed some of their implications. In particular, we examined the LI curve and the modulation response of the laser. We also discussed important large signal effects such as turn-on delay and frequency chirping.

Finally, we reviewed some practical diode laser structures. Practical diode lasers are almost always based on heterostructures. Thus, we discussed the buried heterostructure laser as an example and used it to illustrate several important concepts related to laser structures. We also discussed quantum well and distributed feedback lasers, which are examples of advanced diode lasers with superior performance in many aspects. Finally, we considered VCSELs, which are surface emitting structures. These devices are somewhat different from standard edge emitting lasers and thus provide certain advantages and drawbacks. As a result, they are widely used in cost-sensitive, lower data rate applications. On the other hand, long wavelength telecommunication applications still rely mostly on edge emitting lasers, based on their advantages in terms of wavelength, power, and single-mode operation.

References

[1] S. Boutami et al., "Vertical Fabry-Perot cavity based on single-layer photonic crystal mirrors," Optics Express, Vol. 15, pp.12443–12449, 2007

[2] T. Steinmetz et al., "Stable fiber-based Fabry-Perot cavity" Applied Physics Letters, Vol. 89, Article Number 111110, 2006

[3] Y. D. Jeong et al., "Tunable single-mode Fabry-Perot laser diode using a built-in external cavity and its modulation characteristics," Optics Letters, Vol. 31, pp. 2586–2588, 2006

[4] B. G. Streetman, *Solid State Electronic Devices*, Prentice-Hall, Englewood Cliffs, NJ,1990

[5] W. E. Lamb, "Theory of optical maser", Physical Review A, Vol. 134, pp. 1429–1450, 1964

[6] M. Scully and W. E. Lamb, "Quantum theory of an optical maser," Physical Review Letters, Vol. 16, pp. 853–855, 1966

[7] M. Scully and W. E. Lamb, "Quantum theory of an optical maser, 1. General theory," Physical Review, Vol. 159, pp. 208–226, 1967

[8] M. Scully and W. E. Lamb, "Quantum theory of an optical maser, 2. Spectral profile," Physical Review, Vol. 166, pp. 246–249, 1968

[9] M. Johnsson et al., "Semiclassical limits to the linewidth of an atom laser," Physical Review A, Vol. 75, Article Number 043618, 2007

[10] A. Yariv, "Dynamic analysis of the semiconductor laser as a current-controlled oscillator in the optical phased-lock loop: applications," Optics Letters, Vol. 30, pp. 2191–2193, 2005

[11] S. Stenholm and W. E. Lamb, "Theory of a high intensity laser," Physical Review, Vol. 181, pp. 618–635, 1969

[12] M. Azadeh and L. W. Casperson, "Field solutions for bidirectional high-gain laser amplifiers and oscillators," Journal of Applied Physics, Vol. 83, pp. 2399–2407, 1998

[13] P. Szczepanski, "Semiclassical theory of multimode operation of a distributed feedback laser," IEEE Journal of Quantum Electronics, Vol. 24, pp. 1248–1257, 1988

[14] L. W. Casperson, "Laser power calculations, sources of error," Applied Optics, Vol. 19, pp. 422–434, 1980

[15] S. Foster and A. Tikhomirov, "Experimental and theoretical characterization of the mode profile of single-mode DFB fiber lasers," IEEE Journal of Quantum Electronics, Vol. 41, pp. 762–766, 2005

[16] C. Etrich, P. Mandel, N. B. Abraham, and H. Zeghlache, "Dynamics of a two-mode semiconductor laser," IEEE Journal of Quantum Electronics, Vol. 28, pp. 811–821, 1992

[17] L. A. Coldren and S. W. Corzine, *Diode Lasers and Photonic Integrated Circuits*, John Wiley & Sons, New York, 1995

[18] F. Habibullah and W. P. Huang, "A self-consistent analysis of semiconductor laser rate equations for system simulation purpose," Optics Communications, Vol. 258, pp. 230–242, 2006

[19] J. T. Verdeyen, *Laser Electronics*, 3rd Ed., Prentice Hall, Englewood Cliffs, NJ, 1995

[20] P. V. Mena et al., "A comprehensive circuit-level model of vertical-cavity surface-emitting lasers," Journal of Lightwave Technology, Vol. 17, pp. 2612–2632, 1999

[21] N. Bewtra, et al., "Modeling of quantum-well lasers with electro-opto-thermal interaction," IEEE Journal of Selected Topics in Quantum Electronics, Vol. 1, pp. 331–340, 1995

[22] A. Haug, "Theory of the temperature dependence of the threshold current of an InGaAsP laser," IEEE Journal of Quantum Electronics, Vol. 21, pp. 716–718, 1985

[23] A. Haug, "On the temperature dependence of InGaAsP semiconductor lasers," Physica Status Solidi (B) Basic Solid State Physics, Vol. 194, pp. 195–198, 1996

[24] M. Montes et al., "Analysis of the characteristic temperatures of (Ga,In)(N,As)/GaAs laser diodes," Journal of Applied Physics D-Applied Physics, Vol. 41, Article Number 155102, 2008

[25] K. Lau and A. Yariv, "Ultra-high speed semiconductor lasers," IEEE Journal of Quantum Electronics, Vol. 21, pp. 121–138, 1985

[26] C. Y. Tsai et al., "A small-signal analysis of the modulation response of high-speed quantum-well lasers: effects of spectral hole burning, carrier heating, and carrier diffusion-capture-escape," IEEE Journal of Quantum Electronics, Vol. 33, pp. 2084–2096, 1997

[27] N. Dokhane and G. L. Lippi, "Improved direct modulation technique for faster switching of diode lasers," IEE Proceedings Optoelectronics, Vol. 149, pp. 7–16, 2002

[28] S. Kobayashi, Y. Yamamoto, M. Ito, and T. Kimura, "Direct frequency modulation in AlGaAs semiconductor lasers," IEEE Journal of Quantum Electronics," Vol. 18, pp. 582–595, 1982

[29] S. Odermatt and B. Witzigmann, "A microscopic model for the static and dynamic lineshape of semiconductor lasers," IEEE Journal of Quantum Electronics, Vol. 42, pp. 538–551, 2006

[30] G. Agrawal, "Power spectrum of directly modulated single-mode semiconductor lasers: Chirp-induced fine structure," IEEE Journal of Quantum Electronics, Vol. 21, pp. 680–686, 1985

[31] N. K. Dutta et al., "Frequency chirp under current modulation in InGaAsP injection lasers," Journal of Applied Physics, Vol. 56, pp. 2167–2169, 1984

[32] P. J. Corvini and T. L. Koch, "Computer simulation of high-bit-rate optical fiber transmission using single-frequency lasers," Journal of Lightwave Technology, Vol. 5, pp. 1591–1595, 1987

[33] T. L. Koch and R. A. Linke, "RA effect of nonlinear gain reduction on semiconductor laser wavelength chirping," Applied Physics Letters, Vol. 48, pp. 613–615, 1986

[34] Y. Yoshida et al., "Analysis of characteristic temperature for InGaAsP BH lasers with p-n-p-n blocking layers using two-dimensional device simulator," IEEE Journal of Quantum Electronics, Vol. 34, pp. 1257–1262, 1998

[35] J. Jin, J. Shi, and D. Tian, "Study on high-temperature performances of 1.3-μm InGaAsP-InP strained multiquantum-well buried-heterostructure lasers," IEEE Photonics Technology Letters, Vol. 17, pp. 276–278, 2005

[36] Y. Sakata et al., "All-selectively MOVPE 1.3-μm strained multi-quantum-well buried-heterostructure laser diodes," IEEE Journal of Quantum Electronics, Vol. 35, pp. 368–376, 1999

[37] H. Ghafouri-Shiraz, *Distributed Feedback Laser Diodes and Optical Tunable Filters*, John Wiley and Sons, New York, 2003

[38] E. Garmine, "Sources, modulators, and detectors for fiber optic communication systems" in *Fiber Optics Handbook*, edited by Michael Bass, McGraw-Hill, New York, 2002

[39] F. Koyama, "Recent advances of VCSEL photonics," Journal of Lightwave Technology, Vol. 24, pp. 4502–4513, 2006

[40] K. Iga, "Surface-emitting laser-Its birth and generation of new optoelectronics field," IEEE Journal of Selected Topics in Quantum Electron, Vol. 6, pp. 1201–1215, 2000

Chapter 5
Optical Fibers

5.1 Introduction

The revolution in fiber optic communication has been made possible by technological advancements that have resulted in the availability of low-loss silica fibers. The attenuation in a single-mode fiber can be as low as 0.25 dB/km. This allows for the propagation of optical signals for long distances without the use of repeaters or amplifiers. In this chapter we will discuss different types of fibers commonly used in fiber optic communication, as well as the parameters that affect light coupling into the fiber.

We will also examine the principles behind light propagation in single-mode and multimode optical fibers. Closely related to light propagation is the topic of signal degradation. In general, signal degradation phenomena can be divided into linear and nonlinear. In linear phenomena, the degradation is not a function of the optical intensity. On the other hand, nonlinear degradations generally show up at high optical intensities.

Attenuation and dispersion are the main forms of linear degradation. Signal attenuation is mainly due to absorption and scattering, while dispersion can be a result of factors such as waveguide properties of the fiber or wavelength dependence of the index of refraction.

Nonlinear degradation mechanisms include self and cross-phase modulation, stimulated Raman and Brillouin scattering, and four wave mixing. Although nonlinear effects are usually small, they can become significant for very long distance or high-power links. Such long distances and high intensities were not common in the early days of fiber optics, but are now much easier to achieve because of the availability of high-power lasers and optical amplifiers.

In this chapter we will also examine a very important class of optical devices, i.e., optical fiber amplifiers. In these devices the mechanism for attenuation is reversed, and optical gain is realized through an additional pump signal. The most important class of these devices is the erbium-doped fiber amplifier (EDFA), which we will discuss briefly at the end of this chapter.

5.2 Optical fiber materials, structure, and transmission windows

As we know from the classical electromagnetic theory, electromagnetic waves can propagate freely in a homogenous dielectric at the speed of light. Alternatively, it is possible to guide these waves by confining them to certain geometries. Such a geometry is called a *waveguide*. Waveguides are used extensively in microwave

M. Azadeh, *Fiber Optics Engineering*, Optical Networks,
DOI 10.1007/978-1-4419-0304-4_5,

applications to channel the signals from a source to a destination. Microwave waveguides are commonly made of hollow metal pipes which create the boundary conditions for the electromagnetic waves that propagate inside them. In principle, it is possible to transfer electromagnetic energy of any frequency with the appropriate waveguide. However, the dimensions of the waveguide must be approximately comparable to the wavelength that it transfers, so lower frequencies require a larger waveguide. Thus, it is not practical to build waveguides for, say, radio waves that have wavelengths of tens to hundreds of meters. Since light waves are high-frequency electromagnetic waves, they too could be transferred in a waveguide with appropriate size, and an optical fiber is essentially a waveguide for light waves [1].

Obviously, an optical waveguide must be made from a transparent material. In general the materials used in the construction of optical fibers include plastics and glasses. Therefore, optical fibers can be divided into glass fibers and plastic optical fibers (POFs). Optical attenuation in plastics is much higher than the attenuation in glasses. For instance, materials such as polymethacrylates (PMMA) or polystyrene, which are commonly used in plastic fibers, show losses in the order of 100–1000 dB/km [2]. Nevertheless, plastic fibers are mechanically sturdy and easier to manufacture. As a result, plastic fibers can provide an economic solution for very short-distance applications.

On the other hand, glasses are excellent candidate materials for optical waveguides because of their very low loss. As a result, virtually all communication fibers are glass based. The most common glass used in fabrication of fibers is silica (SiO_2). In particular, the index of refraction of silica can be modified through *doping*. Dopants such as P_2O_5 and GeO_2 slightly increase the index of refraction of silica, while dopants like B_2O_3 and F reduce the index of refraction [3]. With these materials, the achievable loss in optical fibers reduced from 20dB/km in 1970 to almost 0.2 dB/km (at 1550 nm) in 1979 [4]. Although this is close to the theoretical limit of 0.15dB/km, optical fibers have continued to evolve since then in response to the evolution of fiber optic industry [5].

The standard form of optical fibers is the cylindrical structure shown in Fig. 5.1. Because the size of an optical waveguide must be comparable to the wavelength of light, the diameter of optical fibers are typically in the order of few to tens of micrometers.

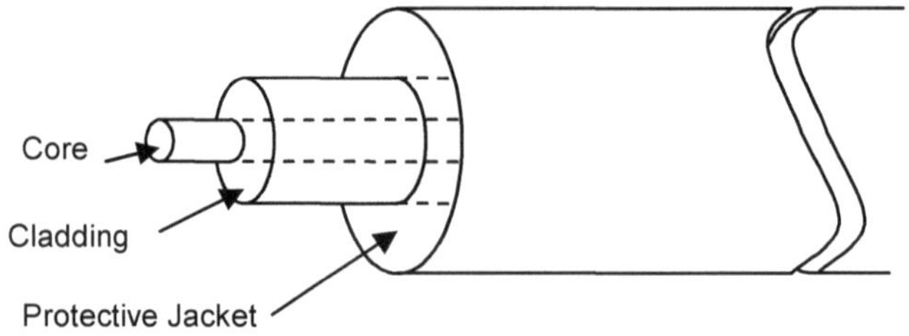

Fig. 5.1. Basic construction of optical fiber

The structure consists mainly of three sections. The center of the structure is called the *core*, and this is the area most of the optical energy is confined to. The core is surrounded by a layer of *cladding*. Both the core and the cladding are made of glass, but the index of refraction of the core is slightly higher than that of the cladding. As we shall see, this slight difference in index of refraction ensures that certain modes of light can propagate along the fiber as guided modes while most of their energy remains confined to the core. The cladding is usually covered by additional layers of protective material. These additional layers do not have a direct role in light confinement. Their role, however, is to provide mechanical support and protection for the inner core and cladding layers. It is also possible to include several optical fibers in optical cables. Such cable assemblies are commonly used in actual optical communication infrastructures.

As noted above, the core must have a slightly larger index of refraction compared to the cladding. The transition of the index of refraction from the core to cladding can take many forms, and several optical properties of the fiber are determined by the radial profile of the index of refraction. Figure 5.2 includes some of the most common profiles used in optical fibers. Single-mode[1] fibers have a much narrower core diameter, usually in the range of 9 μm. Multimode fibers, on the other hand, have a larger core diameter, typically in the range of 50 μm. The typical diameter of the cladding is around 125 μm [6, 7].

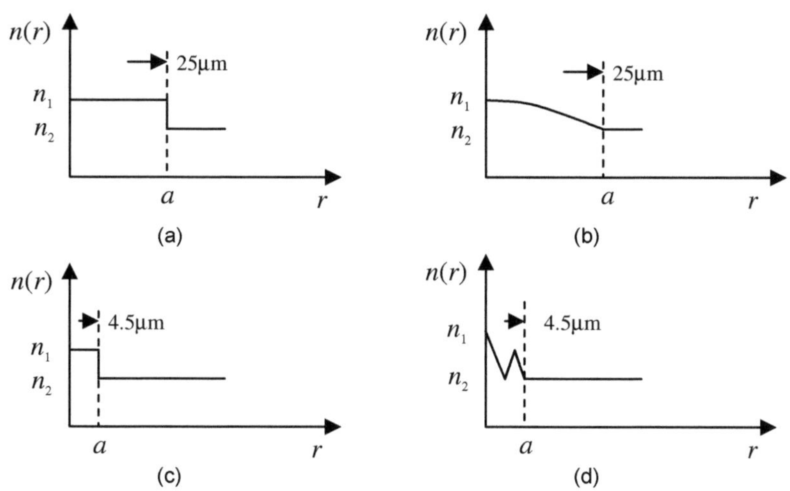

Fig. 5.2. Common profiles for index of refraction, (**a**) step-index multimode, (**b**) graded index (GRIN) multimode, (**c**) step-index single mode, and (**d**) dispersion shifted fiber

[1] We will discuss the single and multimode fibers later in this chapter.

Index profiles are by no means limited to those shown in Fig. 5.2. More complex index profiles have also been proposed that result in different dispersion properties [8–12].

As noted above, achievement of low loss in silica was the main enabling factor in the realization of practical fiber optic links. However, this loss is a function of wavelength, and therefore only those wavelengths where the loss is minimum can be used for practical systems. The wavelengths used in fiber optic communications can be divided into two broad ranges of short wavelengths and long wavelengths. Short wavelengths cover the range of ~800 to 900 nm. As we will see later in this chapter, the attenuation of silica fibers in this range is low enough to allow for short-range links. On the other hand, long-distance links are possible in the range of ~1200 to ~1600 nm, where fiber attenuation reaches its minimum values. Because of the absorption peaks that used to dominate certain wavelength values within this range, it was common to refer to the 800–900 nm range as the first transmission window, and divide the low loss regions in the 1200–1600 range into second, third, and fourth transmission windows. However, it is now possible to build fibers that do not exhibit the local absorption peaks [13]. As a result, it is now common to divide the 1200–1600 nm range into spectral bands. Figure 5.3 shows the designation for these spectral bands.

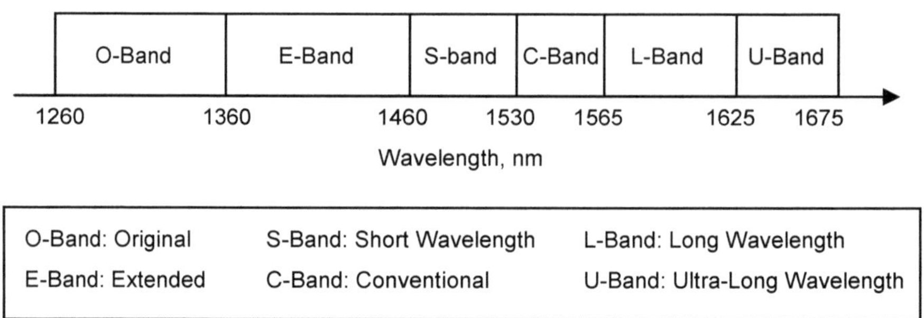

Fig. 5.3. Long-wavelength spectral band designation

As we will see in this chapter, for single-mode fibers, the dispersion becomes zero within the O-band. This means the links operating in this window tend to be *attenuation limited.* On the other hand, attenuation is minimum at 1550 nm, which means links operating in the longer wavelength windows tend to be *dispersion limited.*

5.3 Guided waves in fibers

Intuitively, we tend to think of light as rays propagating in straight lines. Obviously, this intuition is deeply rooted in our daily experiences. The *ray theory* of light formalizes this intuition into a theory that can explain many observations. In reality, however, ray theory can be thought of as the zero wavelength limit of the more general wave theory.

In spite of the wave nature of light, the idea of a ray of light is still a very useful conceptual tool for analyzing cases where the dimensions of the objects that the light interacts with are much larger than the wavelength of light. We can make the connection between a ray and the wave nature of light by thinking of a ray as the direction of light propagation, which is perpendicular to the phase fronts of the light waves, as illustrated in Fig. 5.4 for the case of ideal spherical waves. Thus, the concept of a ray incorporates information about the direction of propagation of optical energy without including any wave behavior.

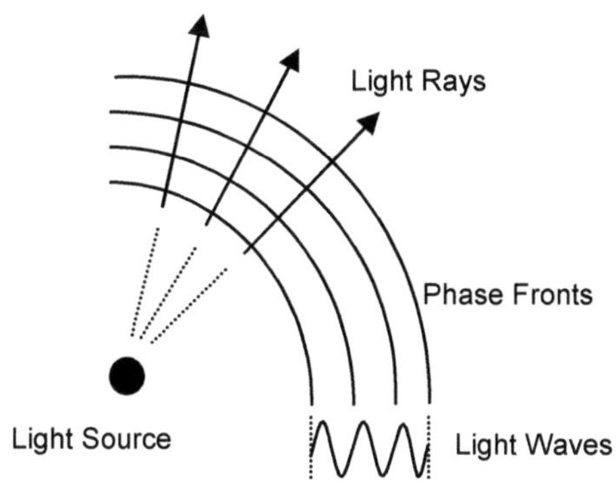

Fig. 5.4. Illustration of light waves, phase fronts, and rays

5.3.1 Guided modes, ray description

The ray theory provides a useful and clear starting point for analysis of light propagation in optical fibers. Although optical fibers have a circular profile, for simplicity we consider only the cross section of a fiber that contains the axis of symmetry of the fiber. Light rays with a small angle relative to the axis of the fiber remain contained within the fiber core due to *total internal reflection*. Figure 5.5 illustrates the situation for one such ray. The indices of refraction of the core, cladding, and the surrounding medium are given by n_1, n_2, and n_0, respectively. The angle of incidence of light from the surrounding medium is ϕ. Part of the light energy is reflected back with the same angle, and part of the light enters the core. The fact that part of the light is reflected back shows that normally it is impossible to couple 100% of the light into the fiber, as there is always some loss associated with the transport of an optical signal between two media with different indices of refraction. The angle of incidence and the angle of refraction are related according to Snell's law:

$$n_0 \sin(\phi) = n_1 \sin(\theta) \tag{5.1}$$

The light ray that has entered the core travels inside the fiber until it reaches the core–cladding interface. The ray has to go through total internal reflection at this point if it is to remain inside the fiber.

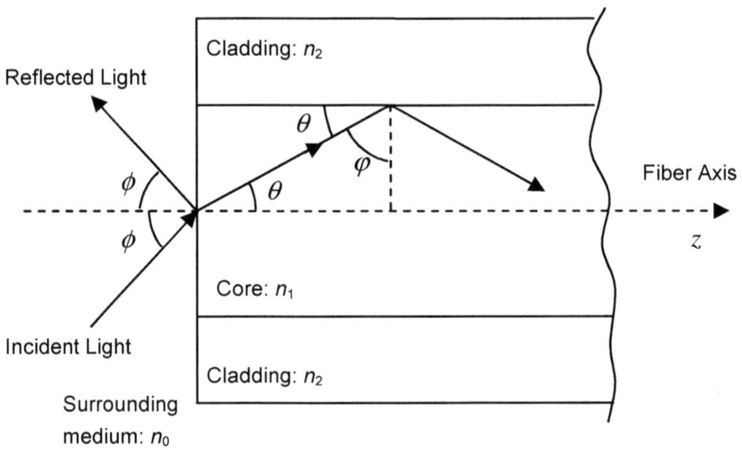

Fig. 5.5. Ray tracing inside a fiber

The *minimum* angle of incidence that will result in total internal reflection is given by

$$\sin(\varphi_M) = \frac{n_2}{n_1} \tag{5.2}$$

Using Eqs. (5.2) and (5.1), and noting that $\sin(\theta)=\cos(\varphi)$, we can relate φ_M to ϕ_M, the maximum possible angle of incidence that will result in a guided ray:

$$\sin(\phi_M) = \frac{\sqrt{n_1^2 - n_2^2}}{n_0} \tag{5.3}$$

Because of the axial symmetry of the fiber, ϕ_M describes a cone around the axis of the fiber. Any ray that enters the fiber and lies in this cone will remain confined inside the fiber due to total internal reflection. This cone describes the *numerical aperture* of the fiber, defined as

$$NA \equiv n_0 \sin(\phi_M) = \sqrt{n_1^2 - n_2^2} \tag{5.4}$$

Note that numerical aperture is a dimensionless quantity that depends only on n_1 and n_2. For commonly used glass fibers, n_2 is very close to n_1, typical values are n_1=1.48 and n_2=1.46. Therefore, Eq. (5.4) predicts that the acceptance angle is very small, in other words, the only rays that remain guided within the fiber are those that are approximately parallel to the axis of the fiber.

Our analysis so far does not put any restriction on the angle of incidence other than the maximum limit set by Eq. (5.3). Therefore, it seems as if any ray that enters the fiber with an angle less than that given by Eq. (5.3) can be guided by the fiber. However, in reality the fiber can only support a finite number of *modes*, which correspond to a finite number of angles of incidence. Here is a case where the ray theory cannot describe all aspects of the physical phenomena involved, and we need to use the wave theory to obtain a more accurate picture.

5.3.2 Guided modes, wave description

As mentioned in Chapter 1, the most comprehensive description of electromagnetic wave propagation is given by Maxwell's equations, given by Eqs. (1.1), (1.2), (1.3), and (1.4), which, combined with the constitutive relations, give rise to the wave equation given by Eq. (1.7). To obtain the behavior of light waves in an optical fiber, the wave equation must be solved in the fiber subject to the boundary conditions imposed by the geometry of the fiber core and cladding. From a mathematical point of view, this is an involved process, and it has been the subject of extensive work [14–17]. A detail analysis is beyond the scope of this book, but the process can be outlined as follows [18]. First, because of the geometry of the optical fiber, the wave equation is written in cylindrical coordinates. Next, it is assumed that the z and t dependence of the waves is given by $\exp j(\beta z-\omega t)$. The radial and azimuthal dependencies are also assumed to be in the form of the product of two separate functions. The azimuthal function must have a periodic dependence on angle, because it must repeat itself after 2π. Substitution of these assumed forms in the wave equation results in a Bessel differential equation for the remaining radial component of the field.

As expected, application of boundary conditions will yield to discrete solutions or *modes*. Although the mathematical description of these modes is complicated, conceptually it is not hard to imagine why such modes may exist. The modes in a dielectric slab waveguide,[2] although not exactly similar to those in a circular

---

[2] A dielectric slab waveguide consists of a layer of waveguide with one index of refraction sandwiched between two dielectric layers with a different index of refraction.

waveguide, can serve as an illuminating case. Figure 5.6 presents a conceptual representation of such modes.

The waves propagate in the z-direction, while they are restricted in the y-direction by the two layers of a dielectric with a lower index of refraction. Because there are no reflections in the *z*-direction, the waves propagate freely along the *z* axis. However, in the y-direction the waves are reflected back from the boundary separating the two dielectrics, as a result, standing waves form in the *y*-direction. Figure 5.6 shows the first few modes in the waveguide. The main mode, designated as k=1, has one peak in the center, while subsequent modes have more peaks. Note that the power profile of these modes does not go to zero at the boundary. Boundary conditions require that fields exist on both sides of the boundary. Solving the wave equation yields evanescent fields that propagate along with the main mode in the *z*-direction, but have an exponentially decaying profile in the *y*-direction.

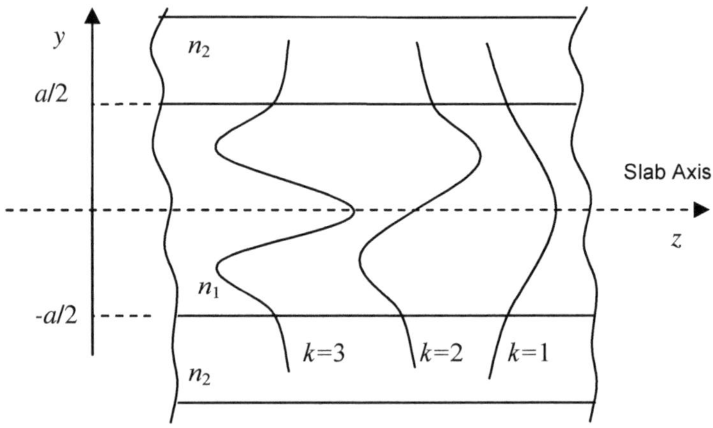

Fig. 5.6. Illustration of guided modes in a dielectric slab waveguide

Figure 5.6 explains why a waveguide supports only a finite number of modes along the direction in which its size is limited.[3] In the *y*-direction, a standing wave is formed when an integer number of half wavelengths fit in the distance separating the two dielectrics ($-a/2$ to $a/2$ in Fig. 5.6). Those waves which do not satisfy this condition will radiate out of the waveguide and will not propagate for long.

It is obvious from Fig. 5.6 that the main mode designated as k=1 has the longest wavelength, and thus the lowest optical frequency, compared to the other modes. Consequently, we can say that the waveguide has a cut-off frequency (or

[3] The situation is very similar to the wave function of an electron in a quantum well. When the size of a potential well gets so small that it becomes comparable to the wavelength of the electron, standing waves form and the electron can occupy only certain discrete energy levels. Each energy level corresponds to one mode, very similar to what is depicted in Fig. 5.6. Moreover, the wave function of the electron decays exponentially outside the potential well.

wavelength). In other words, by choosing the dimensions of the waveguide we can select the modes that it can support. This is indeed what is done in single-mode optical fibers. As we will see, having only a single mode in the fiber significantly enhances the distance that optical signal can propagate without being distorted. As a result, long-distance optical communication is based on single-mode transmission in the fibers.

5.3.3 Signal degradation in optical fibers

As optical signals propagate in the fiber, they experience several forms of degradation which ultimately limit the range of a particular fiber link. The signals cannot be transmitted beyond this limit, unless they go through some form of active reconditioning. For instance, a repeater may be used to detect the signal, convert it to the electrical domain, and then regenerate the signal in the optical domain. Obviously, it is desirable to extend the reach of an optical link by minimizing the deterioration of signals.

Signal degradations can be divided into the two general categories of linear and nonlinear. Linear degradations are those that do not depend on the optical power and affect weak and strong signals equally. In nonlinear degradations, on the other hand, the effects become significant at high-power levels. Thus, nonlinear effects limit the maximum power that can be launched into the fiber.

The most common types of linear degradation are attenuation and dispersion, which we will consider next. Important nonlinear effects include self and cross-phase modulation, four wave mixing, stimulated Raman scattering, stimulated Brillouin scattering, and four wave mixing. The last sections of this chapter deal with these nonlinear effects.

5.4 Attenuation

The most common form of signal degradation is attenuation. In fact, it was only after manufacturing of low loss silica fiber became possible that the field of fiber optics communication took off. As we noted in Chapter 1, attenuation is commonly characterized in units of decibel per kilometer. The rationale behind this choice becomes clear once we recognize that the power in an optical signal attenuates exponentially with distance:

$$I(z) = I_0 e^{-\alpha_F z} \tag{5.5}$$

Here z is the distance along the length of the fiber, I is the optical power, I_0 is power at z=0, and α_F is the attenuation factor. Equation (5.5) can alternatively be written as

$$10\log I(z) = 10\log(I_0) - 4.343\,\alpha_F z \tag{5.6}$$

where we have used $\ln(x)\approx 2.302\log(x)$. Equation (5.6) provides a simple and convenient way of expressing the power loss as a function of distance. If we define $\alpha=4.343\alpha_F$, Eq. (5.6) becomes

$$P_{\text{dBm}}(z) = P_{\text{dBm}}(0) - \alpha z \tag{5.7}$$

where power is now expressed in units of decibel-milliwatt, and α is known as the attenuation coefficient of the fiber in units of decibel per kilometer. To obtain the power at a point z along the fiber, one needs only to multiply the attenuation coefficient α by the distance z and subtract the result from the launching power at z=0.[4]

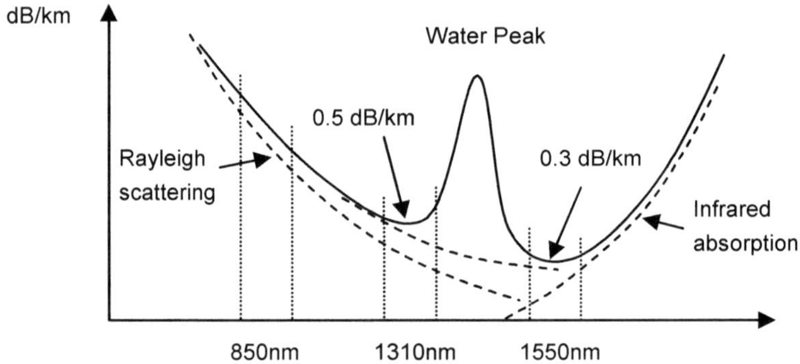

Fig. 5.7. Attenuation profile of silica fibers

Figure 5.7 shows the typical attenuation profile of modern silica glass fibers as a function of wavelength. Traditionally, three wavelength windows have been extensively used in optical communication: the 850, 1310, and 1550 nm windows. Because of higher attenuation, the 850nm window is used for short-distance links. This window corresponds to the bandgap of GaAs. Long-distance links are based on the two other windows, where the nominal loss is 0.5 dB/km at 1310 nm and 0.3 dB/km at 1550 nm. The longer wavelengths correspond to the bandgap of ternary and quaternary semiconductors such as InGaAs and InGaAsP.

Attenuation of light in glass fibers can be contributed to a number of factors. Most important among these are absorption and scattering. We will briefly discuss these mechanisms here.

[4] Note that Eq. (5.7) does not take into account the coupling losses. In reality, there is always some additional loss when light is coupled in and out of fiber.

5.4.1 Absorption

Absorption refers to dissipation of optical energy as the waves propagate in the fiber. Part of the absorption is inherent to the glass substance of the fiber. This component of absorption can be attributed to the interaction of light with the atoms and molecules that make up the bulk of material in a glass fiber. As described in Chapter 4, light can interact with electrons in a material when a photon of light is absorbed and causes transition of an electron to a higher energy level. However, photons interact not only with electrons, but also with molecules. Like electrons, molecules also have energy levels as a result of their various vibration states associated with their internal atomic bonds. In optical fibers, electronic transitions correspond to shorter wavelengths, and the loss associated with these electronic transitions decreases exponentially as wavelength increases. Moreover, these losses are much smaller than scattering losses (to be discussed later). Therefore, for the most part, ultraviolet losses can be ignored. On the other hand, infrared wavelengths interact with molecular transitions, causing infrared absorption. These transitions are the main component of the total loss in the longer wavelength region. These losses increase exponentially with wavelength as shown in Fig. 5.7.

Another component of absorption is due to impurities, that is, materials other than glass that find their way in during the production process. These materials include metals such as copper and iron, and water. In order to have acceptable levels of impurity absorption, the concentration of these impurities must be kept at extremely low levels. For instance, water has an absorption peak around 2.7 μm, which results in harmonics at 1400, 950, and 725 nm. Originally, the transmission windows of glass fibers had to lie between these absorption peaks. Most commercially available fibers still have a water peak at around 1400 nm, as shown in Fig. 5.7. However, it is possible to eliminate water almost completely, and the result will be a fiber which shows low loss over a wide range of wavelengths, starting from 1300 nm all the way to 1600 nm [13].

5.4.2 Scattering

The dominant form of attenuation at shorter wavelengths is scattering. This is in fact the same phenomena known as *Rayleigh scattering*, responsible for the blue color of sky. Shorter (bluer) wavelengths of sun light scatter more as they travel through the atmosphere and encounter the gas molecules in the air, giving rise to the blue color of the sky. Longer (redder) wavelengths scatter less, and therefore tend to travel in straight line. Rayleigh scattering essentially functions the same in fibers. As light propagates in glass, it encounters random local variations of index of refraction resulting from random defects, material composition fluctuations, and various inhomogeneities. Rayleigh scattering scales as λ^{-4}, and therefore drops rapidly at longer wavelengths. In optical fibers, the losses associated with Rayleigh scattering dominate at shorter wavelengths, mainly below 1000 nm.

To summarize, the typical total absorption curve of Fig. 5.7 is the result of at least three mechanisms. At short wavelengths, Rayleigh scattering is the dominant cause of loss, where the λ^{-4} dependence results in the sharp decrease in attenuation as the wavelength increases toward the communication windows. At longer wavelengths, infrared absorption dominates, and this causes the sharp increase in total attenuation as wavelengths increase beyond the communications windows. Finally, around the communications windows, impurity absorption can cause local peaks in the attenuation profile, such as the dominant water peak at 1400 nm.

5.5 Dispersion

An important factor that limits the performance of optical links especially at higher data rates is dispersion. In general, dispersion refers to the spreading of the pulses that make up a signal as it travels along the fiber. Figure 5.8 illustrates pulse spreading due to dispersion for a representative byte of data as it travels along the fiber.

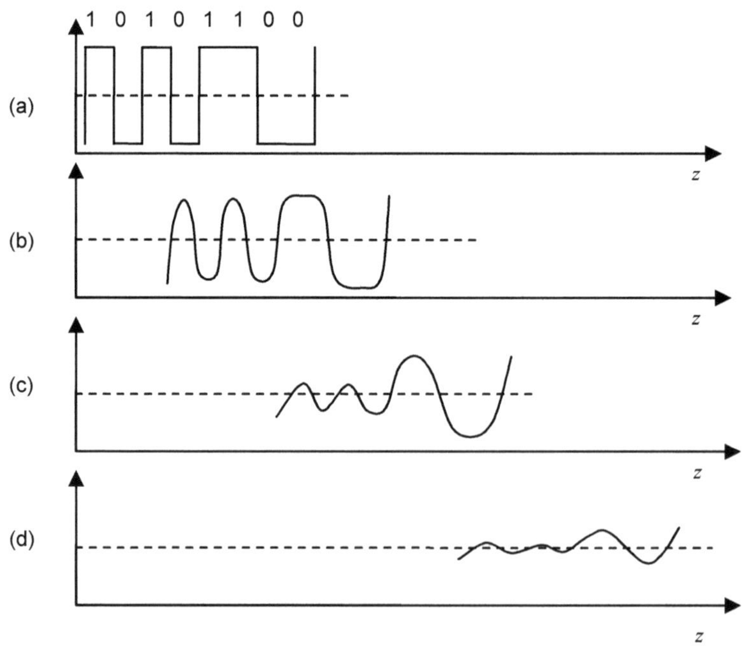

Fig. 5.8. Pulse dispersion as function of the distance traveled within the fiber

Let us assume that the byte is given by the sequence 10101100 and that when it is launched in the fiber it has sharp edges clearly separating individual bits. Note that in this example the first half of the byte is composed of higher frequency components, while the second half consists of lower frequency components. As this byte of data travels, the sharp edges start to round. Initially, all the individual bits are still clearly distinguishable, as shown in Fig. 5.8b. However, as the pulse travels further, the rounding effect continues, and at some point the first four bits, which represented a higher transition rate, start to merge. This is the situation in Fig. 5.8c. Eventually, after sufficient distance, the last four bits of the packet also start to merge, which means soon after this point all this byte of data will be completely lost.

This example demonstrates that dispersion affects higher data rates more severely, because at lower data rates the width of each bit is wider and thus it takes a longer distance before the edges completely collapse as a result of pulse broadening. Therefore, the product of data rate and distance is usually used as a figure of merit to characterize the information capacity of fibers. For instance, a fiber with a capacity of 100 Mbps.km can support a data rate of 100 Mbps over a distance of 1 km, or a data rate of 1 Gbps over a distance of 100 m, etc. Multimode fibers have generally a much lower capacity compared to single-mode fibers, and this is because the dominant mechanism of dispersion is different in a single-mode fiber compared to a multimode fiber, as discussed shortly.

Pulse broadening or dispersion can be due to several mechanisms. Therefore, dispersion can be divided into several categories, the most important of which are modal, chromatic, waveguide, and polarization dispersion. The first mechanism dominates in multimode fibers, whereas the others dominate in single-mode fibers.

5.5.1 Modal dispersion

In multimode fibers, modal dispersion is the dominant form of dispersion. To understand the root of modal dispersion, it is easier to use the ray theory to describe the propagation of light. Although a rigorous treatment requires the wave theory, the ray theory provides a sufficient qualitative description. Let us consider Fig. 5.9 which depicts a pulse of light that is launched into the fiber. The pulse starts as several overlapping modes.

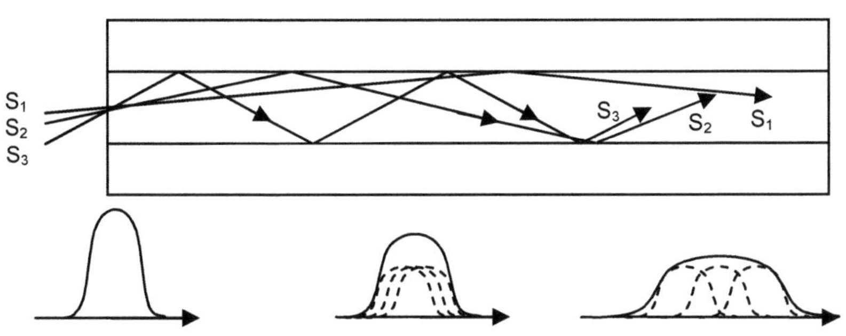

Fig. 5.9. Modal dispersion in a multimode fiber

To keep the figure simple, only three modes are shown which are represented by three rays, S_1–S_3. At the launch point, all the three modes that make up the pulse are aligned. However, as they propagate along the fiber, the mode represented by S_1 travels a shorter distance relative to the other modes, and as a result moves ahead of the others. On the other hand, the mode represented by S_3 travels a longer distance relative to others, and therefore lags behind the others. The result of this difference in effective propagation speeds is that the original pulse spreads out in time, and if it is part of a sequence of pulses, it will merge with other pulses.

Graded index (GRIN) fibers can significantly reduce signal deterioration due to modal dispersion. In a GRIN fiber, the rays that propagate further from the axis of the fiber travel a longer distance in a medium with a lower index of refraction, and therefore travel faster. As a result, they catch up with those rays that propagate closer to the center and travel at a lower speed. Therefore, as shown in Fig. 5.10, the rays that start diverging at the launch point get refocused periodically as they travel along the fiber.

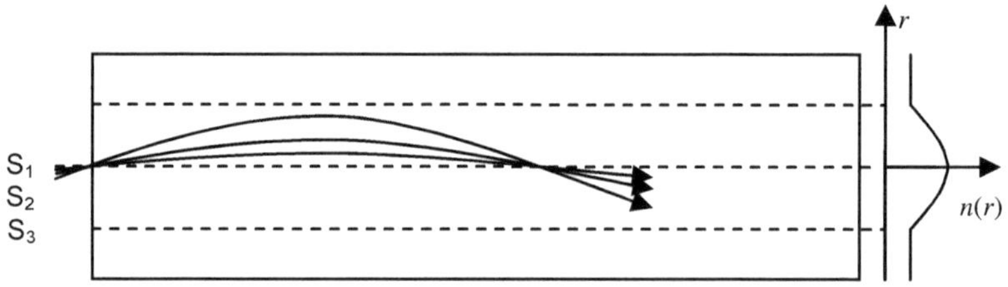

Fig. 5.10. Trajectory of various modes in a graded index fiber

As noted earlier, modal dispersion is the dominant cause of signal distortion in multimode fibers. In step-index fibers, the information capacity is limited to approximately 20 MHz·km. GRIN fibers have a much higher capacity, typically in the range of 2 GHz·km or even higher. On the other hand, single mode fibers do not show modal dispersion precisely because only one mode carries all the optical energy.

5.5.2 Chromatic dispersion

Although in most cases the index of refraction of glass is treated as a constant, in reality it is a weak function of frequency. At the same time, optical signals do not have an infinitely narrow-band spectrum, i.e., their energy consists of a range of optical frequencies. Therefore, various frequency components of an optical signal experience a slightly different index of refraction as they travel through the fiber. Because the speed of propagation is a function of index of refraction, these various components travel at different speeds. The result is that a given pulse of light spreads out as the pulse propagates, giving rise to what is normally called chromatic dispersion. It should be noted that chromatic dispersion (also known as intra-modal dispersion) is different from modal dispersion discussed before, because it affects signals within a single mode. Thus, in a multimode fiber both modal and chromatic dispersion exist, although modal dispersion dominates. On the other hand, a single-mode fiber eliminates modal dispersion, but still shows chromatic dispersion.

Chromatic dispersion is usually characterized in terms of the *dispersion coefficient*, D, expressed in units of picosecond per (nanometer-kilometer). Knowing the dispersion coefficient of the fiber, we can estimate the time spread of a given pulse as

$$\tau = DBL \tag{5.8}$$

where τ is total delay introduced between the frequency components of the pulse in ps, B is the spectral width of the light source in nm, and L is the distance the pulse has traveled.

Equation (5.8) provides a simple tool for analyzing the effects of various parameters in dispersion. First, it is obvious that dispersion is more detrimental at higher data rates, because a given value of pulse spreading that can easily be tolerated at a low data rate can completely distort a signal at a higher data rate. In general, in order for dispersion to be negligible, the pulse spreading τ must be much smaller than the bit period T_B:

$$\tau << T_B \tag{5.9}$$

Moreover, to minimize the effects of dispersion for a given distance, either D or B must be minimized. Reduction of B can be achieved using a source with a narrower spectral width. For example, an LED can have a spectral width of 50 nm, a multimode Fabry–Perot diode laser may have a spectral width of 10 nm, and a DFB laser may have a spectral width of less than 1 nm. Moreover, as explained in Chapter 4, direct modulation can cause spectral broadening. As a result, an externally modulated DFB laser can show the narrowest spectral width achievable, therefore minimizing chromatic dispersion.

On the other hand, the dispersion coefficient D is a function of the structure of the fiber. In standard fibers the dispersion is at a minimum at around 1300 nm. Thus, using a source with a wavelength close to 1300 nm can significantly reduce the dispersion. This coincides with the transmission window of the fiber at the same wavelength. It is also possible to change the dispersion characteristics of fiber by manipulating the profile of the index of refraction. For example, it is possible to shift the zero-dispersion wavelength of a fiber to longer wavelengths. The resulting fiber, known as *dispersion-shifted* fiber, has a zero dispersion wavelength around 1550 nm. The advantage of this fiber is that its zero dispersion wavelength coincides with the transmission window of 1550 nm where the loss reaches the overall minimum value of 0.2 dB/km. However, these fibers have certain shortcomings, for instance, they yield to more crosstalk between adjacent channels in WDM applications [19]. Another kind of fiber, called *dispersion-flattened* fiber, shows a relatively constant dispersion characteristic over a wider wavelength range. Like dispersion-shifted fibers, this can be achieved by careful manipulation of index profile [20]. Figure 5.11 shows the dispersion curves of standard and dispersion-shifted fibers.

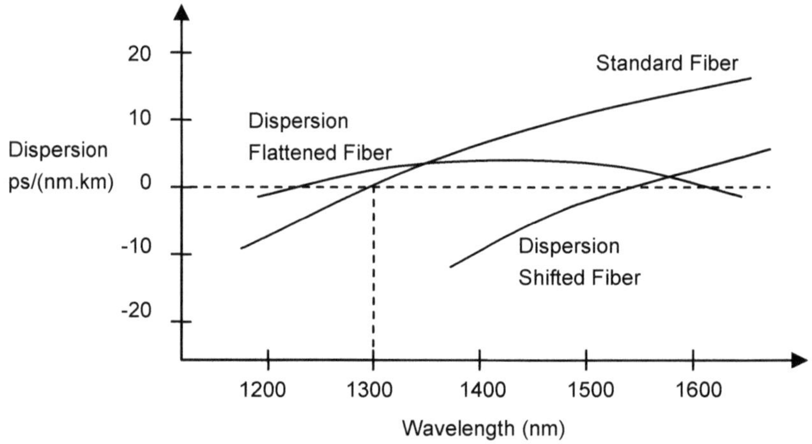

Fig. 5.11. Dispersion coefficient for standard and dispersion-shifted fibers

Another interesting structure is the *dispersion compensating* fiber [21–23] It is possible to design fibers that have a waveguide dispersion with the opposite sign compared to the standard fibers. These fibers are called dispersion compensating fibers (DCFs), and they can be used to offset the effects of accumulation of chromatic dispersion in standard fibers. The overall magnitude of dispersion in a DCF is usually very high, say, 100 ps/nm·km. As a result, a short patch of DCF can be used to counteract the dispersion resulting from much longer signal transmission in a standard fiber. However, DCFs are usually implemented with a very narrow core, which can result in high power densities and thus nonlinear effects.

5.5.3 Waveguide dispersion

Like chromatic dispersion, waveguide dispersion results from non-zero spectral width of an optical source. However, unlike chromatic dispersion which results from wavelength dependence of the index of refraction of glass, waveguide dispersion is a result of the fact that in a single-mode fiber part of the light energy propagates in the cladding. Because the index of refraction in the cladding is slightly less than that of the core, the portion of signal that is in the cladding moves faster, causing the spread of energy in a pulse.

Although the effects of waveguide dispersion are usually small, in some cases they become important. For example, as noted above, in dispersion compensating fibers, waveguide dispersion effects are deliberately utilized and their effects are maximized to achieve an overall dispersion coefficient opposite to that of the standard fibers.

5.5.4 Polarization dispersion

Another form of dispersion results from pulse broadening due to polarization of an optical pulse and the birefringence of fibers. To be more accurate, a given pulse may consist of different polarization components or states, and fibers do not provide an identical index of refraction for these polarization states. Factors such as stress, fiber bends, variations in the composition or geometry, and non-circularity of the fiber can preferentially cause slight changes in the effective index of refraction of fiber along different polarization directions. When different polarization components of an optical signal experience different indices of refraction, they propagate with different speeds, causing pulse broadening and dispersion. This effect is known as polarization mode dispersion (PMD) [24–26].

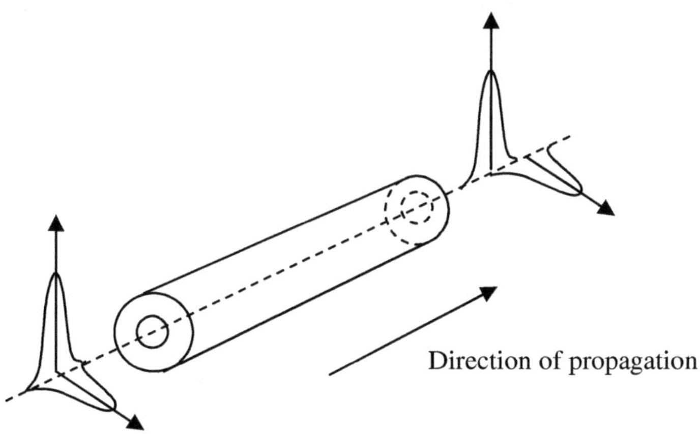

Fig. 5.12. Polarization mode dispersion (PMD) as a result of fiber birefringence

Figure 5.12 is an illustration of PMD. We can think of a given pulse of light as being consisted of two orthogonal polarizations. Because of the birefringence of the fiber, the two modes travel at different group velocities, and therefore they start to separate as the pulse propagates along the fiber. The important point about PMD is that unlike other dispersion phenomena, it is random and in general time varying. This is because environmental causes of birefringence such as temperature or stress variations are generally random. However, we should note that PMD effects are generally small, and they start to become important only at very high data rates, i.e., around 10 Gbps and above.

The random nature of PMD results in a sub-linear accumulation of delay as a function of distance. The delay between the orthogonal polarization states is approximated by [18]

$$\tau = D_{PMD}\sqrt{L} \tag{5.10}$$

where D_{PMD} is known as the polarization dispersion coefficient (in ps·km$^{-1/2}$), and L is distance of pulse propagation. As expected, D_{PMD} is a function of the construction of the fiber, but in general it is in the range of 0.1–1 ps·km$^{-1/2}$.

It should be noted that the random nature of PMD can be controlled by intentionally introducing birefringence in the fiber. The additional birefringence, which can be a result of non-symmetrical fiber designs, can result in a decoupling between polarization states. Moreover, it can introduce preferential attenuation for one polarization state relative to the other [27]. Using such phenomena, it is possible to design *polarization maintaining* fiber structures, which are useful in polarization-sensitive applications [27–30].

5.6 Nonlinear effects in fibers

As noted in the beginning of this chapter, nonlinear effects start to play a role at higher optical intensity levels. Such high intensities are typically encountered in single-mode fibers where the cross-section is small. Moreover, because of optical amplifiers, very high-power signals can now commonly be produced. Nonlinear effects in fibers are mainly due to two causes. One root cause lies in the fact that the index of refraction of many materials, including glass, is a function of light intensity. This dependence gives rise to effects such as four wave mixing, as well as self- and cross-phase modulation. The second root cause is the nonelastic scattering of photons in fibers, which results in stimulated Raman and stimulated Brillouin scattering phenomena [31–35]. In this section we discuss the origin and the effects of these phenomena.

5.6.1 Self- and cross-phase modulation (SPM and XPM)

As noted above, the index of refraction of many materials has a dependence on the intensity of the optical waves that pass through the material. This is in addition to the dependence of the index of refraction on wavelength, which gives rise to dispersion effects. For most cases, a linear expansion of index of refraction with intensity is sufficient for modeling the resulting effects:

$$n(I) = n_0 + n_2 I \tag{5.11}$$

Here n_0 is the usual index of refraction of the material at low-intensity limits, n_2 is the *nonlinear index coefficient*, and I is the optical intensity in W/m^2 [36,37]. For silica fibers, n_0=1.46 and n_2=3.2×10^{-20} m^2/W. [32]

The effects of an index variation in the form of Eq. (5.11) can be explained as follows. The electric field variation in an electromagnetic wave can be described by the following equation [15]:

$$E = E_0(z,t)\exp i\left(\left[n_0 + n_2 I(z,t)\right]k_0 z - \omega_0 t\right) \tag{5.12}$$

where $E_0(z,t)$ is the envelope of the wave, ω_0 is the center frequency of the optical wave, and k_0 is the propagation constant in vacuum. Note that when n_2 is zero, Eq. (5.12) reduces to the standard form of an electromagnetic wave with the phase given by $(n_0 k_0 z - \omega_0 t)$. The time dependence of $I(z,t)$ can then be lumped into the frequency term by taking the time derivative of the phase, and the result can be expressed in terms of an instantaneous frequency, $\omega(z,t)$:

$$\omega(z,t) = n_2 k_0 z \frac{\partial I}{\partial t} + \omega_0 \tag{5.13}$$

Equation (5.13) provides an insight into the mechanism responsible for self-phase modulation (SPM) [38]. As a pulse of light travels through the medium, a change in its intensity causes a change in the optical frequency of the wave. In other words, the intensity of the wave causes chirp or spectral broadening. Moreover, it can be observed from Eq. (5.13) that the instantaneous frequency is a function of the slope of the pulse. The rising and falling edges of the pulse generate the largest frequency shift. Figure 5.13 shows this effect.

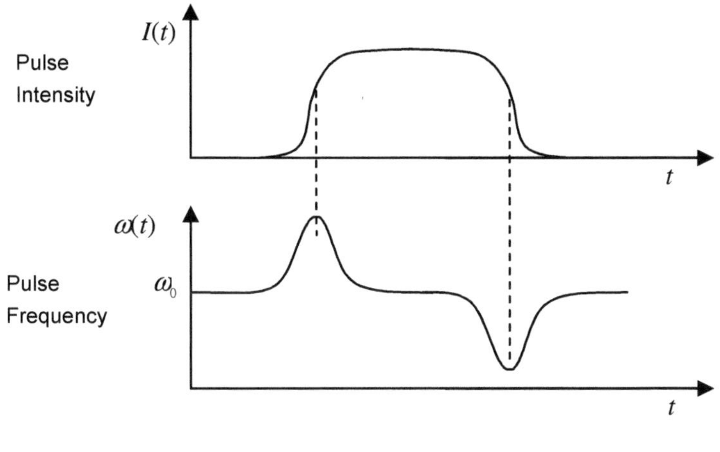

Fig. 5.13. Self-phase modulation as a result of intensity dependence of index of refraction

The addition of self-phase modulation with other dispersion mechanisms can result in interesting effects. If the phase shift due to self-phase modulation has the same sign as the other dispersion effects, the effects will add and the result is faster pulse spread with distance. On the other hand, if the two have the opposite sign, they can cancel each other. This is the basis for *solitons*, pulses that can travel for extraordinarily long distances without suffering from dispersion effects.

The same nonlinear mechanism is the underlying principle behind cross-phase modulation (XPM) [39]. The only difference is that in cross-phase modulation two (or more) pulses propagate simultaneously, and the intensity variation in one of them modulates index of refraction of the medium, which in turn causes phase modulation in the other pulse. The typical example is in WDM systems where several wavelengths propagate along each other. As a result, cross-phase modulation acts like a crosstalk mechanism, where the existence of one signal disturbs the other.

5.6.2 Four Wave Mixing (FWM)

Another nonlinear effect that affects signal propagation in fibers is four wave mixing (FWM) [40]. This happens when three pulses at three different frequencies travel side by side each other for a long distance in the fiber. If the frequencies of these three pulses are f_1, f_2, and f_3, they tend to mix and generate waves at a new frequency, f_{FWM}, given by

$$f_{FWM} = f_1 + f_2 - f_3 \tag{5.14}$$

The energy of this new wave becomes larger when the power of the individual waves is larger. It can be checked easily that if the three frequencies are equally spaced, and if f_2 is between f_1 and f_3, then f_{FWM} will be at the same frequency as f_2. This means that f_1 and f_2 will generate an interference signal for f_3. Moreover, the more these waves propagate along side each other, the more energy is depleted from the signals at f_1 and f_2 and transferred to the newly generated signal at f_3. Figure 5.14 is an illustration of this situation, where three pulses P_1, P_2, and P_3 are shown at the beginning of the fiber and after they have traveled together. Also shown are the spectral components of these three pulses, f_1, f_2, and f_3. As a result of FWM, P_1 and P_3 are depleted and additional signal component have been created at f_2, causing distortion to P_2. Obviously FWM can create problems for WDM systems, causing crosstalk between channels.

The key to manage FWM lies in the fact that the FWM process is efficient when pulses get a chance to travel side by side each other for a long distance. However, this happens only close to the zero dispersion wavelength of a fiber, because it is only at this point that the group velocity of different wavelengths is the same and therefore different wavelengths get a chance to interact with each other for a long distance.

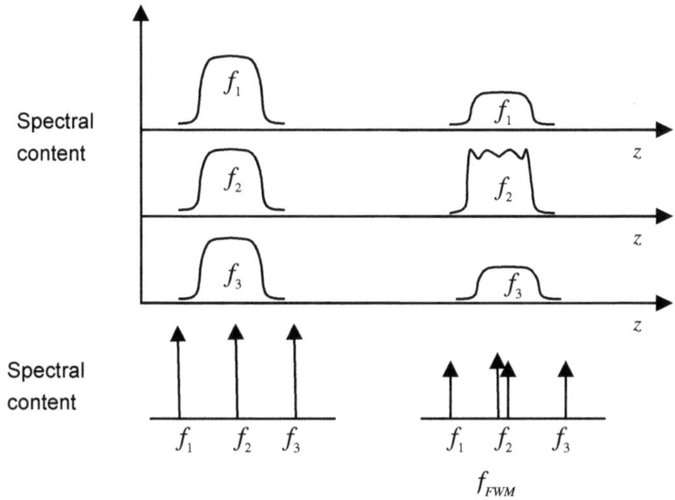

Fig. 5.14. Illustration of four wave mixing

Therefore, eliminating dispersion completely is not very beneficial for WDM systems. This is why dispersion-shifted fibers that have zero dispersion at the 1550 nm range are not very suitable for WDM applications. Instead, non-zero dispersion shifted fibers in which dispersion is reduced, but not totally eliminated show the optimal WDM performance.

5.6.3 Stimulated Raman scattering (SRS)

When we were discussing the principles behind light amplification in lasers we described the phenomenon of stimulated emission. Stimulated Raman scattering (SRS) is a somewhat similar process, except that it involves the interaction of photons with phonons which represent the vibrational modes of molecules in silica fiber. Figure 5.15 shows an overview of the photon–phonon interaction involved in SRS. A photon with energy hf_1 is absorbed by the molecular structure of the silica, causing it to transfer from the initial energy state of E_I to E_R. The molecule then decays down to energy state E_F, releasing a photon with energy hf_2. This photon, whose frequency (and therefore energy) is less than the incident photon, is called the *Stokes* photon. The difference between hf_1 and hf_2 is the energy of the phonon produced by the process. Because part of the energy of the incident photon is absorbed in the form of a phonon, the energy of the scattered photon will be less. In other words, the scattered light will be red-shifted relative to the incident light.

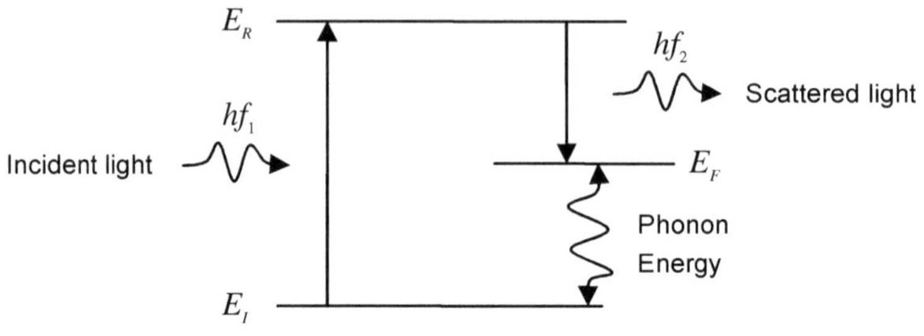

Fig. 5.15. Stimulated Raman scattering

For silica fibers, the spectrum of Raman-generated light increases almost linearly below f_1 until it reaches a maximum power at ~15 THz (a red shift of ~125 nm). After that it drops suddenly, but a tail end extends well beyond 15 THz [41].

As expected, this can be a significant degradation factor for WDM systems, because energy from shorter wavelengths transfers to longer wavelengths, interfering with signals that already exist in those longer wavelengths. The effects of SPS can be somewhat mitigated by careful choice of wavelengths. Nevertheless, SRS is a serious limiting factor in WDM systems.

5.6.4 Stimulated Brillouin scattering (SBS)

The scattering of light waves from acoustic waves (low-frequency phonons), known as stimulated Brillouin scattering (SBS), is another nonlinear effect in fibers [42,43]. The scattered light undergoes a frequency change as a result of the Doppler shift caused by the speed of sound waves (Fig. 5.16).

The phenomenon is called stimulated because the sound waves can be generated by the light waves, once the energy of the light wave increases beyond a critical level. For single-mode fibers this threshold is around 10 mW. In silica fibers, the scattered light is red-shifted by approximately 11 GHz. This is a very small frequency shift relative to optical frequencies.

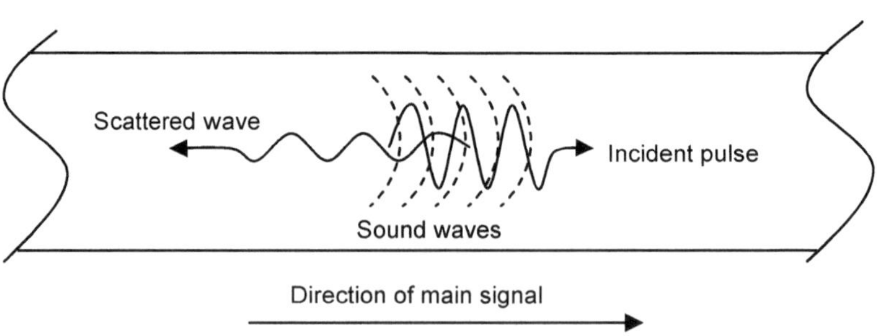

Fig. 5.16. Stimulated Brillouin scattering. The scattered wave is red-shifted as a result of Doppler shift.

As a result, SBS does not introduce crosstalk between WDM channels, which are usually spaced farther apart. Instead, SBS depletes the energy out of an optical signal and converts it into scattered waves that travel in the opposite direction relative to the incident wave. Therefore, the optical signal effectively experiences a higher attenuation as a result of SBS, and the extra attenuation increases as the launched power is increased. As a result, SBS effectively acts like a saturation mechanism for single-mode fibers. Moreover, the reflected light can cause problems in the source, because certain optical sources (such as DFB lasers) are very sensitive to back scattered light. In such cases optical isolators must be used to prevent the scattered light from reaching sensitive devices.

5.7 Fiber amplifiers

So far in this chapter we have discussed several phenomena related to propagation of light in an optical fiber. In particular, the propagation of light in a normal fiber always involves attenuation, a factor that limits the range of an optical link. In principle, there are two ways to address signal degradation. One way is to detect the optical signal before it is degraded too much, convert it into the electrical domain, and regenerate a new copy of the optical signal. The advantage of this approach is that once the signal is converted into electrical domain, additional high level processing tasks can also be performed on the signal. An alternative approach is to use an optical amplifier. An optical amplifier can amplify the optical signals in the optical domain without the need to convert them back to the electrical domain. In fact, just like their electrical counterparts, optical amplifiers can be used in a variety of applications. They can be used as inline amplifiers to compensate the attenuation of optical signals, they can be used as pre-amplifiers in optical detectors to improve the sensitivity of the receiver, or they can be used as power boosters to increase the output power of an optical transmitter.

In general, amplifiers are divided into two categories: semiconductor optical amplifiers (SOAs) and fiber amplifiers, also known as doped fiber amplifiers (DFAs). SOAs can provide gain in both 1310 and 1550 nm low attenuation windows. We discussed the principles behind optical gain in a semiconductor junction in Chapter 4. To summarize, optical gain in semiconductors is based on a forward-biased junction. If the reflections from the facets of the cavity that encloses the junction are minimized, the system will not self-oscillate, and therefore a semiconductor amplifier is realized. Fiber amplifiers, on the other hand, are based on optical gain in an optical fiber. Because of the importance of fiber amplifiers, we briefly review the principals behind their operation [44–49].

In fiber amplifiers, an optical fiber is doped with a rare earth element, such as neodymium (Nd) and praseodymium (pr) in fluoride-glass fibers for 1310 nm operation, and ytterbium (Yb) or erbium (Er) in silica-glass fibers for the 1550 nm window. Just like a semiconductor amplifier, realization of optical gain in a fiber amplifier requires population inversion. The introduction of dopants generates new energy bands (corresponding to new transition wavelengths) within the fiber. The required population inversion is achieved through *optical pumping* at a wavelength corresponding to these new energy levels. This process is schematically shown in Fig. 5.17.

The pump signal, at a frequency f_p, interacts with the dopant ions and moves them to a higher energy state, E_P. The lifetime of these ions in this state (τ_P) is relatively short, thus, they quickly decay to a lower energy band, E_1, where they have a longer lifetime (τ_1). The difference in energy between E_1 and E_0 corresponds to the frequency of the main signal which needs to be amplified. Because $\tau_1 >> \tau_P$, a condition of population inversion is created. Photons of the main signal with frequency f_s interact strongly with the inverted system, and as a result of stimulated emission the signal is amplified. Just like lasers, there is also spontaneous decay from the E_1 band into E_0 band which results in *amplified spontaneous emission* (ASE) adding to the noise at the output of the amplifier.

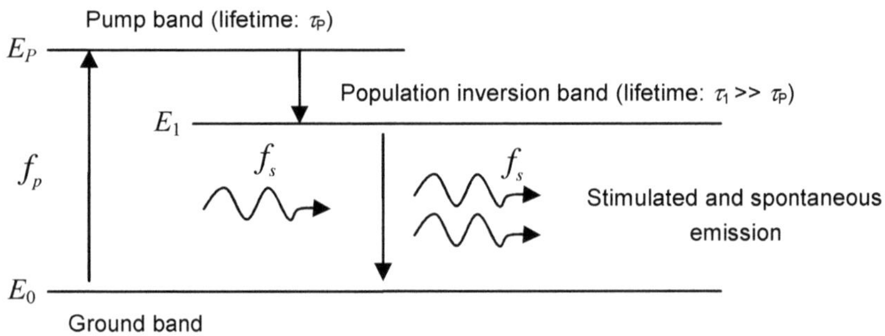

Fig. 5.17. Band diagram of a doped fiber amplifier (DFA)

Among fiber amplifiers, the erbium doped fiber amplifier (EDFA) is the most popular type, because E_1–E_0 corresponds to the wavelength of 1550 nm. Thus, EDFAs can amplify signals in the window of lowest attenuation in silica fibers, and therefore are widely used in long-distance communication [46–49]. In EDFAs the pump signal is usually at 980 nm, although a lower energy transition corresponding to 1480 nm is also available. In reality, the bands that are shown as sharp lines in Fig. 5.17 are broadened, which means an EDFA can provide gain across a range of wavelengths. Typical EDFA amplifiers can provide gain across a 30 nm band (~1530–1560 nm) with a gain of 20–30 dB. As a result, they are ideal for signal amplification in WDM systems, because a whole range of channels can be amplified simultaneously without the need for any electronic processing.

Figure 5.18 shows a possible EDFA configuration. The main signal which needs to be amplified enters the setup from left bottom side, where it is coupled to the pump signal through a 2×2 coupler (for a discussion of this and other passive devices see Chapter 7). The main output from the coupler enters the doped fiber. Here the main and the pump signals travel together in the fiber for a typical length of several meters. Along the way, the energy from the pump signal goes into creating population inversion in the fiber, which in turn causes amplification of the main signal. The other output branch of the coupler can be used for signal monitoring purposes. At the end of the doped fiber, another coupler is used to transfer the signal to the rest of the network.

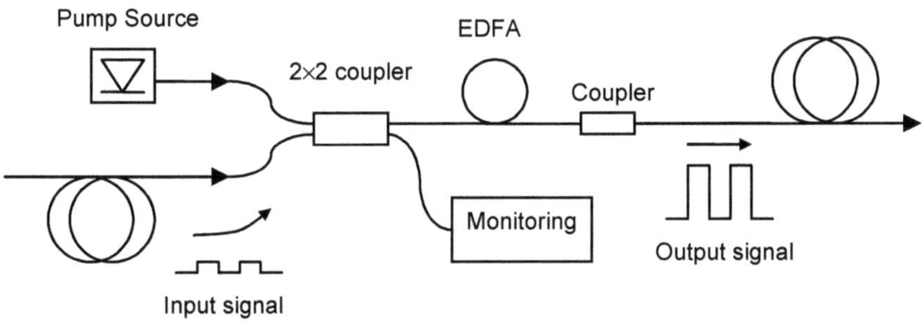

Fig. 5.18. Sample EDFA configuration

The setup illustrated in Fig. 5.18 is an example of *co-directional* pumping, because the main and the pump signals travel in the same direction. It is also possible to have *counter-directional* pumping, where the pump signal is applied from the other side of the EDFA and travels in the opposite direction to the main signal. Alternatively, the pump signal can be applied in a *bi-directional* configuration, from both sides of the fiber. A bi-directional setup results in a more steady pump power (and therefore gain profile) along the fiber at the price of additional complexity and price.

EDFAs are crucial components in long-distance fiber optic communication. They have enabled signal transmission for thousands of kilometers without the need for optical to electrical conversion. Therefore, improving their performance and extending their bandwidth is an active research area [50,51].

5.8 Summary

In this chapter we examined both the propagation and the degradation of signals in optical fibers. Technological advancements have made the production of extremely pure glass fibers possible. Fiber optic communication is based on the low loss transmission windows in glass silica fibers, which are essentially cylindrical optical waveguides.

An optical fiber consists of a core and a cladding. Because the index of refraction of the core is slightly higher, certain light modes, once launched into the fiber, for the most part remain confined within the core, and thus can travel for long distances with relatively small attenuation. If the diameter of the core is large, the fiber can support many such modes. Such fibers are called multimode fibers and are usually used in short-distance links, around the wavelengths of 700–800 nm. If the diameter of the core is made smaller, eventually a point is reached where the fiber can support only one mode. The result is a single-mode fiber, commonly used in the wavelength range of 1200–1600 nm.

As an optical signal travels along the fiber, it suffers from several degradation effects. In general, these effects can be divided into linear and nonlinear, depending on whether they are intensity dependent or not. Linear mechanisms can further be divided into attenuation and dispersion. On the other hand, major nonlinear mechanisms include stimulated Raman and Brillouin scattering, self-and cross-phase modulation, and four wave mixing.

Attenuation can be attributed to several factors. At shorter wavelengths, say, below 1000 nm, Rayleigh scattering is the dominating factor. However, at longer wavelengths, inherent infrared absorption of glass starts to dominate. The wavelength range of 1200–1600 nm is the mid-range where both Rayleigh scattering and infrared absorption are low, enabling the possibility of long distance links. In most fibers, the existence of impurities introduce additional peaks within this range. The most important example of such impurities is water which causes several absorption peaks, mainly one at around 1400 nm.

Dispersion, the other major linear cause of signal degradation, can be attributed to a host of phenomena that cause pulse broadening. In multimode fibers modal dispersion dominates, which is mainly due to different group velocities for various modes that make up a given pulse. In single-mode fibers chromatic dispersion is a major factor, which is due to the fact that the effective index of refraction of glass is a function of wavelength. As a result, the frequency components of a given pulse travel at different group velocities, again causing the pulse to spread. Other dispersion mechanisms include polarization mode dispersion and waveguide dispersion.

Among nonlinear degradation mechanisms, self-phase modulation, cross-phase modulation, and four wave mixing result from the dependence of index of refraction on optical intensity, also known as Kerr effect. Self-phase modulation (SPM) occurs when a pulse modulates the index of refraction and hence causes a shift in its own frequency. In cross-phase modulation (XPM) a given pulse, by modulating the index of refraction, affects another pulse which is propagating in the medium simultaneously. Kerr effect can also cause four wave mixing (FWM) when three (or more) waves with different frequencies propagate together. In this case the nonlinearity creates harmonics or beat frequencies.

Nonelastic scatterings are the root cause of another class of nonlinear effects, most notably stimulated Raman scattering (SRS) and stimulated Brillouin scattering (SBS). In SRS, photons from the incident wave interact with high-frequency phonons. From the optical point of view, the scattered wave is red-shifted by roughly 12–16 GHz. In SBS, a high-power pulse bounces back off the low-frequency phonons or acoustic waves that it generates. The reflected waves are red-shifted by approximately 11 GHz. These nonlinear effects are especially important in WDM systems, as they increase the crosstalk between channels and limit the maximum power of each channel.

Although signals generally deteriorate as a result of propagation in fibers, it is also possible to use fibers as optical amplifiers to partly counteract the effects of attenuation. Doped fiber amplifiers, which are based on optical fibers doped with certain rare earth elements, can provide optical gain in certain wavelength ranges. The most important example of these is the erbium doped fiber amplifier (EDFA), which can provide gain in the low loss 1550 nm window. EDFAs are an essential part of long-distance fiber optic technology and constitute the basis for high data rate uninterrupted fiber transmission for distances in excess of thousands of kilometers.

References

[1] A. W. Snyder and J. D. Love, *Optical Waveguide Theory*, Kluwer Academic Publishers, Dordrecht, 1983

[2] P. Polishuk, "Plastic optical fibers branch out," IEEE Communications Magazine, Vol. 44, pp. 140–148, 2006

[3] P. C. Schultz, "Vapor phase materials and processes for glass optical waveguides," in *Fiber Optics: Advances in Research and Development*, Edited by B. Bendow and S. S. Mitra, Plenum, New York, 1979

[4] T. Miya et al., "Ultimate low-loss single-mode fibers at 1.55 mm," Electronics Letters, Vol. 15, pp. 106–108, Feb. 1979

[5] M. J. Li and D. A. Nolan, "Optical transmission fiber design evolution," Journal of Lightwave Technology, Vol. 26, pp. 1079–1092, 2008

[6] G.651.1, "Characteristics of a 50/125 μm multimode graded index optical fiber cable for the optical access network," ITU-T, 2007

[7] G.652, "Characteristics of a single-mode optical fiber and cable," ITU-T, 2005

[8] G.653, "Characteristics of a dispersion-shifted single-mode optical fiber and cable," ITU-T, 2006

[9] P. L. Liu and S. D. De, "Fiber design-from optical mode to index profile," Optical Engineering, Vol. 42, pp. 981–984, 2003

[10] T. Miya, K. Okamoto, Y. Ohmori, and Y. Sasaki, "Fabrication of low dispersion single-mode fibers over a wide spectral range," IEEE Journal of Quantum Electronics, Vol. 17, pp. 858–861, 1981

[11] X. P. Zhang and X. Wang, "The study of chromatic dispersion and chromatic dispersion slope of WI- and WII-type triple-clad single-mode fibers," Optics and Laser Technology, Vol. 37, pp. 167–172, 2005

[12] B. J. Ainsle and C. R. Day, ''A review of single-mode fibers with modified dispersion characteristics,'' Journal of Lightwave Technology, Vol. 4, pp. 967–979, 1986

[13] G.656, "Characteristics of a fiber and cable with non-zero dispersion for wide-band optical transport," ITU-T, 2006

[14] R. Olshansky, "Propagation in glass optical waveguides," Review of Modern Physics, Vol. 51, pp. 341–367, 1979

[15] J. A. Buck, *Fundamentals of Optical Fibers*, 2nd Ed., John Wiley & Sons, New York, 2004

[16] A. Kapoor and G. S. Singh, "Mode classification in cylindrical dielectric waveguides," Journal of Lightwave Technology, Vol. 18, pp. 849–852, New York, 2000

[17] H. Shu and M. Bass, "Calculating the guided modes in optical fibers and waveguides," Journal of Lightwave Technology, Vol. 25, pp. 2693–2699, 2007

[18] G. Keiser, *Optical Fiber Communication*, McGraw-Hill, New York, 2000

[19] C. DeCusatis and L. Guifang, "Fiber optic communication links, telecom, datacom, and analog," in *Fiber Optics Handbook*, edited by M. Bass and E. W. Van Stryland, McGraw-Hill, New York, 2002

[20] X. Tian and X. Zhang, "Dispersion-flattened designs of the large effective-area single-mode fibers with ring index profiles," Optics Communications, Vol. 230, pp. 105–113, 2004

[21] S. Dasgupta, B. P. Pal, and M. R. Shenoy, "Design of dispersion-compensating Bragg fiber with an ultrahigh figure of merit," Optics Letters, Vol.30, pp. 1917–1919, 2005

[22] F. Gerome et al., "Theoretical and experimental analysis of a chromatic dispersion compensating module using a dual concentric core fiber," Journal of Lightwave Technology, Vol. 24, pp. 442–448, 2006

[23] B. P. Pal and K. Pande, "Optimization of a dual-core dispersion slope compensating fiber for DWDM transmission in the 1480–1610 nm band through G.652 single-mode fibers," Optics Communications, Vol. 201, pp. 335–344, 2002

[24] N. Boudrioua et al., "Analysis of polarization mode dispersion fluctuations in single mode fibers due to temperature," Optics Communications, Vol. 281, pp. 4870–4875, 2008

[25] M. Brodsky et al., "Polarization mode dispersion of installed fibers," Journal of Lightwave Technology, Vol. 24, pp. 4584–4599, 2006

[26] J. Cameron et al., "Time evolution of polarization mode dispersion in optical fibers," IEEE Photonics Technology Letters, Vol. 10, pp. 1265–1267, 1998

[27] A. Kumar, R. K. Varshney, and K. Thyagarajan, "Birefringence calculations in elliptical-core optical fibers" Electronics Letters, Vol. 20, pp.112–113, 1984

[28] K. Saitoh and M. Koshiba, "Single-polarization single-mode photonic crystal fibers," IEEE Photonics Technology Letters, Vol. 15, pp.1384–1386, 2003

[29] Y. Liu, B. M. A. Rahman, and K. T. V. Grattan, "Analysis of the birefringence properties of optical fibers made by a preform deformation technique," Journal of Lightwave Technology, Vol. 13, pp. 142–147, 1995

[30] D. Marcuse, "Simplified analysis of a polarizing optical fiber," IEEE Journal of Quantum Electronics, Vol. 26, pp. 550–557, 1990

[31] E. H. Lee, K. H. Kim, and H. K. Lee, "Nonlinear effects in optical fiber: advantages and disadvantages for high capacity all-optical communication application," Optics and Quantum Electronics, Vol. 34, pp. 1167–1174, 2002

[32] S. P. Singh and N. Singh, "Nonlinear effects in optical fibers: origin, management and applications," Progress in Electromagnetics Research, Vol. 73, pp. 249–275, 2007

[33] A. Carena et al., "A time-domain optical transmission system simulation package accounting for nonlinear and polarization-related effects in fiber," IEEE Journal on Selected Areas in Communications," Vol. 15, pp. 751–765, 1997

[34] N. Shibata et al., "Transmission limitations due to fiber nonlinearities in optical FDM systems," IEEE Journal on Selected Areas in Communications, Vol. 8, pp. 1068–1077, 1990

[35] A. R. Chraplyvy, "Limitations on lightwave communications imposed by optical-fiber nonlinearities," Journal of Lightwave Technology, Vol. 8, pp. 1548–1557, 1990

[36] T. Brown, "Optical fibers and fiber-optic communications," in *Fiber Optic Handbook*, edited by M. Bass and E. W. Van Stryland, McGraw-Hill, New York, 2002

[37] E. Lichtman, A. A. Friesem, S. Tang, and R. G. Waarts, "Nonlinear Kerr interactions between modulated waves propagating in a single-mode fiber," Journal of Lightwave Technology, Vol. 9, pp. 422–425, 1991

[38] R. H. Stolen and C Lin, "Self phase modulation in silica fibers," Physical Review A, Vol. 17, pp. 1448–1453, 1978

[39] N. Kikuchi, K. Sekine, and S. Saski, "Analysis of XPM effect on WDM transmission performance," Electronics Letters, Vol. 33, pp. 653–654, 1997

[40] A. V. Ramprasad and Meenakshi, "Performance of NRZ, RZ, CSRZ, and VSB-RZ modulation formats in the presence of four-wave mixing effect in DWDM optical systems," Journal of Optical Networking, Vol. 6, pp. 146–156, 2007

[41] A. R. Sarkar, M. N. Islam, M. G. Mostafa, "Performance of an optical wideband WDM system considering stimulated Raman scattering, fiber attenuation and chromatic dispersion," Optical and Quantum Electronics, Vol. 39, pp. 659–675, 2007

[42] R. Y. Chiao, C. H. Townes, and B. P. Stoicheff, "Stimulated Brillouin scattering and coherent generation of intense hypersonic waves," Physical Review Letters, Vol. 12, pp. 592–595, 1964

[43] V. I. Kovalev and R. G. Harrison, "Means for easy and accurate measurement of the stimulated Brillouin scattering gain coefficient in optical fiber," Optics Letters, Vol. 33, pp. 2434–2436, 2008

[44] *Rare Earth Doped Fiber Lasers and Amplifier*, 2nd Ed., Edited by M. J. F. Digonnet, Taylor & Francis, London, 2001

[45] G.662, Generic characteristics of optical amplifier devices and subsystems, ITU-T, 2005

[46] E. Desurvire, *Erbium-Doped Fiber Amplifiers, Device and System Developments*, John Wiley, New York, 2002

[47] B. Pedersen et al., "Detailed theoretical and experimental investigation of high-gain erbium-doped fiber amplifier," IEEE Photonics Technology Letters, Vol. 2, pp. 863– 865, 1990

[48] A. R. Bahrampour et al., "A theoretical analysis of the effects of erbium ion pair on the dynamics of an optical gain stabilized fiber amplifier," Optics Communications, Vol. 265, pp. 283–300, 2006

[49] J. J. Jou and C. K. Liu, "Equivalent circuit model for erbium-doped fibre amplifiers including amplified spontaneous emission," IET Optoelectronics, Vol. 2, pp. 29–33, 2008

[50] M. Pal et al., "Study of gain flatness for multi-channel amplification in single stage EDFA for WDM applications," Optical and Quantum Electronics, Vol. 39, pp. 1231–1243, 2007

[51] H. Masuda H and Y. Miyamoto, "Low-noise extended L-band phosphorus co-doped silicate EDFA consisting of novel two-stage gain-flattened gain blocks," Electronics Letters, Vol. 44, pp. 1082–1083, 2008

Chapter 6
PIN and APD Detectors

6.1 Introduction

As mentioned in Chapter 1, in a fiber optic link an optical source, such as semiconductor laser, converts an electrical signal to an optical signal. The optical signal, once coupled properly into an optical fiber, can travel as a guided wave for relatively long distances. At destination, the optical signal must be converted back from the optical domain to the electrical domain. This conversion is accomplished by using a photodetector, which is a light-sensitive device that converts the received photons into electrons.

There are a wide variety of photodetectors that can be used for different purposes. In fiber optics, two types of photodetectors are of primary interest: PIN diodes and APD diodes. Almost all practical fiber optic receivers use one of these two devices at their front end. Therefore, this chapter is dedicated to these two detector structures. Later in Chapter 9, we discuss complete optical receiver circuits, in which the electrons generated by the detector are converted into a useful electrical signal that represents the original data sent from the transmitter.

6.2 The PIN diode and photon–electron conversion

As noted in Chapter 1, the generation and detection of light is a phenomenon that is most properly described by quantum physics [1]. The photoelectric effect provides a clear example where the absorption of photons results in the release of free electrons. If high-energy photons hit a metal, they knock out electrons from the surface of the metal. If an external electric field is applied, these electrons can be collected and generate a photocurrent. This is the operating principle behind vacuum tube devices such as vacuum photodiodes and vacuum photomultipliers. However, such devices are not very useful for fiber optic applications, because they are bulky, require high voltage, and hard to operate. Semiconductor photodetectors provide several advantages in terms of size, cost, operating voltage, responsivity, reliability, and integration with other optoelectronic devices. That is why virtually all photodetectors used in fiber optic receivers are semiconductor based.

In many ways, the process of conversion of photons to electrons in a semiconductor is the opposite of the process of electron to photon conversion that takes place in a semiconductor laser. We start by examining the process of photodetection in PIN diodes.

M. Azadeh, *Fiber Optics Engineering*, Optical Networks,
DOI 10.1007/978-1-4419-0304-4_6, © Springer Science+Business Media, LLC 2009

6.2.1 PIN diode, static characteristics

The most widely used semiconductor detector is a revered biased p-i-n (PIN) junction [2–10]. Figure 6.1 outlines the basic operating principles of a PIN photodetector. As can be seen from the figure, the PIN diode consists of a p-doped and an n-doped semiconductor separated by an intrinsic material. The junction is reverse biased by an external source. The bias voltage, V_s, appears almost entirely across the intrinsic section in the middle. This is because the intrinsic section is almost devoid of free charges, and therefore is much more resistive compared to the *p* and *n* sections. As a result, a strong electric field is formed in the intrinsic section.

When photons hit this section, they cause valence band electrons to jump into the conduction band, leaving a positive charge, or hole, behind. Thus, a population of photo-generated carriers is created in the intrinsic region. These carriers drift out of the intrinsic region because of the present electric field. The electrons move toward the *n*-region, and the holes move toward the *p*-region. This causes a photocurrent to flow in the circuit.

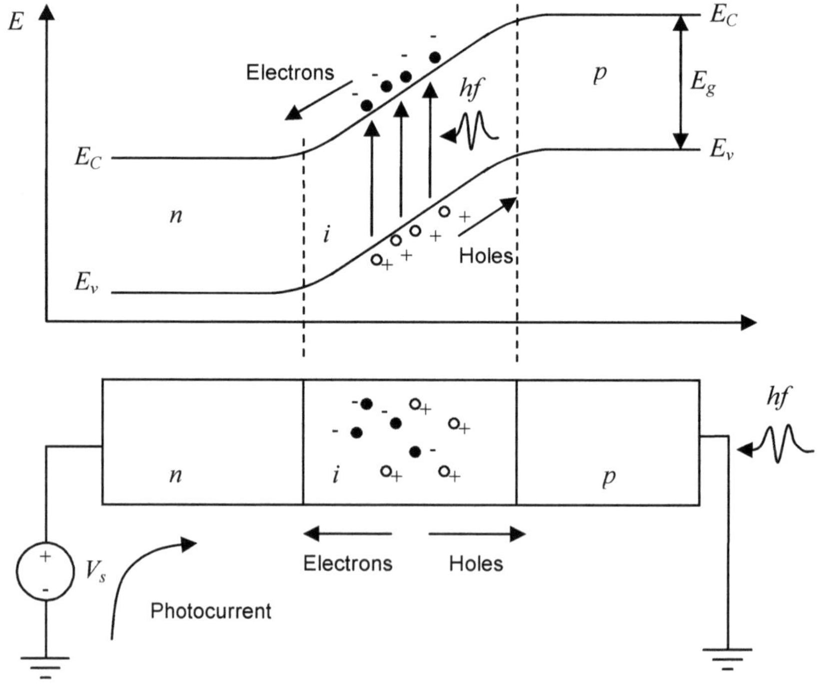

Fig. 6.1. Energy band diagram and schematic of a reverse bias PIN photodetector

From this description, it is obvious that a photon can knock an electron out of the valence band and into the conduction band if its energy exceeds the bandgap energy of the semiconductor:

$$hf \geq E_g \tag{6.1}$$

where $E_g = E_c - E_v$ is the bandgap energy, f is the photon's optical frequency, and h is Planck's constant. The efficiency of a photodetector is characterized by its *responsivity*, which is defined as

$$\rho = \frac{I}{P} \tag{6.2}$$

where I is the photocurrent, P is the optical power incident on the detector, and ρ is responsivity in units of A/W.

Equation (6.2) provides the definition of responsivity in terms of measurable quantities. It is also instructive to express Eq. (6.2) in terms of quantum mechanical parameters. This is rather straight forward. If P is the optical power and $E=hf$ is the energy of a photon, then P/E or P/hf is the number of photons that hit the photodetector. Because not every absorbed photon generates an electron/hole pair, we need to define a *quantum efficiency*, η, which represents the fraction of photons that generate an electron–hole pair and thus contribute to the photocurrent. Therefore, the number of generated electrons (or holes) in a second is given by $\eta P/hf$. Multiplying by e, the charge of an electron, we get the value of photocurrent I as $e\eta P/hf$. Finally, using the definition given by Eq. (6.2), we get

$$\rho = \frac{e\eta}{hf} \tag{6.3}$$

which is an expression relating responsivity (which is a measurable parameter) to the quantum mechanical parameters and frequency.

From a circuit point of view, a PIN detector can be thought of as a light-controlled current source. Figure 6.2 shows the typical IV characteristic of a PIN diode. As long as the junction is reverse biased, application of optical power generates a photocurrent proportional to the optical power.[1] It can also be seen that even in the absence of light application of voltage results in a small current. This is because the intrinsic region also includes thermally generated free carriers which contribute to a current even in the absence of light. This is known as the dark current.

---

[1] The diode can also generate photocurrent when it is forward biased. This is called the photovoltaic mode of operation, where light is converted to forward bias current. Photocells work in the photovoltaic mode. However, PIN detectors work in the reverse-biased region, also known as photo-conductive mode.

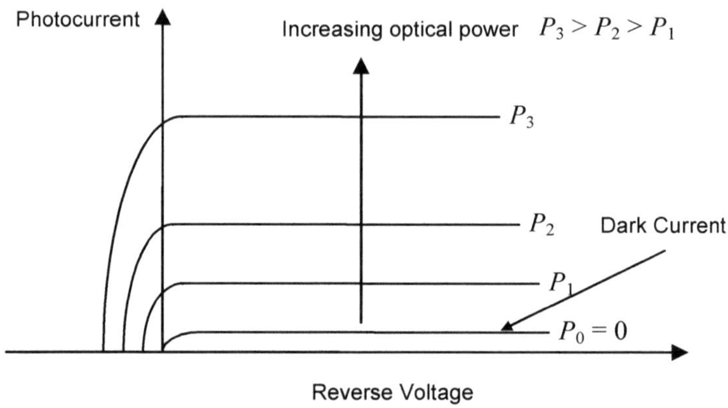

Fig. 6.2. Current vs. voltage characteristics of a reverse-biased PIN diode

The characteristic of a current source is that its voltage must be determined by other elements in the circuit. Thus, as shown in Fig. 6.3, if the diode is connected to a load resistance of R_L, the output voltage of the circuit is given by

$$V = P\rho R_L \tag{6.4}$$

This equation is valid as long as $V < V_S$. Figure 6.3 shows the simplest way to bias a PIN diode to convert the optical signal into a voltage signal. However, as a practical circuit, this scheme suffers from certain shortcomings, which we will discuss further in Chapter 9.

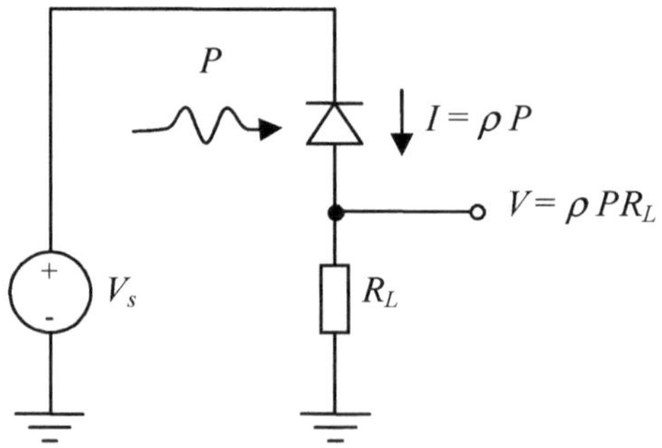

Fig. 6.3. Biasing a PIN photodetector

6.2.2 PIN diode, dynamic characteristics

The static response of a PIN diode provides a sufficient description when the input optical signal varies slowly compared to the internal time constants of the device. However, when the signal has high-frequency components, it is also important to take into account the dynamic response of a PIN diode [11].

The dynamic response of a PIN diode is determined by several factors, some of which are a result of the structure of the diode and others result from the external circuits. In general, carriers can move as a result of diffusion and drift. Drift is the main mechanism behind carrier movement in the intrinsic depletion region, because it is where the electric field is strong. On the other hand, as a result of the high conductivity in the *p* and *n* regions, the voltage drop and thus the electric field is small, and carrier movement takes place as a result of diffusion. Among the limiting factors, the most important are

• Transit time: The drift of carriers across the intrinsic region is not instantaneous, and this is a major factor that limits the response time of the detector. The transit time is a function of the speed of the carriers and the width of the depletion region. To reduce the transit time, either the speed of carriers will have to be increased by increasing the reverse voltage or the width of the intrinsic region must be reduced. Increasing the reverse voltage is effective as long as long as carrier velocities do not saturate. On the other hand, the intrinsic region cannot be made very thin because then the photons may pass through it without being absorbed [11]. Carrier speeds can be in the order of thousands of meters per second, and the width of depletion region can be around a few microns. This can result in sub-nanosecond transits times.

• Diffusion time: Diffusion is a slower process compared to drift, and therefore to reduce the response time, it is desirable to minimize the amount of carriers that are generated in the *p* and *n* regions. One way to achieve this is to reduce the length of the *p* or *n* regions. For instance, in Fig. 6.1 the photons have to pass through the *p* region before they can reach the depletion region. Therefore, by making the *p* region narrower, we can minimize the number of carriers that are generated there.

• Capacitance: Another limiting factor is the capacitance of the junction as well as the parasitic capacitance of the package. The reverse-biased PIN junction acts like a capacitor because charge can be stored in the *p* and *n* sides with the depletion layer acting as a dielectric. Notice that although reducing the length of the depletion region helps transit time, it cannot be made very narrow, because it increases the junction capacitance. The package capacitance includes the capacitance of the pads provided for wirebonding to the detector. These capacitances, in conjunction with the external load connected to the detector, form a low-pass filter.

To analyze the operation of a photodetector, it is useful to develop models that can predict its behavior based on the detector's parameters [12–15]. Figure 6.4 shows a simple linear ac equivalent circuit of a PIN diode along with the load resistor attached to it.

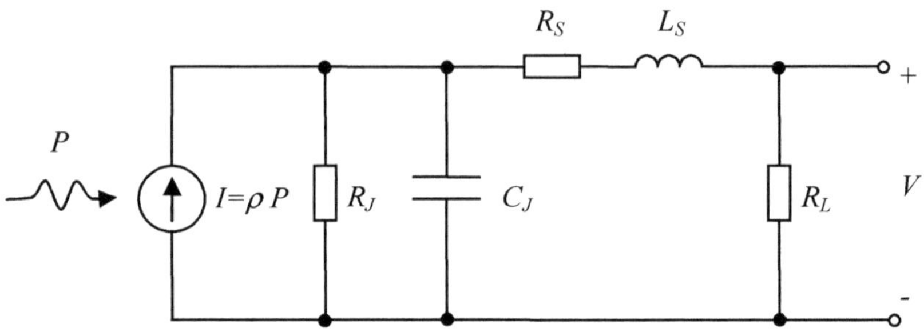

Fig. 6.4. ac model of a PIN detector

The photodetector is represented by a current source, with the junction capacitance C_J and the internal resistance of the depletion layer, R_J parallel to it. Typically there is some extra series resistance associated with the *p* and *n* regions, which is shown as R_S. The inductance of the wire bonding used to connect the diode to the load is shown by L_S, and the load itself is shown by R_L.

It is obvious that this circuit acts as a low-pass filter, with a possible resonance created by L_S and by C_J. Normally, the resistance of the depletion layer R_J is high, while the series resistance of the *p* and *n* regions is relatively low. This means R_J and R_S may be ignored for approximate analysis. Also, assuming very short wire-bondings, L_S can be ignored. With these simplifications, the circuit's behavior can be approximated as a regular low-pass RC filter, with the 3dB frequency given by

$$f_{3dB} = \frac{1}{2\pi R_L C_J} \tag{6.5}$$

Thus, in order to maximize the frequency response of the diode, both its capacitance and the resistance of the load it is attached to must be minimized. Minimizing the junction capacitance can be achieved by careful design of the diode structure, and practical highspeed diodes can have capacitances in the order of sub-pF. Minimizing the load resistance is not a function of the diode itself, but is related to the receiver's electrical design. This is a subject we will consider in Chapter 9.

6.3 Avalanche photodiode (APD)

The PIN diode described in the previous section is the most commonly used detector in fiber optics. However, in some cases the responsivity of a PIN diode is not sufficient to achieve the needed performance, in which case an avalanche photodiode structure (APD) is typically used. The main difference between an APD and a regular PIN detector is that an APD provides inherent current gain. This is a clear advantage, because in a receiver, if the signal is amplified in an earlier stage, the signal-to-noise ratio will be improved. In an APD, the reverse bias voltage applied to the junction is much higher. As a result, once a photon generates an electron–hole pair, the intense electric field present in the depletion region greatly accelerates the pair. These high-energy carriers then collide with other atoms. As a result of these collisions the atoms are ionized, releasing additional pairs of electron holes. This process is known as *impact ionization* or *avalanche multiplication*. Therefore, a single photo-generated carrier pair can ultimately generate many more carriers, resulting in a very high current gain.

Figure 6.5 shows a popular APD structure known as a *reach-through* APD [16,17]. Also shown are the electric field and energy diagrams along the junction under reverse bias. Note that the band gap voltage is small compared to the reverse bias.

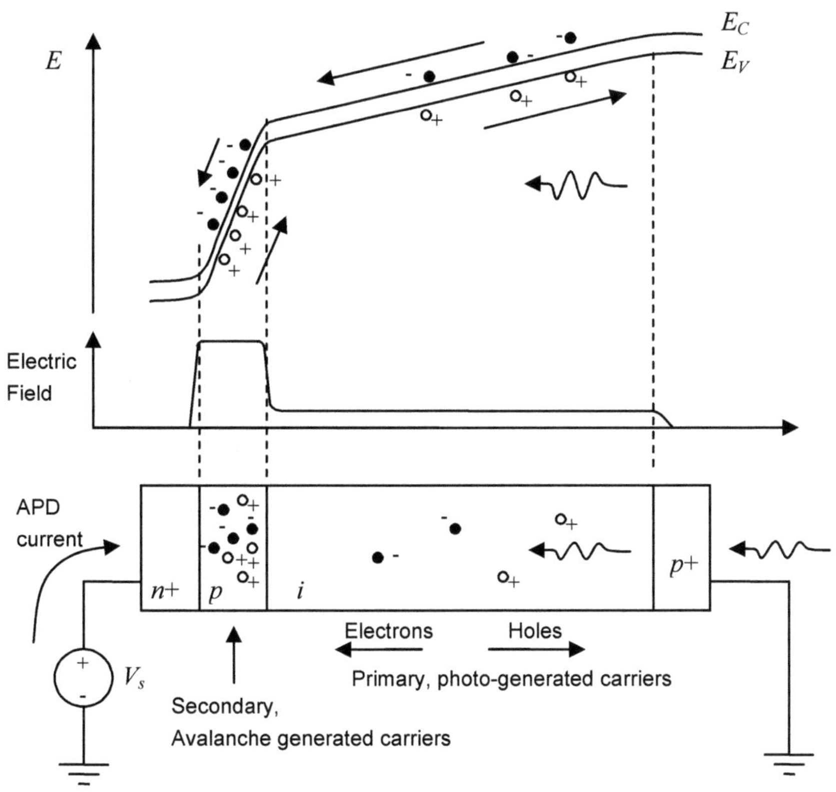

Fig. 6.5. The structure of a reach-through APD as well as the energy and electric field diagrams along the junction

The structure consists of highly doped $p+$ and $n+$ regions on either side of a lightly doped, almost intrinsic region in the middle. The intrinsic region is sometimes referred to as the π region. There is also an additional p layer sandwiched between the $n+$ and p regions. When the structure is reverse biased, most of the voltage drop appears across the $n+p$ junction. The photons enter from the $p+$ side, and are absorbed in the intrinsic region where they generate electrons and holes. These are primary photo-generated carriers. The holes drift back toward the $p+$ side. But the electrons drift toward the reverse bias junction where the electric filed is intense. In a reverse bias junction, the width of the depletion region on either side of the junction is inversely proportional to the doping level. Therefore, in this structure, the depletion region extends in the p side, and with enough reverse bias, "reaches through" to the intrinsic region. Once the electrons enter the high electric field region, they accelerate and cause avalanche multiplication through impact ionization. These *secondary carriers* are the main contributors to the overall APD current. Note that in this device only electrons migrate toward the high field area and initiate the avalanche multiplication. This is beneficial for noise purposes, because the noise performance of devices that rely on one primary carrier is better than the performance of those relying on both carriers.

The current gain of an APD, denoted by M, is a function of the reverse bias voltage. As the reverse bias voltage approaches the *breakdown voltage*, V_{BR}, the gain starts to increase. An empirical description of this behavior can be given by the following formula [18]:

$$M = \frac{1}{1-\left(V/V_{BR}\right)^{n}} \tag{6.6}$$

where V is the reverse bias voltage and n is an empirical parameter that varies between 2 and 7. For long wavelength detectors based on InGaAs, typical breakdown voltages can range from 30 to 100 V, and gains of 10–30 are achievable. Gains of 100 and higher can be achieved by silicon-based APDs, but the breakdown voltage is higher, and they cannot be used for long-wavelength applications. We will discuss the wavelength response of different semiconductors later in this chapter.

It may seem from Eq. (6.6) that it would be better to bias the APD as close to the breakdown voltage as possible to maximize the APD gain. In practice, however, there is a relatively narrow optimal range to bias an APD. Beyond that range, the additional gain of the APD is quickly offset by excess noise.

Figure 6.6 shows a typical plot of M and signal-to-noise ratio as a function of bias voltage. The optimal bias range may lie up to few volts below the breakdown voltage. Biasing the APD beyond that point will result in a lower signal-to-noise ratio, thus deteriorating the sensitivity of the receiver.

We should also note that the gain of an APD decreases at higher temperatures. This is because at higher temperatures the mean free path between carrier-atom impacts decreases.

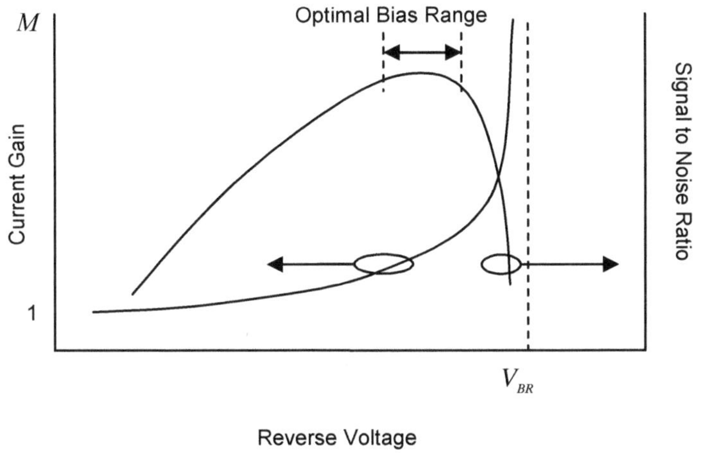

Fig. 6.6. Signal-to-noise ratio and current gain vs. reverse bias voltage

In other words, as a result of more thermal vibration, the carriers are scattered more frequently and do not reach high-enough velocities needed to initiate secondary ionization. As a result, the breakdown voltage generally increases with temperature. To first order, this dependence is linear, and can be described by

$$V_{BR}(T) = V_{BR}(T_0) + K_{BR}(T - T_0) \tag{6.7}$$

where T and T_0 are two temperatures, $V_{BR}(T)$ and $V_{BR}(T_0)$ are the breakdown voltages at those two temperatures, and K_{BR} is the breakdown voltage temperature coefficient in volts/°C. Typical values for K_{BR} are in the range of 0.1–0.3 volts/°C.

Another phenomenon that one needs to consider when using APDs is *gain saturation.* The current gain expression given by Eq. (6.6) holds at low optical powers. As the received optical power increases, the differential current gain reduces. Thus, an optical power–photocurrent curve starts linearly at low powers but becomes sub-linear at high optical power. This behavior can be attributed to a field screening effect of very high concentrations of free carriers, so that at very high powers, the effective electric field inside the device is reduced [19].

The dynamic performance of the APD is also crucial for highspeed applications. Here the common figure of merit is APD's *gain-bandwidth product.* We expect that biasing an APD at a higher gain should result in a reduction of its bandwidth. In reality, the gain-bandwidth product is a complex parameter and is a function of many factors, including quantum efficiency, material, device geometry, and even the gain itself. Bandwidths in the range of 20–30 GHz, and gain-bandwidth products as high as 150 GHz, have been achieved [20].

6.4 Noise in photodetectors

In discussing the gain of an APD we alluded to the fact that merely increasing the current gain in an APD does not necessarily improve the performance of a receiver. Beyond a certain point, the noise generated by the APD starts to dominate, causing the detected signal to deteriorate. Noise always accompanies physical signals, and anytime a signal is detected or amplified, the accompanying noise is also amplified. Moreover, any electronic device adds some additional noise of its own to the signal. As a result, it is common to keep track of the signal-to-noise ratio (SNR) as a signal propagates through a system. To reconstruct a signal at the receiver faithfully, the noise power must be small compared to the signal level. In other words, the SNR is a figure of merit in characterizing a receiver circuit.

In a photodetector, noise can be attributed primarily to two factors. The *shot noise*, also known as *quantum noise*, originates from the statistical nature of photon to electron conversion. *Thermal noise*, on the other hand, is an intrinsic property of any electrical circuit that is connected to the photodetector.

6.4.1 Shot noise

Photon to electron conversion is fundamentally a quantum mechanical process. When a photon is absorbed, a pair of electron–holes is generated. Therefore, the photo-generated current is not truly continuous, but has a discrete nature. It fluctuates around some average value as a result of the discrete charge of the carriers that contribute to it.

Because of the random nature of the current fluctuations, the noise current must be characterized in a statistical manner. It is common to describe the noise current by its mean square value. For a PIN detector, the mean square value of the shot noise is [18]

$$\left\langle i_N^2 \right\rangle = 2eI_P B \tag{6.8}$$

where I_P is the photocurrent, e is the electron charge, and B is the bandwidth within which the noise is being measured. Equation (6.8) implies that shot noise has a constant spectral density, an assumption that holds for all frequencies of interest. Normally, B is set by the bandwidth of the receiver. This shows that one way to minimize the effects of shot noise is to keep the bandwidth of circuit as narrow as possible.

The current flowing through a PIN diode is not just photo-generated. Any reverse bias junction has a leakage current. For photodetectors the leakage current is called dark current, because it exists even when these is no optical power. As a result, the mean square value of the total shot noise is given by

$$\langle i_N^2 \rangle = 2e(I_P + I_D)B \tag{6.9}$$

For APDs, both the photocurrent and the dark current are amplified by the inherent current gain of the device. It can be shown that the mean square value of the total shot noise for an APD is given by

$$\langle i_N^2 \rangle = 2e(I_P + I_D)BM^2 F(M) \tag{6.10}$$

where *F(M)* is the excess noise factor, and describes the statistical nature of the avalanche multiplication [21]. An expression for *F(M)* is given by [22]

$$F(M) = M - (1-k)M^3 (M-1)^2 \tag{6.11}$$

In Eq. (6.11), k is the ratio of the ionization coefficient of the holes to that of the electrons, assuming that the electrons initiate the multiplication.

Figure 6.7 shows an equivalent circuit for analysis of noise in a photodetector. This figure is a simplification of Fig. 6.4 when $R_J >> R_L$, $R_s << R_L$, and $L_S \cong 0$.

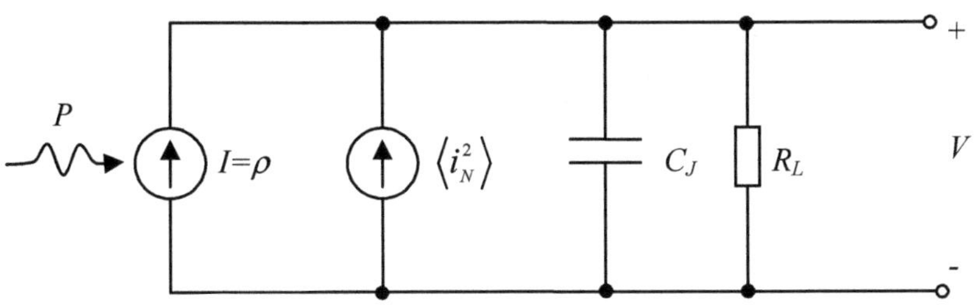

Fig. 6.7. Modeling shot noise as an extra current source

As noted earlier, the junction capacitance and the load resistance form a low-pass filter. As a result, the noise bandwidth term B in Eqs. (6.9) and (6.10) is determined by the bandwidth of this low-pass filter as defined by Eq. (6.5).

6.4.2 Thermal noise

Shot noise is a consequence of the quantum nature of light detection. Therefore, it is a fundamental property of the photodetector and sets a maximum limit on the value of SNR. In such a case, the SNR is said to be *quantum limited.* In reality, however, almost always there are other sources of noise present. Chief among these is *thermal noise*, also known as *Johnson noise*, associated with the electric circuits connected to the detector.

The source of this noise is the thermal motion of electrons in the load resistor R_L. The mean squared of the thermal noise in the load resistor is given by

$$\langle i_T^2 \rangle = \frac{4kTB}{R_L} \tag{6.12}$$

where k is Boltzmann's constant, T is the absolute temperature, and B is bandwidth. Like shot noise, thermal noise has a constant spectral density. This is another reason to keep the bandwidth of a receiver as low as possible, i.e., just sufficient to pass the signals of interest. We can incorporate the thermal noise as an additional current source into our circuit model, with the result shown in Fig. 6.8. This figure is similar to Fig. 6.7 with the difference that all band-limiting effects including those of the junction capacitance are lumped together in the form of a low-pass filter (LPF) with bandwidth B.

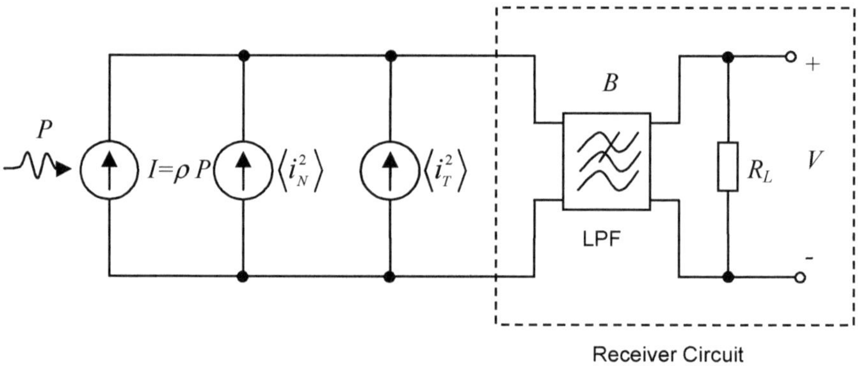

Fig. 6.8. Circuit schematic of the thermal and shot noise

If the photocurrent drives an amplifier instead of a resistor, B will be determined by the bandwidth of the amplifier. In either case, the total noise current at the input of an optical receiver is limited by the effective bandwidth of the receiver. This is why receivers with a higher bandwidth generally have a lower sensitivity.

6.4.3 Signal-to-noise ratio (SNR)

Once we have characterized the noise level at the input of a receiver, it is possible to analyze the SNR. The SNR is an important parameter because it determines the performance of a receiver. In digital receivers, SNR can be related to the bit error rate. In analog receivers, SNR is the main figure of merit and characterizes the quality of the analog link.

Let us assume that the optical power is given by a signal $P(t)$. Assuming the responsivity is given by Eq. (6.3), the detector current is given by

$$I(t) = M\frac{e\eta}{hf}P(t) \tag{6.13}$$

Note that a gain current gain term M is added to account for APD detectors. For a PIN detector, M=1. The electrical signal power associated with this current is given by

$$P_E(t) = R_L I(t)^2 = R_L M^2 \left(\frac{e\eta}{hf}\right)^2 P(t)^2 \tag{6.14}$$

where $P_E(t)$ is the *electrical* power delivered to the load.

On the other hand, the total noise power delivered to the load, consisting of shut and thermal noise, can be expressed as

$$P_N = \left[\langle i_N^2\rangle + \langle i_T^2\rangle\right]R_L = 2e\left[\frac{e\eta}{hf}P(t) + I_D\right]BM^2\,F(M)R_L + 4kTB \tag{6.15}$$

where P_N is the total noise power delivered to the load. Again, the two terms M and $F(M)$ relate to APD detectors, and for a PIN diode they are both 1. From Eqs. (6.14) and (6.15) we can calculate SNR, defined as $P_E(t)/P_N$. After simplification, we obtain

$$SNR = \frac{\eta}{2Bhf} \times \frac{P(t)^2}{\left[P(t) + \dfrac{hf}{e\eta}I_D\right]F(M) + \dfrac{2hfkT}{\eta e^2 M^2 R_L}} \tag{6.16}$$

This equation provides several insights into the SNR behavior of a receiver. Notice that as expected, increasing the optical power $P(t)$ increases the SNR. On the other hand, increasing the bandwidth of the receiver, B, reduces the SNR. The denominator of Eq. (6.16) shows the contribution of thermal noise and shot noise to SNR. An interesting point is that as R_L increases, the effects of thermal noise decrease. In a practical circuit, however, R_L cannot be increased too much because it will reduce the bias headroom of the photodetector diode (refer to Fig. 6.3). As we will see in Chapter 9, using a transimpedance amplifier removes this limitation. Another point to notice is the effect of current gain, M. Higher M values reduce the effects of thermal noise, and therefore improve the SNR. That is why an APD detector provides a higher sensitivity compared to a PIN diode. However, beyond a point, the $F(M)$ term in the denominator starts to take over and will cause SNR degradation. This means that when an APD is used, there is an optimal biasing point for best SNR. We discussed this point qualitatively in a previous section, where we noted that usually the best point to bias the APD is a few volts below the breakdown voltage (see Fig. 6.6).

Equation (6.16) can be simplified in several cases. For example, we can consider a case where the optical power is not very low, in which case the dark current I_D can be ignored. Now, consider a case where thermal noise power is much less than the shut noise. In this case, known as the *quantum* or *shut noise limit*, Eq. (6.16) simplifies to

$$SNR = \frac{\eta}{2hfBF(M)} P(t) \tag{6.17}$$

which represents the best possible SNR attainable from a detector. Note that in this case using an APD does not provide a clear advantage, except possibly in reducing the effects of thermal noise by diminishing the thermal noise term in the denominator of Eq. (6.15).

On the other hand, we can consider the case where thermal noise dominates over shut noise. In this case, Eq. (6.16) simplifies to

$$SNR = \left(\frac{e\eta M}{hf} \right)^2 \frac{R_L}{4B} P(t)^2 \tag{6.18}$$

Note that in this case, which usually corresponds to low optical powers, the SNR is more sensitive to both optical power and the current gain M. As expected, this implies that using an APD is very effective in improving the SNR at low optical powers.

6.5 Photodetector materials and structures

So far we have concentrated on the operational principles behind PIN and APD diodes. In this section we briefly review the different materials and structures that are used in building these detectors.

6.5.1 Photodetector materials

We mentioned before that in order for a photon to be absorbed and generate carriers, its energy must be larger than the bandgap of the semiconductor material. As a result, the responsivity of a detector to light is a function of the material and wavelength. This dependence is characterized by the quantum efficiency, η. This makes responsivity a function of wavelength and material.

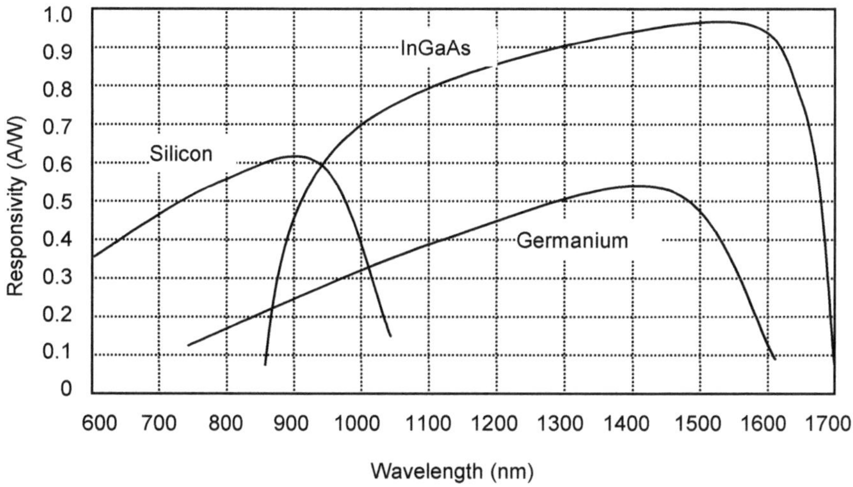

Fig. 6.9. Typical responsivity curves for silicon, InGaAs, and germanium

Figure 6.9 shows the responsivity of the common semiconductors used in photodetectors as a function of wavelength. It can be seen that the responsivity of a detector varies widely with wavelength and is maximum at a particular wavelength for each material. At longer wavelengths responsivity drops sharply because the photons do not have enough energy to knock electrons from the valence band to the conduction band. At lower wavelengths the responsivity also reduces because of increased absorption. For long wavelength applications at 1300 and 1550 nm, InGaAs is an excellent choice, as its response extends beyond the 1550 nm window. Moreover, it provides the highest responsivity, which approached 1A/W at 1550 nm. Figure 6.9 shows the inherent responsivity of the materials, without any additional current gain. In an APD, as a result of the avalanche gain *M*, the overall device responsivity can be much higher.

Table 6.1. Typical parameters for silicon, germanium, and InGaAs detectors

	Wavelength range (nm)	Responsivity (A/W)	Dark current (nA)	Avalanche gain
Silicon	400–1100	0.4–0.6	0.1–10	20–400
Germanium	800–1600	0.4–0.5	50–500	50–200
InGaAs	1100–1700	0.7–1	0.5–50	10–50

Table 6.1 provides a summary of typical values for the critical receiver parameters for these three materials.

6.5.2 PIN diode structures

Practical APD and PIN diode structures are based on the same principles we discussed before. Figure 6.10 shows an example of a practical InGaAs PIN diode structure. The light enters from the top and goes through an aperture surrounded by a metal contact. The aperture is anti-reflection coated to minimize light reflection. Because of the high index of refraction of semiconductors, without anti-reflection coating a large percentage of incident light will be reflected, causing a substantial reduction in the efficiency. Typical anti-reflection coatings may consist of quarter-wavelength single or multi-layers.

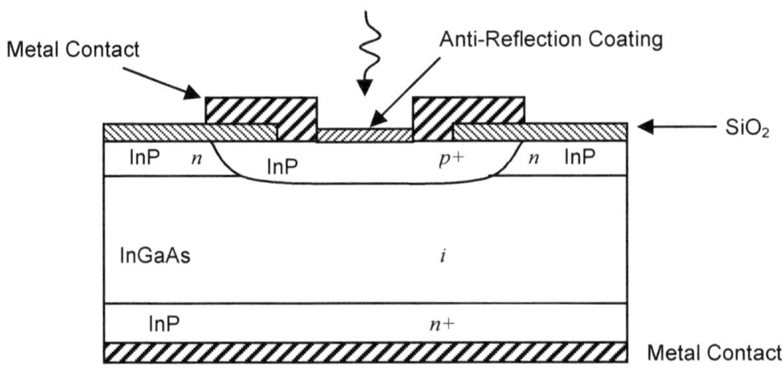

Fig. 6.10. Structure of a practical InGaAs PIN diode

The diode itself consists of a lightly doped, almost intrinsic InGaAs surrounded by two *n*-doped InP layers. On the top side, an additional diffused *p* layer makes up the *p* side of the junction. The top InP layer has a wider bandgap and therefore allows photons to pass through and reach the internal InGaAs layer. Therefore, the wavelength response of the detector is determined by the intrinsic layer, as photons are mostly absorbed in this region. Metal contacts on both sides provide the electrical connections to the junction.

6.5.3 APD structures

A practical APD structure is in many ways similar to a PIN diode. Figure 6.11 shows an example of an InGaAs APD structure similar to the PIN structure of Fig. 6.10. The difference is the addition of a *p* layer between the intrinsic region and the bottom *n* contact. When the APD is reverse biased, most of the voltage is dropped across the n+p junction. Because of the higher level of doping in the n+ layer, the depletion region extends into the p layer, and with enough reverse bias reaches the intrinsic region. This is the basis of the "reach-through" APD structure that we discussed previously. Like the PIN structure, the InP has a wider bandgap and is almost transparent for the long wavelengths that pass through it and are absorbed by the intrinsic InGaAs layer. These photons generate the primary carriers. However, the avalanche gain occurs in the p-doped InP layer, which is where the electric field is much higher because of the reverse voltage drop in this layer.

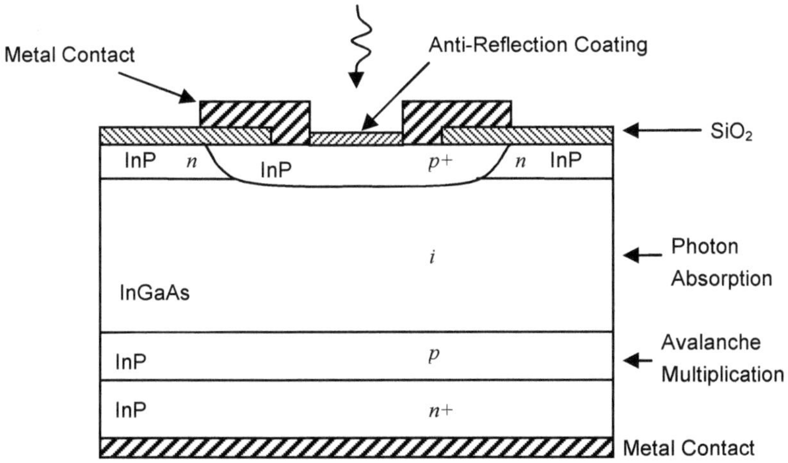

Fig. 6.11. Structure of a practical InGaAs PIN diode

In effect, absorption and avalanche multiplication are separated in this structure. This is advantageous, because each layer can be optimized for a different purpose. For instance, the InP layer is more suitable for impact ionization, because as a result of its higher bandgap tunneling current is reduced, and therefore higher electric fields can be supported.

6.6 Summary

As the front end of optical receivers, photodetectors convert photons to electrons. Although a variety of devices can accomplish this function, for fiber optic communications almost exclusively PIN and APD diodes are used. These devices provide very good responsivity in the wavelengths used in fiber optics. Moreover, they are robust, cheap, small in size, and mechanically compatible with other optoelectronic devices and optical subassemblies.

A PIN diode consists of an intrinsic semiconductor layer sandwiched between *p*- and *n*-doped materials. Typically, the device is reverse biased as a photodetector. The photons are absorbed in the intrinsic region, as long as the photon's energy is larger than the bandgap of the material. Photons are absorbed in the intrinsic region and generate carrier pairs. Because of the electric field in the intrinsic region, these carriers are swept out of the junction, making up the output current of the device. This process is characterized by the quantum efficiency, defined as the number of generated electrons for every 100 photons absorbed. Related to this is the responsivity of a detector, defined as the ratio of photocurrent to the optical power. Responsivity is a parameter that can easily be measured and is a key factor in the performance of a receiver. Typical responsivities vary from 0.5 to 1 A/W.

An APD detector's primary advantage over a PIN detector is internal current gain, which can substantially improve the sensitivity of an optical receiver. In an APD, the junction is reverse biased at a much higher voltage. Therefore, photogenerated carriers are accelerated to very high speeds and, as a result of impact ionization, generate additional secondary carriers. These secondary carriers themselves accelerate to high speeds and generate more carriers. This is known as avalanche multiplication. In the way very high current gains can be achieved. Usually, the optimum biasing point for an APD is a few volts below the breakdown voltage of the junction.

Noise is a key parameter to consider in photodetectors. There are two primary mechanisms for noise: shot noise or quantum noise, and thermal or Johnson noise. While there are some ways to reduce the effects of thermal noise, shot noise is always present and sets a limit on the maximum achievable signal-to-noise ratio. Simple mathematical results allow for the study of the effects of these two noise sources.

The most common materials used in building PIN and APD detectors are Si, Ge, and InGaAs. These materials have different operating wavelengths. For long-wavelength applications, InGaAs provides an excellent choice both in terms of responsivity and in terms of optical bandwidth. Its operating wavelength spans from almost 1000–1700 nm and its responsivity peaks around the 1550 nm. Consequently, in practical PIN and APD structures it is common to use InGaAs as the absorption layer.

References

[1] H. Zenk, "Ionization by quantized electromagnetic fields: the photoelectric effect," Review in Mathematical Physics, Vol. 20, pp. 367–406, 2008

[2] S. M. Sze and K. K. Ng, *Physics of Semiconductor Devices*, 3rd Ed., John Wiley & Sons, Hoboken, NJ, 2007

[3] A. J. Seeds and K. J. Williams, "Microwave photonics," Journal of Lightwave Technology, Vol. 24, pp. 4628–4641, 2006

[4] G. Wang et al., "Highly reliable high performance waveguide-integrated InP/InGaAs pin photodiodes for 40 Gbit/s fibre-optical communication application," Electronics Letters, Vol. 39, pp. 1147–1149, 2003

[5] L. Y. Lin et al., "High-power high-speed photodetectors-design, analysis, and experimental demonstration," IEEE Transactions on Microwave Theory and Techniques, Vol. 45, pp. 1320–1331, 1997

[6] Y. J. Chiu et al., "GaAs-based, 1.55 μm high speed, high saturation power, low-temperature grown GaAs p-i-n photodetector," Electronics Letters, Vol. 34, pp. 1253–1255, 1998

[7] Y. J. Chiu et al., "Ultrafast (370GHz bandwidth) p-i-n traveling wave photodetector using low-temperature-grown GaAs," Applied Physics Letters, Vol. 71, pp. 2508–2510, 1997

[8] Y. G. Wey et al., "110-GHz GaInAs/InP double heterostructure PIN photodetectors," Journal of Lightwave Technology, Vol. 13, pp. 1490–1499, 1995

[9] A. R. Williams, A. L. Kellner, and P. K. L. Yu, "Dynamic range performance of a high speed, high saturation InGaAs/InP pin waveguide photodetector," Electronics Letters, Vol. 31, pp. 548–549, 1995

[10] J. E. Bowers, C. A. Burrus, and R. J. McCoy, "InGaAs PIN photodetectors with modulation response to millimeter wavelengths," Electronics Letters, Vol. 21, pp. 812–814, 1985

[11] A. Bandyopadhy and M. J. Deen, "Photodetector for optical fiber communications," in *Photodetectors and Fiber Optics*, Edited by H. S. Nalwa, pp. 307–368, Academic Press, New York, 2001

[12] M. Lazovic, P. Matavulj P, and J. Radunovic, "The few SPICE models of ultra fast P-i-N photodiode," Journal of Optoelectronics and Advanced Materials, Vol. 9, pp. 2445–2448, 2007

[13] G. Wang et al., "A time-delay equivalent-circuit model of ultrafast p-i-n photodiodes," IEEE Transactions on Microwave Theory and Techniques, Vol. 51, pp. 1227–1233, 2003

[14] J. J. Jou et al., "Time-delay circuit model of high-speed p-i-n photodiodes," IEEE Photonics Technology Letters, Vol.14, pp. 525–527, 2002

[15] K. J. Williams, R. D. Esman, and M Dagenais, "Nonlinearities in p-i-n microwave photodetectors," Journal of Lightwave Technology, Vol. 14, pp. 84–96, 1996

[16] T. F. Refaat, H. E. Elsayed-Ali, and R. J. DeYoung, "Drift-diffusion model for reach-through avalanche photodiodes," Optical Engineering, Vol. 40, pp. 1928–1935, 2001

[17] H. Ando, Y. Yamauchi, and N. Susa, "Reach-through type planar InGaAs/InP avalanche photodiode fabricated by continuous vapor phase epitaxy," IEEE Journal of Quantum Electronics, Vol. 20, pp. 256–264, 1984

[18] E. Garmire, "Sources, modulation, and detectors for fiber-optic communication systems," in *Fiber Optics Handbook*, Edited by M. Bass, pp. 4.1–4.78, McGraw-Hill, New York, 2002

[19] J. W. Parks et al., "Theoretical study of device sensitivity and gain saturation of separate absorption, grading, charge, and multiplication InP/InGaAs avalanche photodiodes," IEEE Transactions on Electron Devices," Vol. 43, pp. 2113–2121, 1996

[20] J. Campbell et al., "Recent advances in avalanche photodiodes," IEEE Journal of Quantum Electronics, Vol. 10, pp. 777–787, 2004

[21] R. J. McIntyre, "The distribution of gains in uniformly multiplying avalanche photodiodes: Theory," IEEE Transactions on Electron Devices, Vol. 19, pp. 703–713, 1972

[22] C. Yeh, *Applied Photonics*, Academic Press, New York, 1994

Chapter 7
Light Coupling and Passive Optical Devices

7.1 Introduction

In electrical circuits, passive components refer to resistors, capacitors, and inductors; elements that overall consume power. On the other hand, active components deliver power to a system. In fiber optic systems, passive components typically refer to those that are not involved in opto-electric conversion, i.e., they neither generate nor detect light. Instead they are involved in guiding or manipulating the light without adding energy to it.

The most obvious example of a passive optical element is the optical fiber itself. Because of the importance of the fiber, we dedicated a complete chapter to it. But there are many other areas where passive components play an important role.

We start this chapter by discussing two critical problems. The first deals with methods of coupling light from a laser source into a fiber. The importance of this problem lies in the fact that ultimately the integrity and performance of a fiber optic link depends and relies on our ability to couple the light emitted by an optical source into the fiber. The second problem has to do with coupling light between two fibers. Connecting fibers together, both on a temporary basis (connectors) and on a permanent basis (splicing), is critical for all practical and industrial applications.

Both of these problems are passive in nature and require passive components such as lenses, fiber guides, and connectors. However, there are a wide variety of other passive components that are widely used to address other problems, components such as beam splitters, couplers, attenuators, and filters. We will provide an overview of some these devices and their applications in the remainder of the chapter.

7.2 Coupling light to and from a fiber

An optical link is based on generation and modulation of light by a source, propagation of light in an appropriate medium, and finally detection of light by a proper detector. In fiber optics, these functions are typically carried out by semiconductor lasers, optical fibers, and PIN or APD detectors. So far we have examined each of these areas separately in various chapters. However, we have not explicitly addressed the question of coupling light from a laser to the fiber. This is an important problem, because the light emerging from a typical semiconductor laser diverges rapidly as a result of its original small size, while the core of an optical

M. Azadeh, *Fiber Optics Engineering*, Optical Networks,
DOI 10.1007/978-1-4419-0304-4_7,

fiber has a very small diameter. Therefore, directing the divergent light into a small fiber core becomes a nontrivial, but crucial problem.

The coupling between a source and a fiber is expressed in terms of coupling efficiency or coupling loss. The *coupling efficiency* is simply the ratio of the power that is coupled into the fiber in the form of guided modes, P_T, to the total power of the source, P_S

$$\eta = \frac{P_T}{P_S} \tag{7.1}$$

The most efficient coupling methods approach a coupling efficiency of 95%, although much lower efficiencies are more common in normal applications. Another common way of characterizing coupling efficiency is in terms of *coupling loss*, expressed in dBs:

$$L = -10\log(\eta) = -\log\left(\frac{P_T}{P_S}\right) \tag{7.2}$$

For example, a coupling efficiency of 79% corresponds to a coupling loss of 1 dB. As we shall see, achieving low coupling loss between a source and a fiber is a challenging task. In principle, we may have the same problem on the other side of the fiber too, where the light exits the fiber facet. However, the problem is not as severe there, because photodetectors typically have a much larger diameter and wider acceptance angle compared to optical fibers. As a result, we will limit our discussion to the problem of light coupling from a source into the fiber, although many of the solutions are applicable to light detection as well.

7.2.1 Direct coupling

The most straightforward way of coupling light from an optical source to the fiber is to bring the two in close physical proximity of each other, as shown in Fig. 7.1. For simplicity, we use the ray description of light, while remembering that in reality each ray represents a wave propagating in the direction of the ray.

The coupling efficiency in direct coupling is a function of the beam size at the source, the angle of divergence of the beam, the fiber core size, and the physical distance between the fiber and the source. Ideally, if the size of the fiber core is large compared to the beam size, and if the beam divergence angle is within the acceptance angle of the fiber, and if the source can be brought to very close proximity of the fiber, a high coupling efficiency can be achieved. Unfortunately, in most cases, these conditions cannot be met.

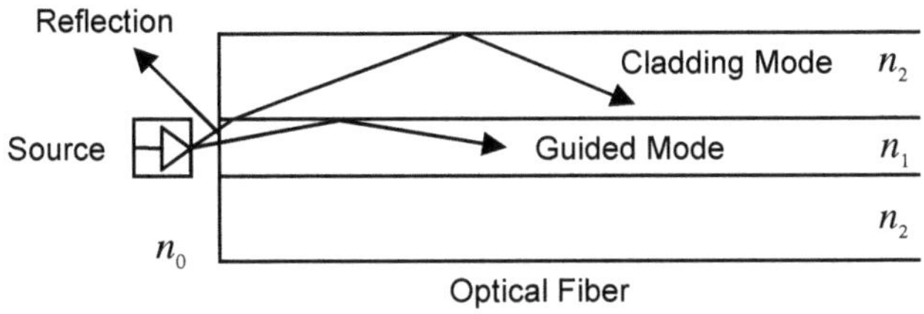

Fig. 7.1. Direct coupling between a light source and fiber

As an example, let us consider the problem of reflection and refraction from the surface of a flat fiber. Let us assume that the indices of refraction of the medium surrounding the source and the fiber core are n_0 and n_1, respectively. For normal incident angle, the reflectance at the fiber surface is given by

$$R = \left(\frac{n_1 - n_0}{n_1 + n_0} \right)^2 \tag{7.3}$$

And the transmittance is given by

$$T = \frac{4n_1 n_0}{(n_1 + n_0)^2} \tag{7.4}$$

As can be seen from Fig. 7.1, not all light waves arrive at the surface of the fiber with a normal angle of incidence. Thus, Eq. (7.4) sets the maximum limit of transmitted power into the fiber.

However, not all the light waves that enter the fiber will end up as guided waves. As shown in Fig. 7.1, some of the light will scatter into the cladding, causing radiative modes that attenuate rapidly with distance. Only those rays that fall within the numerical aperture[1] of the fiber will be guided by the fiber. Thus, the more divergent the light rays coming out of the source are, the less efficient they are coupled into the fiber. A consequence of this result is that light from smaller sources is generally harder to couple into the fiber, because light from a smaller source diffracts faster with distance.

As a result of these shortcomings, direct coupling between a source and a fiber is used wherever the coupling efficiency is not very critical and low cost and manufacturing ease are primary factors. This can be the case for short-distance multimode fiber links, especially because the larger core of multimode fibers

---

[1] See Chapter 5 for a discussion on numerical aperture and guided modes in a fiber.

makes direct coupling easier. However, for high coupling efficiency single-mode fiber applications, more elaborate schemes are needed.

7.2.2 Lensed fibers

One of the most efficient methods of increasing coupling efficiency is by using lensed fibers. These are fibers with tips that are tapered in the form of a microlens [1–8]. Figure 7.2 shows two examples of lensed fibers. Because the tip of the fiber acts as lens, the effective numerical aperture of the fiber increases and this improves coupling efficiency. In other words, light diffracting at wider angles which otherwise would be lost can now be gathered and guided into the fiber.

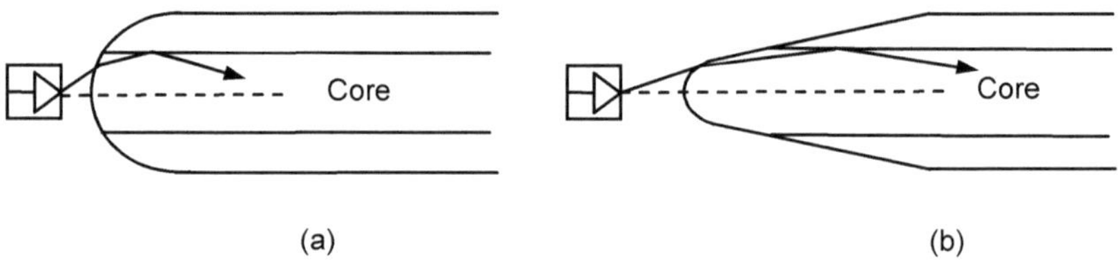

Fig. 7.2. Examples of lensed fibers: (**a**) Spherical tip and (**b**) tapered tip

One of the main problem with lensed fibers is their sensitivity to alignment. Because of the small size of the tip, a slight lateral movement can cause a large change in the coupling efficiency. On the other hand, the fiber tip must be brought to very close physical proximity of the laser. This proximity has some disadvantages. For instance, if a manual alignment process is used, the fiber can touch the laser chip and cause damage either to the laser facet or the fiber tip itself. Moreover, it becomes difficult to incorporate other optical elements such as filters, beam splitters, and isolators. Therefore, in general, the physical proximity and tight alignment requirements can make lensed fibers less attractive.

7.2.3 Fiber coupling via lens

In cases where using lensed fibers is not practical, a separate lens can be utilized [9–13]. In general, using a separate lens provides more degrees of freedom in the design. A separate lens is generally large compared to both the laser and the fiber. The large size of the lens is a plus because it provides a wider area for light collection. Moreover, the lens can be placed at some distance from the source and the fiber, because the lens now acts as an imaging device, with the source and the fiber located at the imaging planes of the lens. Therefore, if needed, additional devices can be placed in the path of the light. Figure 7.3 shows several commonly used lens structures.

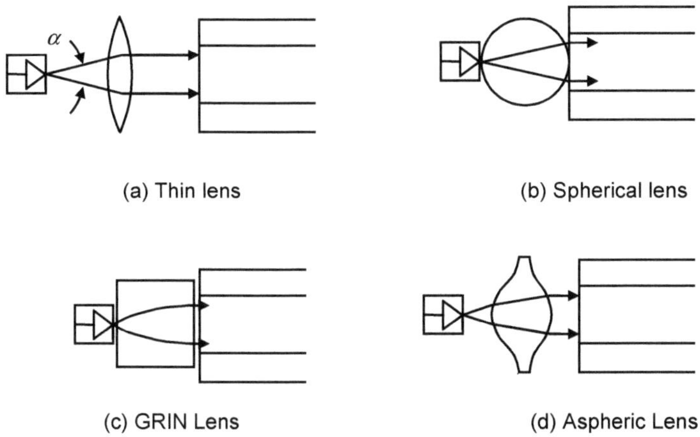

Fig. 7.3. Commonly used lens structures to improve coupling efficiency

Figure 7.3a shows a standard *thin lens* used to calumniate the diffracting light from a light source. For thin lenses, the position of the object, image, and the focal length are related by

$$\frac{1}{D_O}+\frac{1}{D_I}=\frac{1}{F} \tag{7.5}$$

where D_O and D_I are the distances of the object and image to the lens and F is the focal length.

From Eq. (7.5) it can be seen that if the source is placed in the focal plane ($D_O = F$) then the image will form at infinity, which means the light rays are parallel and calumniated at the other side of the lens. In this case, we can say that the lens has reduced the diffraction angle of the source, α, to zero.[2] More generally, the diffraction angle of the source can be mapped to a narrower diffraction angle of the image. The image's diffraction angle can then be controlled to match the acceptance angle of the fiber, thus improving the coupling efficiency.

Figure 7.3b shows a *spherical lens*. The interesting point about a spherical lens is that by choosing the right index of refraction for lens, it is possible to move the focal point of the lens to the surface of the lens. This allows the lens to be placed very close to the source. In some cases this results in easier alignment procedures, while capturing a larger percentage of the emitted light.

Figure 7.3c shows a graded index or *GRIN lens*, also known as a GRIN rod. Unlike the other lenses, the GRIN lens is made of an inhomogeneous material. Just like a GRIN fiber, in a GRIN lens the index of refraction is a function of radius.

[2] Of course, in realty, any finite size column of light traveling in free space will eventually diffract.

For a parabolic index of refraction profile with the maximum value of profile at the center, light rays traveling along a GRIN rod move along sinusoidal paths, periodically refocusing at the center. Therefore, by cutting the rod at a length equal to half of the distance between two subsequent refocusing points, it is possible to make the rod behave like a lens. Diverging rays entering from one side will exit as parallel rays from the other side. The advantage of GRIN lenses is in their small size and short focal lengths. Moreover, the operation of a GRIN lens is independent of the difference between the index of refraction of the medium and the lens. Therefore, it can operate without an air gap between it and the source or the fiber, or when the surrounding medium is not air.

Finally Fig. 7.3d shows an *aspheric lens*. In standard lenses, the surfaces are normally spherical, i.e., they are part of a sphere. Spherical geometries are easy to manufacture. However, when it comes to focusing light, spherical lenses suffer from aberration, i.e., the focusing point of an optical ray depends on the angle that it makes with respect to the optical axis. To eliminate aberration, the surfaces must follow more complex geometries, such as hyperboloids or ellipsoids. In general, it is possible to improve the optical properties of a lens at the expense of more complex geometries. Such lenses in which one or both surfaces deviate from spherical are called aspheric lenses. Because they are harder to manufacture, aspheric lenses are more expensive and typically used in high-end applications [14].

Lenses are used not only in coupling optical sources to a fiber, but also in coupling the light coming out of the fiber to a detector. When light exits the fiber, it diffracts, and unless the detector is placed very close to the fiber facet, part of the light will be lost. Using a lens, the light can be calumniated as a wider beam or refocused into a detector. This is very important in optical receivers, because any improvement in coupling efficiency directly results in improving the signal-to-noise ratio at the receiver.

7.3 Fiber-to-fiber coupling

In electrical circuits an electrical connection can be established between two conductors by simply creating a physical contact between them. As a result, a wide range of electrical connectors are available for all sorts of electrical applications.

In Fiber optics, just like electrical circuits, we need to connect various optical elements together, and this often requires using fiber jumpers and coupling light from one fiber patch to another. Unfortunately, coupling light between fibers or other optical elements is much more difficult compared to an electrical connection and requires great care in terms of mechanical alignment and accuracy. This brings us to the problem of coupling light between two fibers. We will consider fiber-to-fiber coupling in two sections: temporary connections or connectorized coupling, and permanent connections or splicing.

7.3.1 Connectorized couplings

Quite often it is desirable to have a means of connecting two fibers together through a temporary mating device or connector [15–19]. Figure 7.4 shows a common way to implement such a connector. Each fiber is placed in a ferrule whose function is to provide the mechanical support for the fiber and hold it in place tightly. The ferrule can be made out of plastic, metal, or ceramic materials. The central piece of the connector itself is an alignment sleeve. The two ferrules are inserted in the sleeve, and proper alignment between the cores is ensured because of the tight mechanical tolerances of the ferrules and the sleeve. The gap between the two fibers can be controlled by a mechanical stop which determines the exact stopping positions of the fibers. In some variations, the alignment sleeve is tapered to improve connector mating and demating.

Well-designed connectors provide low coupling loss, in the order of 0.1 dB or less. However, as shown in Fig. 7.5, a number of undesirable situations can reduce the coupling efficiency. Figure 7.5a shows a case of two fibers with different core diameters. In general, whenever the numerical apertures of two fibers are different, the potential for power loss exists. In this case, light coupling from a narrower fiber core to a wider fiber core is easier and more efficient compared to coupling in the other direction. Figure 7.5b shows an example of poor concentricity. Fibers that do not provide a tight concentricity tolerance may show large coupling variations depending on the orientation or from one pair of fibers to the next.

A large air gap, shown in Fig 7.5c, is another reason for loss of power. An air gap can result from incomplete insertion of the fiber or from mechanical problems inside the sleeve. It is also common for microscopic dust particles to get into fiber connectors, preventing them from making proper contact, or even scratching and damaging the fiber facets.

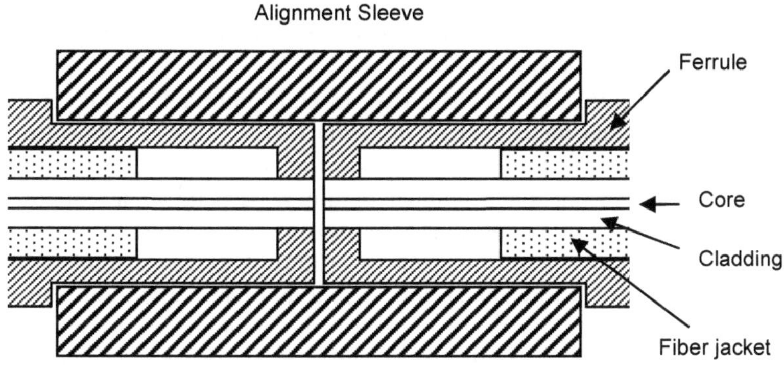

Fig. 7.4. Connectorized coupling between two fibers

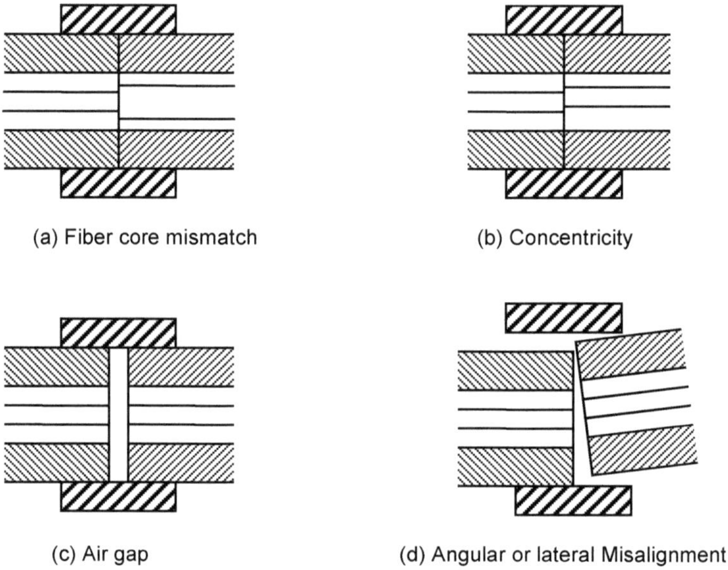

Fig. 7.5. Coupling loss due to (**a**) fiber core mismatch, (**b**) concentricity, (**c**) air gap, and (**d**) angular and/or lateral misalignment

More dramatic power reduction results when dust particles land on the fiber core, blocking the light path. As a result, constant monitoring and cleaning of fiber facets are important to prevent such problems. Angular or lateral displacement, shown in Fig. 7.5d, is another potential cause of low coupling. This happens when the mechanical tolerances are not tight enough or when the dimensions of the sleeve and the ferrule do not match.

A fiber connector is characterized by several important parameters. As noted before, the most important factor is *insertion loss*, or simply connector loss. Another important factor is *repeatability*. If the same two fibers are connected through the same connector a number of times, each time the coupling will be slightly different. A good connector assembly provides a small standard deviation for coupling efficiency across multiple insertions. Another desirable specification of a fiber connector is low *return loss*, i.e., a low back reflection. Return loss is defined as the ratio of the reflected power from the connector to the input power. For example, a return loss of 30 dB means 0.001 of the input power is reflected back from the connector. A connector must also be resistant and show a minimal coupling variation in the presence of normal mechanical forces such as axial and lateral forces. This is a practical requirement because in a normal environment it is likely for the connector to encounter a range of mechanical forces.

A wide range of connectors have been designed and are in use in the industry. Here we give an overview of some of the most popular types.

Straight tip or ST connectors are one of the more common types of connectors and in wide use in many applications. The ferrule diameter in an ST connector is 2.5 mm. ST connectors are spring loaded and engaged by a twist-and-lock mechanism [20].

Fixed connector or FC connectors use an alignment key and a threaded (screw-on) socket and are similar to the popular SMA connectors used in electronics [21]. They are in wide use in single-mode applications and provide low insertion loss and high repeatability.

Subscriber connector, or SC, is another common type of connector [22]. The advantage of SC connectors is that they are engaged by a push-and-snap mechanism, without the need for any rotation. This makes plugging and unplugging them very easy and also reduces wear out. Moreover, a higher connector density is achieved. Many transceivers provide either an SC receptacle connector, or a pigtail SC connector, as their optical interface. The push-and-snap feature of SC connectors thus provides a very convenient and easy way of connecting to optical transceivers. SC connectors are available in simplex and duplex variations. The ferrule diameter in an SC connector is 2.5 mm.

The LC or small form factor connector is similar to the SC, but with half the size [23]. The diameter of the ferrule in an LC connector is 1.25 mm, vs. 2.5 mm for most other connectors. This allows for twice the connector density for a given space. Because of their compactness, LC connectors have become more popular and are used in many high-end transceivers such as SFPs and XFPs.

Other common connectors that are used in industry include Biconic [24], MTP/MPO [25], MTRJ [26], and MU [27].

7.3.2 Fiber finish

So far we have not specified the exact shape of the fiber facets, implying a flat shape as the standard. Figure 7.2 showed two cases where a non-flat shape is used to improve light coupling into the fiber. In fiber connectors, the shape of fiber facet is usually referred to as facet finish, and it impacts not only the coupling loss, but also the back reflection or return loss. A desirable characterization of a fiber connector is low return loss.

The most common type of finish for fiber-to-fiber coupling is known as *physical contact* or PC. In a PC connector, the facets have a slightly convex shape, as shown in Fig. 7.6a. This allows the two cores to make physical contact, thus eliminating the air gap and improving the coupling efficiency.

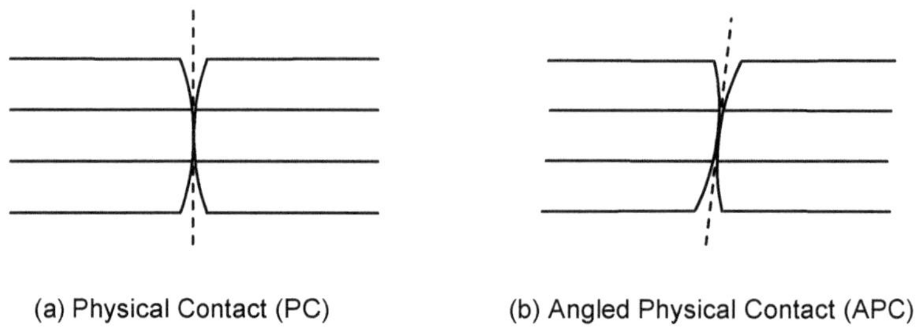

Fig. 7.6. Standard and angled physical contact fiber facet finish

Elimination of air gap also reduces the back reflection. Typical PC connectors have around 0.2 dB of loss and better than 40 dB of return loss.

PC connectors offer acceptable performance for most applications and are the standard finish on most connectors. However, some applications require even lower return loss. For instance, DFB lasers are notoriously sensitive to even minute amounts of back reflection. A standard PC connector may not be able to provide this level of low back reflection, especially in cases where imperfect mating or residual air gap can increase the back reflection. An angled physical contact, or APC finish, shown in Fig. 7.6b solves this problem. Because of the angle of the surface, any reflected light generates cladding modes, which are highly attenuated with distance. The standard angle used in APC connectors is 8 degrees, yielding a typical return loss of 60 dB or more.

Because of the angle of the surface, the two fibers can only mate in a specific orientation. This requires additional care to ensure that the angular orientation of the two fibers is matched. This is achieved by using keyed ferrules which allow the two sides to make contact only in the right angle. Because of this additional complexity, APC connectors are typically used only in single-mode applications.

7.3.3 Fiber splicing

A fiber connector provides a way to couple two fibers temporarily. However, sometimes it is necessary to connect two fibers permanently. For instance, when installing long-distance fiber cables, it may become necessary to connect two fiber cables because a single cable is not long enough to support the needed distance. A permanent connection between two fibers is achieved through *fiber splicing* [28, 29].

Fiber splicing can be divided into two categories, mechanical splicing and fusion splicing. In mechanical splicing, typically an external guide such as a sleeve or a V-groove is used to align the two fibers. To improve the coupling between the two fibers, an index matching gel can be applied to fill the air gap between the two fiber facets. Figure 7.7 shows these schemes.

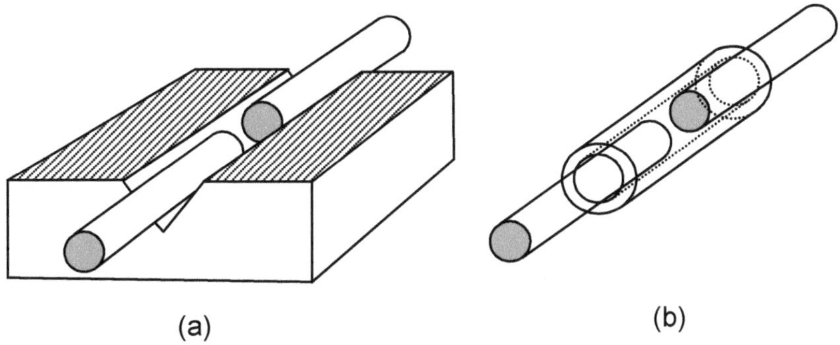

Fig. 7.7. Examples of mechanical splicing, (**a**) V-groove alignment and (**b**) guiding sleeve alignment

The advantage of mechanical splicing is that it does not require expensive splicing equipment. Typical loss is around 0.5–0.7 dB.

Another from of permanent splicing is through fiber fusion, which essentially involves welding the two fibers together. The fiber facets must be cleaned and prepared. They are then aligned accurately by positioning fixtures before welding. The welding is usually done by automatic equipment which uses an electric arc (or flame or a laser) to fuse the two fibers. Figure 7.8 illustrates this process. The advantage of fusion splicing is that it results in better matching and lower loss. In fact, losses as low as 0.02 dB are attainable [30].

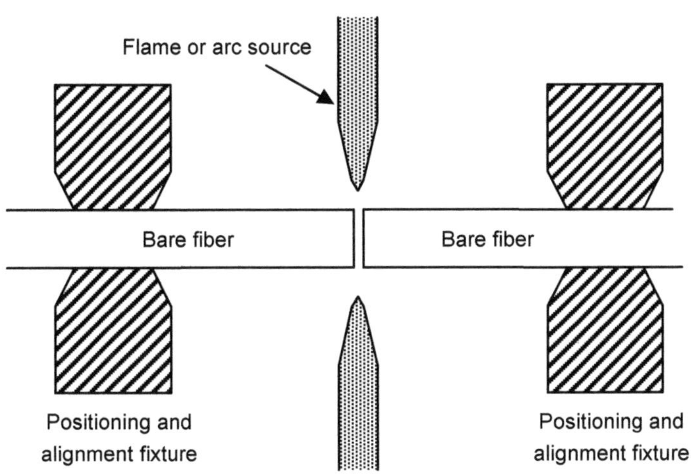

Fig. 7.8. Fiber splicing through fusion

Modern fusion splicing equipments are relatively expensive but easy to operate and highly automated. This makes fusion splicing the method of choice for single-mode long-telecom applications where the cost can be justified. Fusion splicers can also be used to introduce a specified amount of loss in the link, in which case the splice acts as an in-line fixed attenuator.

7.4 Passive components

In addition to fibers and the components used to couple light to and between the fibers, a variety of other passive devices are used in fiber optics. These include a vast array of elements, far beyond what can be covered in a chapter. In the remainder of this chapter we will examine some of the important devices that are widely used.

7.4.1 Splitters and couplers

Quite often it becomes necessary to combine two or more optical signals into one stream or to divide a single optical signal into several branches for different purposes. An obvious example of combining several signals is in the case of WDM systems, where several signals of different wavelengths need to be combined into one WDM signal which can then be transmitted on a single fiber. An example of dividing a single signal into multiple signals is when the received signal at an optical receiver is split into two paths, one path goes to the receiver and the other path goes to an optical power meter to monitor the power level at the receiver. A coupler is a device that can achieve this function.

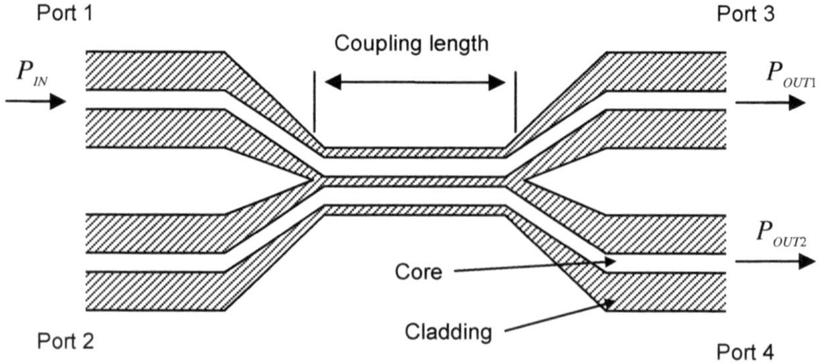

Fig. 7.9. A 2×2 coupler

One of the most basic forms of couplers is the 2×2 coupler [31–33]. The concept behind this device is shown in Fig. 7.9. The coupler has four ports, numbered from 1 to 4 in the figure. This device is usually made by twisting two pairs of fibers together and then pulling them while heating the twisted area. As a result, the two claddings fuse and the thickness of cladding at the twisted area reduces, allowing modes to couple from one fiber to the other.

When light propagates in one fiber, the evanescent fields in the cladding generate guided waves in the other fiber. The energy oscillates back and forth between the two fibers as a function of the coupling distance and the radius of the fibers. Therefore by controlling these parameters the coupling ratio can be controlled. In general, the power entering from one side splits between the two outputs on the other side.

Let us assume a signal is applied to port 1 as the input. It will then split between the two outputs, ports 3 and 4. If the optical power associated with the input and outputs are P_{IN}, P_{OUT1}, and P_{OUT2}, for an ideal lossless coupler we have

$$P_{IN} = P_{OUT1} + P_{OUT2} \tag{7.6}$$

In reality, there is a tiny amount of back reflection in port 1 and port 2, as well as some power loss within the device itself. The coupler's *insertion loss* L_{INS} between the input port 1 and the output port 3 (or 4) is defined as

$$L_{INS} = 10 \log \frac{P_{OUTi}}{P_{IN}} \tag{7.7}$$

where i=1,2. If the power is split equally between ports 3 and 4, the power for each of the outputs is half of the input power. In this case, the device is known as 50/50 splitter or a 3 dB coupler. Obviously the insertion loss for both outputs is 3 dB in this case. On the other hand, a highly asymmetric splitting ratio can be used to tap off a small portion of the signal, for instance, for monitoring purposes.

The *directionality* L_{DIR} of the coupler is defined as

$$L_{DIR} = 10 \log \frac{P_{OUT1} + P_{OUT2}}{P_{R1} + P_{R2}} \tag{7.8}$$

where P_{R1} and P_{B2} are any reflected powers from ports 1 and 2. Well-designed couplers have negligible return loss and thus very high directionality. This is why this device is also known as a directional coupler.

The same concepts hold if two signals are applied to the two inputs of the device at ports 1 and 2. In that case, each of the signals will split, and therefore ports 3 and 4 will each include a mixture of the two inputs.

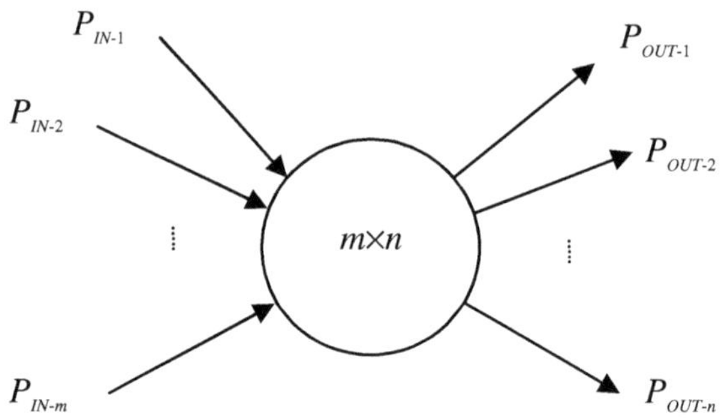

Fig. 7.10. A star coupler

In general, a coupler/splitter can have several inputs and outputs, in which case it is referred to as a star coupler [34–35]. Figure 7.10 shows a representation of an $m\times n$ star coupler. The device has m inputs, P_{IN-1} to P_{IN-m}, and n outputs, P_{OUT-1} to P_{OUT-n}. It acts as both a coupler and splitter. All the inputs mix together and then divide between the outputs. In the most general form, the Sth output is given by

$$P_{OUT-S} = \frac{1}{K_S} \sum_J F_J P_{IN-J} \tag{7.9}$$

where K_S are the coefficients representing the splitting ratio between the outputs, and F_J are the mixing ratios between the inputs. For most practical devices, the input signals are mixed equally and the power is split equally between the outputs. In this configuration, each output power is given by

$$P_{OUT-S} = \frac{1}{n}\left(P_{IN-1} + P_{IN-2} + \ldots + P_{IN-m}\right) \tag{7.10}$$

Because couplers are passive elements, they are highly reliable and easy to use. As a result, they are widely utilized for mixing, splitting, and routing of signals in optical systems such as PON and WDM networks.

7.4.2 Attenuators

Sometimes it becomes necessary to attenuate an optical signal by a known amount. For example, the power output from a transmitter may be more than what a receiver can handle. Another common case is when an optical signal needs to be analyzed by a piece of equipment, but the power is beyond the optimum input range of the equipment. In such cases, an optical attenuator can be used to reduce the signal power.

Optical attenuators can be either fixed or variable. A fixed attenuator is used in cases where the required amount of attenuation is constant. A fixed amount of attenuation can be achieved in a variety of ways. One way is to use a splitter. For instance, as we noted in the previous section, a 50:50 splitter yields 3 dB of attenuation in each of its output branches. There are also devices specifically designed to work as an attenuator. These are typically coupling sleeves that incorporate a stop that ensures a predetermined air gap between the fiber ferrules. Alternatively, a metal-doped fiber with high absorption coefficient can be inserted in the path of the signal. As a result, the light experiences a fixed amount of attenuation as it passes through the device. These attenuators are available for different kinds of connectors and in various attenuation values. Standard connector types are SC, FC, ST, and LC, and typical attenuation values are from 1 to 20 dB.

Variable attenuators are somewhat more sophisticated. We noted in the beginning of this chapter that several factors can reduce the coupling between two fibers. In principle, by controlling any of those factors the attenuation can be varied. For instance, by changing the air gap between the two fibers, a variable attenuator can be achieved. However, because of the very high sensitivity of coupling loss to mechanical parameters (such as air gap or alignment), controlling and calibrating the attenuation is challenging. As a result, mechanical variable attenuators tend to be rather bulky, expensive, and more like a piece of equipment rather than a simple device. Microelectromechanical systems (MEMS) technology can also be used to build variable attenuators. MEMS devices are faster, smaller in size, and consume less supply power. As a result, these devices are becoming increasingly popular as MEMS technology matures [36–38].

7.4.3 Isolators

As light travels along fibers and connectors, there are many opportunities for back reflection. These reflections usually happen when light passes through an interface perpendicular to the direction of light propagation. An angled interface, such as in an APC connector, significantly reduces such reflections. However, in most cases one does not have control over all the link parameters, and it is necessary to reduce the effects of any potential reflection on sensitive devices such as DFB lasers. In such cases, an isolator is used to reduce the amount of light reaching back to the laser.

Isolators normally work on the principle of rotation of the light polarization by a Faraday rotator [39–41]. As Fig. 7.11 shows, the isolator consists of two polarizer plates that are at 45° angle with respect to each other with a Faraday rotator placed in between them.

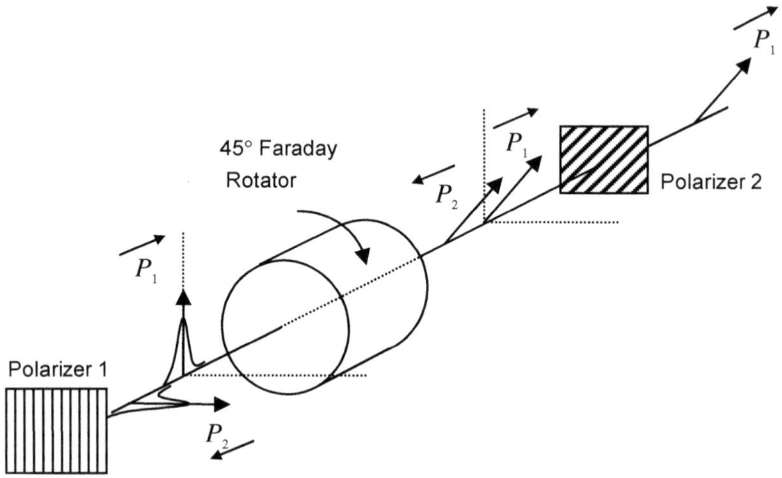

Fig. 7.11. Optical isolator based on a Faraday rotator

Let us first consider a pulse of light, P_1, which is propagating from left to right. After passing through polarizer 1, the light is polarized in the vertical direction. Then the light enters the Faraday rotator which is subject to an external magnetic field usually provided by a permanent magnet. The Faraday rotator rotates the polarization plane of the light in proportion to its length and the strength of the applied magnetic field. For this application, the length of the Faraday rotator is chosen such that the plane of polarization rotates by 45°. Thus, once P_1 goes through the rotator its polarization plane has rotated by 45°. At the output plane, a second polarizer plate is inserted whose axis of polarization is also at 45° with respect to polarizer 1. As a result, P_1 can go through polarizer 2 with little attenuation. It can be seen, therefore, that light can travel through the system from left to right with relatively little attenuation, except possibly for the attenuation due to the filtering of unpolarized light by the first polarizer (if the input light is unpolarized before it enters the isolator).

Now consider a pulse traveling from right to left. Once it passes through polarizer 2, it will be polarized at 45°, as represented by P_2 in Fig. 7.11. Then it goes through the Faraday rotator, which rotates the polarization plane by another 45°. Thus, once P_2 arrives at polarizer 1, its polarization plane is perpendicular to that of polarizer 1. Therefore, it is almost completely absorbed by polarizer 1, and very little of it gets through the system. This means any light propagating through the system from right to left experiences a large amount of attenuation. As a result, the isolator passes light in one direction and blocks it in the opposite direction.

Practical isolators can provide around 40 dB of isolation, with an insertion loss of around 1 dB. As noted before, they are widely used on the output of DFB lasers to prevent any back reflected light from entering the laser cavity and causing instability in the laser output.

7.4.4 Optical filters

In electronic systems filters play a crucial role in numerous signal processing applications. Optical filters play an equally critical role in optical domain [42,43]. Optical filters are essentially frequency domain devices that selectively attenuate or pass certain wavelengths. For example, a bandpass optical filter can be used to select a particular channel in a WDM application from several channels that are stacked next to each other. Another example of optical filter application is the grating in a DFB laser, which acts as a single wavelength bandpass filter by greatly attenuating all but one wavelength, causing DFB lasers to operate as single-mode devices.

The common factor in all optical filters is their dependence on some wavelength-sensitive phenomena. In general, optical filters are based either on interference or diffraction effects. A Fabry–Perot cavity is a common example of an interferometric filter. As discussed in Chapter 4, the cavity has a set of resonance wavelengths (frequencies) at which the transmittivity of the cavity is maximum. Thus, close to the wavelength peaks, the cavity acts as an optical filter with a strong wavelength-dependent transmission coefficient. We will encounter an example of a diffraction-based filter in Chapter 11, where we discuss optical spectrum analyzers (OSA). In an OSA, a diffraction grating filter spatially separates a signal into its various wavelength components, thus providing the frequency spectrum of the signal. Other types of filters, such as arrayed waveguide gratings (AWGs), are very useful tools for wavelength-dependent multiplexing and demultiplexing of signals in WDM applications [44].

7.5 Summary

In addition to generation and detection of light, a wide range of functions must be performed on optical signals. Most common among these functions is coupling the light to and from the fiber, as well as light coupling between two fibers.

Coupling of light from a source into a fiber can be done through a variety of means. The simplest method involves simply positioning the tip of the fiber in close proximity to the source. This technique works well for power-insensitive applications where significant coupling loss can be tolerated. However, in most cases, an imaging device such as a lens is used to increase the coupling efficiency. The role of the lens is to increase the numerical aperture of the fiber or to match the numerical aperture of the fiber to the radiation pattern of the source.

Fiber-to-fiber coupling can be divided into two categories of permanent and temporary connections. Temporary connections are achieved by standard connector types that provide accurate alignment and almost eliminate the air gap between the two cores. The most popular types of connectors include ST, SC, LC, and FC connectors. All of these connectors have special alignment sleeves and adaptors that allow for connecting two fibers together. Because of their simple push-and-snap connection design, SC and LC connectors are also widely used in receptacle form in optical transceivers.

Temporary fiber connectors can also be divided based on the type of their surface finish. A physical contact or PC connector uses a slightly convex finish that allows the cores to make physical contact, thus further reducing the insertion loss and back reflection. Angled physical contact (APC) connectors use an angled fiber facet to significantly reduce any back reflection due to the fiber-to-fiber interface.

Permanently connecting two fibers is known as fiber splicing. Fiber splicing can be by either mechanical means or fiber fusion. In mechanical methods, the two fibers are aligned with the aid of some external mechanical guide such as a V-groove or alignment sleeve, and then a permanent epoxy is used to maintain the structure. In fusion splicing, the two fibers are essentially melted and fused together. Fusion splicing is typically superior to other types of splicing, but needs more expensive specialized equipment.

In addition to the functions that involve fiber coupling, a wide range of passive optical components exist. In this chapter, we examined couplers, attenuators, and isolators.

Couplers are passive devices that allow for mixing or splitting light signals. One of the simplest of such devices is a 2×2 coupler which has two input ports and two output ports. As a coupler, it can mix two optical signals that are applied to its inputs and provide a portion of the result at each of its output ports. As a splitter, it can split a signal that is applied to one of its input ports into two portions and provide each portion at one of the outputs. A more general version of a coupler is an $m \times n$ coupler, also known as a star coupler, which mixes and splits m optical signals. These devices are widely used in optical networking applications.

An attenuator is a device that attenuates the optical power by a known amount. Attenuators are either fixed or variable. A fixed attenuator is an in-line or pluggable device that provides a fixed amount of optical attenuation. Fixed attenuators are simple devices that usually operate on the basis of a fixed air gap or misalignment between two fibers. Variable attenuators, on the other hand, can set the amount of attenuation over a wide dynamic range. These devices tend to be bulkier and more complex.

An optical isolator is a device that allows for passage of light in only one direction. It is usually used to isolate a sensitive source, such as a DFB laser, from any potential back reflection. Optical isolators are usually based on a Faraday rotator and two polarizing plates aligned at 45° with respect to each other. Isolators can provide as much as 40 dB of attenuation for back-reflected light.

Another class of very useful passive elements is optical filters. These devices perform a task similar to their electronic counterparts and allow for discrimination of various components in a signal based on wavelength.

Passive elements, along with optical sources, detectors, and fibers, constitute the backbone of physical layer in fiber optic links. In the next two chapters, we focus our discussion on optical transmitters and receivers, which act as an interface between the optical signals and the electronic systems in which the higher level signal processing takes place.

References

[1] A. Christopher et al., "Ideal microlenses for laser to fiber coupling," Journal of Lightwave Technology, Vol. 11, pp. 252–257, 1993

[2] Z. Jing et al., "Design and characterization of taper coupler for effective laser and single-mode fiber coupling with large tolerance," IEEE Photonics Technology Letters, Vol. 20, pp. 1375–1377, 2008

[3] K. Shiraishi, H. Yoda, T. Endo, and I. Tomita, "A lensed GIO fiber with a long working distance for the coupling between laser diodes with elliptical fields and single-mode fibers," IEEE Photonics Technology Letters, Vol. 16, pp. 1104–1106, 2004

[4] R. A. Modavis and T. W. Webb, "Anamorphic microlens for laser diode to single-mode fiber coupling," IEEE Photonics Technology Letters, Vol. 7, pp. 798–800, 1995

[5] J. Sakai and T. Kimura, "Design of a miniature lens for semiconductor laser to single-mode fiber coupling," IEEE Journal of Quantum Electronics, Vol. 16, pp. 1059–1067, 1980

[6] H. M. Presby and A. Benner, "Bevelled-microlensed taper connectors for laser and fibre coupling with minimal back-reflections," Electronics Letters, Vol. 24, pp. 1162–1163, 1988

[7] S. Mukhopadhyay, S. Gangopadhyay, and S. Sarkar, "Misalignment considerations in a laser diode to monomode elliptic core fiber coupling via a hyperbolic microlens on the fiber tip: efficiency computation by the ABCD matrix," Optical Engineering, Vol. 46, Article No. 095008, 2007

[8] K. Shiraishi et al., "A fiber lens with a long working distance for integrated coupling between laser diodes and single-mode fibers," Journal of Lightwave Technology, Vol. 13, pp. 1736–1744, 1995

[9] K. Kato et al., "Optical coupling characteristics of laser diodes to thermally diffused expanded core fiber coupling using an aspheric lens," IEEE Photonics Technology Letters, Vol. 3, pp. 469–470, 1991

[10] Y. Fu et al., "Integrated micro-cylindrical lens with laser diode for single-mode fiber coupling," IEEE Photonics Technology Letters, Vol. 12, pp. 1213–1215, 2000

[11] T. Sugie and M. Saruwatari, "Distributed feedback laser diode (DFB-LD) to single-mode fiber coupling module with optical isolator for high bit rate modulation," Journal of Lightwave Technology, Vol. 4, pp. 236–245, 1986

[12] K. Kurokawa and E. E. Becker, "Laser fiber coupling with a hyperbolic lens," IEEE Transactions on Microwave Theory and Techniques, Vol. 23, pp. 309–311, 1975

[13] M. Saruwatari and T. Sugie, "Efficient laser diode to single-mode fiber coupling using a combination of two lenses in confocal condition," IEEE Journal of Quantum Electronics, Vol. 17, pp. 1021–1027, 1981

[14] K. Kato and I. Nishi, "Low-loss laser diode module using a molded aspheric glass lens," IEEE Photonics Technology Letters, Vol. 2, pp 473–374, 1990

[15] J. K. Myoung et al., "Lens-free optical fiber connector having a long working distance assisted by matched long-period fiber gratings," Journal of Lightwave Technology, Vol. 23, pp. 588–596, 2005

[16] Y. Abe et al., "16-fiber fiber physical contact connector with MU connector coupling mechanism, compact shutter and fiber clamping structure," IEEE Transactions on Electronics, Vol. E87-C, pp. 1307–1312, 2004

[17] K. Shibata, M. Takaya, and S Nagasawa, "Design and performance of high-precision MT-type connector for 1.55-mu m zero-dispersion-shifted fiber-ribbon cables," IEEE Photonics Technology Letters, Vol. 13, pp. 136–138, 2001

[18] K. M. Wagner, D. L. Dean, and M. Giebel, "SC-DC/SC-QC fiber optic connector," Optical Engineering, Vol. 37, pp. 3129–3133, 1998

[19] K. Kanayama et al., "Characteristics of an SC-type optical fiber connector with a newly developed pre-assembled ferrule," IEEE Photonics Technology Letters, Vol. 7, pp. 520–522, 1995

[20] TIA-604-2-B (FOCIS-2) Fiber Optic Connector Intermateability Standard, Type ST, Telecommunication Industry Association (TIA), 2004

[21] TIA-604-4-B (FOCIS-4) Fiber Optic Connector Intermateability Standard, Type FC and FC-APC, Telecommunication Industry Association (TIA), 2004

[22] TIA-604-3-B (FOCIS-3) Fiber Optic Connector Intermateability Standard, Type SC and SC-APC, Telecommunication Industry Association (TIA), 2004

[23] TIA/EIA-604-10A (FOCIS-10) Fiber Optic Connector Intermateability Standard-Type LC, Telecommunication Industry Association (TIA), 2002

[24] TIA-604-1 (FOCIS 1) Fiber Optic Connector Intermateability Standard, Telecommunication Industry Association (TIA), 1996

[25] TIA-604-5-B (FOCIS 5) Fiber Optic Connector Intermateability Standard-Type MPO, Telecommunication Industry Association (TIA), 2002

[26] TIA/EIA-604-12 (FOCIS 12) Fiber Optic Connector Intermateability Standard Type MT-RJ, Telecommunication Industry Association (TIA), 2000

[27] TIA-604-17 (FOCIS 17) Fiber Optic Connector Intermateability Standard, Type MU, Telecommunication Industry Association (TIA), 2004

[28] A. D. Yablon, *Optical Fiber Fusion Splicing*, Springer, Heidelberg, 2005

[29] Application Note AN103, "Single Fiber Fusion Splicing," Corning, 2001, available at www.corning.com

[30] V. Alwayn, *Optical Network Design and Implementation*, Cisco Press, Indianapolis, IN, 2004

[31] K. S. Chiang, F. Y. M. Chan, and M. N. Ng, "Analysis of two parallel long-period fiber gratings," Journal of Lightwave Technology, Vol. 22, pp. 1358–1366, 2004

[32] S. J. Hewlett, J. D. Love, and V. V. Steblina, "Analysis and design of highly broad-band, planar evanescent couplers," Optical and Quantum Electronics, Vol. 28, pp. 71–81, 1996

[33] A. Ankiewicz, A. Snyder, and X. H. Zheng, "Coupling between parallel optical fiber cores-Critical examination," Journal of Lightwave Technology, Vol. 4, pp. 1317–1323, 1986

[34] M. Tabiani and M. Kavehrad, "An efficient N×N passive optical star coupler," IEEE Photonics Technology Letters, Vol. 2, pp. 826–829, 1990

[35] A. A. M. Saleh and H. Kogelnik, "Reflective single-mode fiber-optic passive star couplers," Journal of Lightwave Technology, Vol. 6, pp. 392–398, 1988

[36] B. Borovic et al., "Light-intensity-feedback-waveform generator based on MEMS variable optical attenuator," IEEE Transactions on Industrial Electronics, Vol. 55, pp. 417–426, 2008

[37] A. Unamuno and D. Uttamchandani, "MEMS variable optical attenuator with Vernier latching mechanism," IEEE Photonics Technology Letters, Vol. 18, pp. 88–90, 2008

[38] H. Cai et al., "Linear MEMS variable optical attenuator using reflective elliptical mirror," IEEE Photonics Technology Letters, Vol. 17, pp. 402–204, 2005

[39] K. Shiraishi, F. Tajima, and S Kawakami, "Compact faraday rotator for an optical isolator using magnets arranged with alternating polarities," Optics Letters, Vol. 11, pp. 82–84, 1986

[40] J. F. Lafortune and R. Vallee, "Short length fiber Faraday rotator," Optics Communications, Vol. 86, pp. 497–503, 1991

[41] D. J. Gauthier, P. Narum, and R. W. Boyd, "Simple, compact, high-performance permanent magnet Faraday isolator," Optics Letters, Vol. 11, pp. 623–625, 1986

[42] H. A. Macleod, *Thin Film Optical Filters*, 3rd Ed., Institute of Physics Publishing, Bristol, 2003

[43] V. Kochergin, *Omnidirectional Optical Filters*, Kluwer Academic Publishers, Dordrecht, 2003

[44] K. Okamoto, *Fundamentals of Optical Waveguides*, 2nd Ed, Academic Press, New York, 2006

Chapter 8
Optical Transmitter Design

8.1 Introduction

In this chapter we discuss design issues related to optical transmitters. An optical transmitter acts as the interface between the electrical and optical domains by converting electrical signals to optical signals. For digital transmitters, the optical output must conform to specifications such as optical power, extinction ratio, rise and fall time, and jitter. In analog transmitters, the optical output must faithfully regenerate the input in terms of linearity, bandwidth, phase delay, etc. It is the responsibility of the designer to ensure that the transmitter meets all the relevant requirements for the intended application of the design.

This task, in general, represents challenges in several fronts. Because of the universal trend toward higher data rates, the transmitter must be able to modulate light at these higher data rates. Currently, state-of-art transmitters can run at speeds around 40 Gbps and beyond, and 10 Gbps transmitters are commonly available. These high data rates require very careful design practices both in electrical and optical domains. On the other hand, demand for additional functionalities is constantly on the rise, and this, coupled with shrinking mechanical dimensions, presents new challenges to the designers.

From a practical perspective, this means packing more and more circuits in a shrinking board space. For example, a small form pluggable (SFP) transceiver, in widespread use across industry, has only a few square centimeters of board space, in which many circuits such as a laser diver, optical receiver, and control and monitoring functions must all fit. In such a small space, problems such as heat generation, crosstalk, layout limitations, manufacturability, and reliability should be considered carefully. Moreover, it is oftentimes necessary to trade off performance in these areas, in which case the designer must clearly understand the interplay of various parameters in order to come up with an optimal design.

Because most optical transmitters are laser based, we will focus our discussion in this chapter on laser transmitters, although many of the discussed concepts can easily be applied to LED-based transmitters as well. We will start by a review of optical subassemblies, which are used widely in all fiber optic transmitters. Next, we examine the LI curve and biasing of a laser. We then proceed with schemes for control of power and modulation in standard laser transmitters. We will also discuss burst mode transmitters, which are becoming more common thanks to the proliferation of passive optical networks.

The common trend in fiber optic transmitters is the migration to ever-increasing data rates. We will therefore conclude the chapter by discussing topics related to the design of highspeed circuits that are at the heart of modern optoelectronic systems.

M. Azadeh, *Fiber Optics Engineering*, Optical Networks,
DOI 10.1007/978-1-4419-0304-4_8,

8.2 Transmitter optical subassembly (TOSA)

Semiconductor lasers used in fiber optics industry are small, sensitive devices. They are typically a few hundred microns long, with tiny pads for cathode and anode that need wire bonding for electrical connection. Moreover, to couple the light generated by them into an optical fiber, focusing lenses with tight alignment tolerances are needed. Because of these delicacies, proper packaging is a crucial aspect for the performance of these devices. For fiber optic applications, diode lasers are often packaged in the form of transmitters optical subassemblies, or TOSAs [1–4]. The idea behind a TOSA is illustrated in Fig. 8.1.

An optical subassembly fulfills several functions. First and foremost, it provides a stable mechanical platform for the laser chip along with the necessary electrical interconnects. Inside the TOSA, the interconnects are wirebonded to the laser's cathode and anode. For simplicity, Fig. 8.1 only shows a laser chip as the active electrical component inside the TOSA. But practical TOSAs may include a number of other electronic parts, such as power monitoring diodes, TEC coolers, and external modulators. The laser diode (and any additional device) is mounted on a substrate. The electrical connections for all these devices enter through the TOSA's pins.

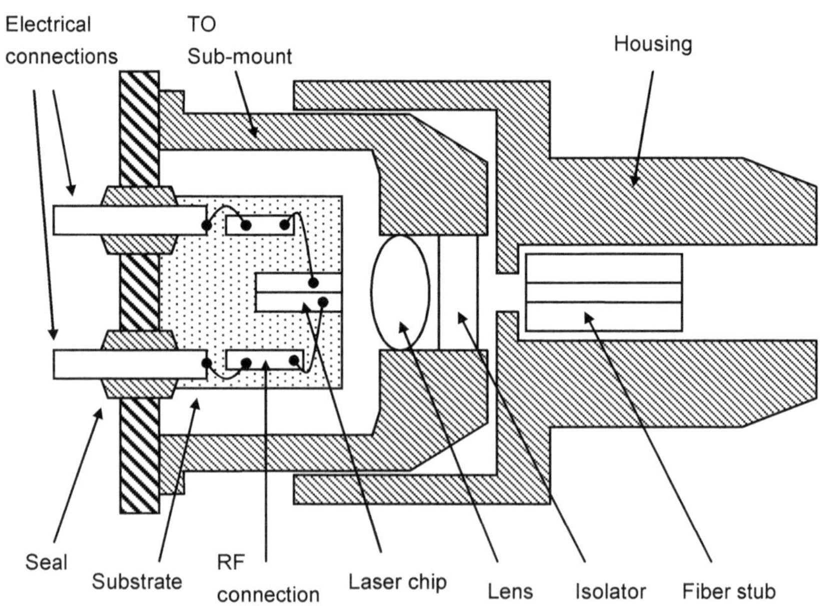

Fig. 8.1. A transmitter optical subassembly (TOSA)

In Fig 8.1, these pins provide the necessary interconnections to the cathode and anode of the laser through internal wirebondings and RF transmission lines. In cases where highspeed performance is crucial, additional matching circuits such as terminations or bias isolators may also be included inside the TOSA. On the other hand, for low-speed applications, the laser may simply be wirebonded to the pins directly.

The TOSA also provides optical alignment and optical interface for external fiber connections. In Fig. 8.1, a lens gathers the diverging light and focuses it at the center of a fiber stub that is placed in the housing. The housing forms a standard connector type, such as SC and LC. As a result, the TOSA can be used with standard fiber connectors commonly utilized in the industry. Figure 8.1 also shows an optical isolator in front of the laser. This is a requirement especially for DFB lasers, because DFB lasers are very sensitive to back reflection. Depending on the application, other optical components such as optical filters and beam splitters may also be included.

The TOSA also provides a hermetically sealed environment for the laser. Such a sealed enclosure is needed to protect the laser from environmental factors such as humidity. From the point of view of heat management, the TOSA provides a heat sinking path for the laser diode, reducing the risk of overheating and damage to the laser. Because of all these functionalities, transmitter optical subassemblies are a critical component of optical transmitters and are available and widely used in a wide variety of forms and functions.

8.3 Biasing the laser: the basic LI curve

From an electrical perspective, a semiconductor laser is a diode, and in order to generate light it must be forward biased. So simply connecting a laser diode to a current source is sufficient to turn on the laser and bias it at some operating point. Figure 8.2 shows the standard LI curve of a typical laser which consists of a subthreshold region, a linear region above threshold, and a saturation region. The laser is normally biased in the linear region.

Also shown is a voltage-controlled current source as the biasing circuit. The input of the driver consists of two components: a bias signal and a modulation signal, provided by two digital to analog converters (DACs). The reason for using DACs is that in almost all modern circuits the bias and the modulation must be controlled very accurately, and this is most commonly done through DACs. Therefore, we use the schematic diagram of Fig. 8.2 as the basis of our analysis.

The LI curve of a laser identifies the useful range of biasing a diode laser. This range starts from just above the threshold current and extends up to the saturation level. This is the linear part of the LI curve. If the laser is biased too close to threshold, spontaneous emission starts to dominate, and therefore spectral characteristics of the laser deteriorate.

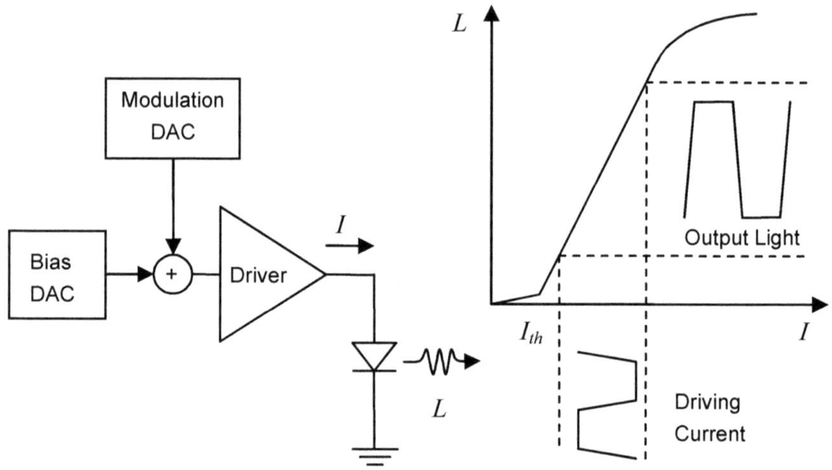

Fig. 8.2. Schematic of a laser driver and the typical LI curve

If the laser is biased too close to saturation, any change in temperature may cause significant power degradations. In practice, the biasing point must be well below the saturation to allow room for modulation around the biasing point as well as quantum efficiency degradation over temperature and life of the device.

As Fig. 8.2 suggests, the driving current of a laser can typically be divided into two parts, a bias current and a modulation current. The bias current can be thought of as a dc current that sets the operating point in the target region of the LI curve and corresponds to the dc optical power. The modulation current corresponds to the variations in the drive current around the bias point in accordance with the data pattern to be transmitted. We will consider these two parts separately, and start with the bias portion of the circuit first.

As noted in Chapter 1, the power of an optical transmitter is measured in units of decibel-milliwatt, which provides a "natural" basis for power because power loss (and gain) is additive in the logarithmic scale. Optical power is a result of the bias current, and because the bias current is ultimately set by a DAC with limited resolution, care must be taken as to how the full range of the DAC is mapped on the range of the optical power.

For instance, suppose the DAC is n bits wide, providing 2^n steps in the setting of power. Also assume the laser has a threshold current of I_{th} (mA) and a saturation current of I_{sat} (mA) with a slope efficiency of K (mW/mA). The optical power in dBm, corresponding to a bias current of I, is then given by

$$L_{\text{dBm}} = 10\log\left[K\left(I - I_{TH}\right)\right] \qquad I > I_{TH} \tag{8.1}$$

If the transconductance gain of the circuit from the DAC voltage to the bias current I is given by A, and if the DAC's value is given by m ($m = 1–2^n$), the power in dBm as function of the DAC's value is given by:

$$L_{dBm}(m) = 10\log\left[K\left(m\frac{A}{2^n}V_{DAC} - I_{TH}\right)\right] \qquad m > \frac{2^n}{AV_{DAC}}I_{TH} \tag{8.2}$$

where V_{DAC} is the full-range voltage of the DAC. We also need to add another constraint, i.e., that the full range of DAC should generate a current less than the saturation current: A× $V_{DAC} < I_{sat}$. This is a design constraint that ensures the utilization of DAC's resolution is maximized. Equation (8.2) maps the value of the DAC to the optical power and determines the minimum step size in power setting. Another way to characterize this is by considering the change in power that results from a single-step change in the DAC, which follows directly from Eq. (8.2):

$$\Delta L_{dB}(m) = L_{dBm}(m) - L_{dBm}(m-1) = 10\log\left[\frac{mI_{MAX} - 2^n I_{TH}}{(m-1)I_{MAX} - 2^n I_{TH}}\right] \tag{8.3}$$

where $I_{MAX} = A\times V_{DAC}$ is the maximum bias current that the circuit can support and which corresponds to the full range of the DAC. Note that the step size in not a function of the slope efficiency of the laser. Equation (8.3) can be written more compactly if we define a ratio $R=I_{TH}/I_{MAX}$:

$$\Delta L_{dB}(m) = 10\log\left[\frac{m - 2^n R}{m - 1 - 2^n R}\right] \tag{8.4}$$

We see that at lower powers, especially close to threshold ($m\approx 1+2^nR$), the step size becomes larger. As a result, when we want to select the width of the power control DAC, we must ensure that enough resolution is attained over the entire range of power settings.

This is often an important issue when the circuit is used in a design that must support various applications with a wide range of target powers. For instance, as we will see in Chapter 12, the SONET standard defines several classes of physical layer link budgets such as SR, IR, and LR, with different min/max optical power specifications.

8.4 Average power control (APC)

Simply biasing the laser with a constant current is not sufficient for most applications. The reason is that the LI characteristic of a diode laser is strongly affected by temperature [5–8]. Even at a given temperature, the LI curves varies over time as a result of aging [9,10]. In general, an optical transmitter must operate over some given range of temperatures and over an extended period of time. Thus, unless provisions are made to compensate the effects of temperature variation and/or aging, the output power of the diode laser will change significantly.

In general, higher temperature causes a lowering of quantum efficiency in a semiconductor laser as well as an increase in the threshold current. As a result, the circuit must ensure that optical power stays the same regardless of these changes. This is typically done with an average power control (APC) scheme. In general, APC schemes can be divided into two main categories of open and closed loop control. Closed loop schemes rely on some sort of optical feedback to maintain the power. Open loop schemes, on the other hand, do not use an optical feedback loop for power control. In the following sections we will discuss the main approaches used in the implementation of APC circuits.

8.4.1 Open loop average power control schemes

In an open loop temperature compensation scheme, the circuit attempts to correct for the effects of temperature variations by changing the biasing point of the laser. In its simplest form, a single temperature-sensitive device is added to the circuit of Fig. 8.2. For example, a negative temperature coefficient (NTC) element, such as a thermistor, can be added to the circuit of Fig. 8.2. The resistance of a thermistor drops when the temperature increases. Thus, it can be used as a means of making the bias point sensitive to the temperature [11,12].

This is a very simple but crude way of controlling the power of a diode laser. The reason is that the LI curve of each individual laser is different from another laser, and this holds even for laser diodes from the same vendor which are even built from the same batch. Moreover, it is very hard to "match" the characteristics of any given laser to a temperature-sensitive device such as a thermistor over a reasonable range. As a result, the circuit can easily under-compensate or over-compensate the temperature variations. Another drawback of this approach is that any variations not due to temperature (such as degradations due to aging) will not be compensated.

A more accurate method of controlling the biasing point of a diode laser as a function of temperature is to use a lookup table. This approach, like the previous one, also uses a temperature-sensitive device such as a sensor. However, the difference is that the temperature sensor does not control the biasing point of the diode laser directly. Instead, the sensed temperature is first digitized through an analog to digital converter (ADC) and then used as an index into a lookup table. A lookup table is essentially a table consisting of two columns and a number of rows. The two entries in each row include a temperature and the corresponding bias point at that temperature. In this way, a state machine or a microcontroller can read the temperature from the temperature sensor through an analog to digital converter, look up that temperature in the table, and find the corresponding biasing point for that temperature. This scheme is illustrated in Fig. 8.3.

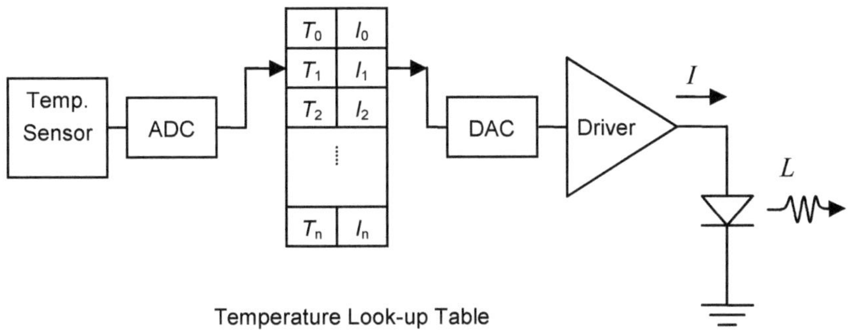

Fig. 8.3. Power control using a temperature look-up table

The advantage of a lookup table approach is that each circuit can be individually tuned to the temperature behavior of the particular laser diode the circuit is driving. In practice, this means each laser diode must be characterized individually over temperature, and the data be used to calculate the lookup table. This approach lends itself well to a production process where the transmitter is tested over a number of temperature points. At each point, the bias current is tuned until the desired optical power is achieved. The temperature point and the corresponding bias current constitute one data point. The test is repeated at several temperature points, and the lookup table can be calculated as a curve that is fit to these collected points. Essentially, any temperature behavior can be accommodated, because a lookup table can implement any function. However, there are several additional points that need to be considered.

One problem arises from the fact that the benefit of accurate power control is achieved at the price of the need to individually characterize each and every laser. This could be a costly process, especially after the laser is assembled into an optical transmitter module. Another point is the resolution or "granularity" of the ADC and the DAC in the lookup table. This is closely related to the power setting error discussed before. As the temperature varies, the pointer jumps from one entry in the table to the next, causing a "jump" in the laser power. This jump can be reduced by increasing the resolution of the ADC and the DAC and by using a larger lookup table. In general, both the quantum efficiency and the threshold of a laser degrade almost exponentially with temperature. This calls for more temperature granularity at higher temperatures. Finally, all open loop schemes, including the lookup table approach, cannot compensate for laser aging. For these reasons, approaches based on closed power control are frequently used.

8.4.2 Closed loop power control

In a closed loop power control approach, a feedback loop is utilized to stabilize the power. Like any feedback loop, a feedback signal must be tapped off the quantity that needs to be controlled, in this case the optical power of the diode laser. In edge emitting lasers, this is achieved through using the back facet optical power from the laser diode. As discussed in Chapter 4, a semiconductor diode laser typically has two optical outputs from the two sides of the optical cavity. While one output can be coupled into the fiber and used as the main output, the other output can be used as a feedback signal to measure the optical power of the diode laser. However, this output cannot be used directly by the control circuits in the form of light. Therefore, it must be converted back into an electrical signal. This is done by placing a photodetector close to the laser diode to monitor the optical power output of the laser diode. The current from this photodetector is known as back-facet current, photodetector monitoring current, power monitoring current, or simply monitoring current. This current, then, can be used as a means of measuring the optical power by the electronic control circuits.

In a closed loop APC scheme, the circuit attempts to keep the optical power constant by keeping the power monitoring current constant. For most diode lasers used in telecommunication, the back-facet photodetector is integrated with the laser as part of the optical package, and either its cathode or anode (or both) is provided as a pin on the package, as shown in Fig. 8.4.

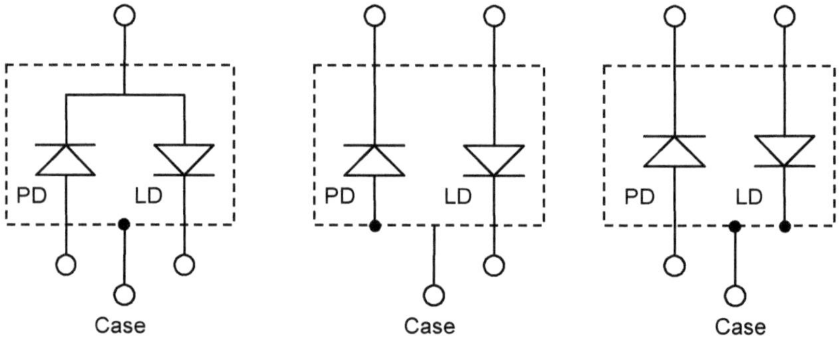

Fig. 8.4. Examples of commonly used packages for laser diodes and power monitoring photodetectors

Once a feedback signal is available, the power control can be implemented in a closed loop scheme, as illustrated in Fig. 8.5.

In this circuit, the photodetector generates a current that is proportional to the optical output power of the laser. If the laser is being modulated, the current goes through a low pass filter (LPF) to remove higher frequency components. After that, it is compared to a set point corresponding to the target optical power. The circuit attempts to minimize the difference between the feedback signal and the set point by controlling the optical power through varying the bias current. Thus, the user controls the power by setting the photodetector current. The feedback circuit will adjust the bias so as to reach the target photocurrent. The magnitude of the bias current depends on the power requirements, type of laser, and temperature.

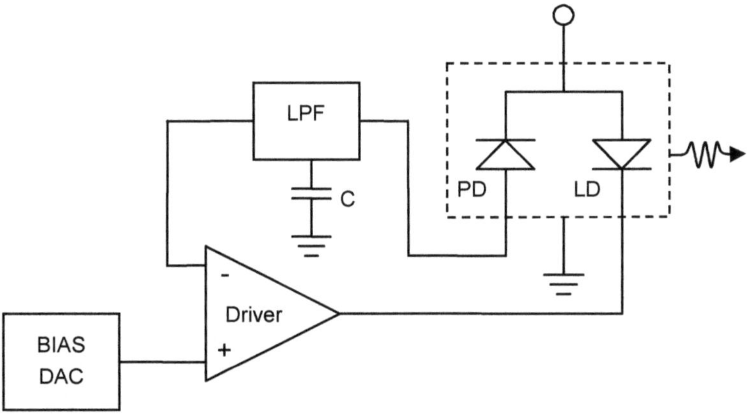

Fig. 8.5. Closed loop APC scheme

For a VCSEL working at room temperature the bias current may be as low as a few milliampere. On the other hand, for a DFB laser operating at high temperature (~70–80°C) and high power (3–5 dBm) the bias current may be as high as 100 mA. Therefore, it must be ensured that the circuit can handle the expected maximum bias current with adequate margin.

The loop time constant is usually set by a capacitor, C, and determines the response time of the transmitter. A slow time constant yields more stability against pattern dependence, an important issue in direct modulated lasers. If the time constant of the loop is too fast, the loop will start to respond to various bit patterns. This would be an undesirable effect because ideally we want the loop to maintain the *average power* constant, while allowing the instantaneous power to vary according to the desired modulation. We will come back to this issue shortly. On the other hand, a slow time constant means it will take longer for the loop to stabilize to a new set value. This could be an issue for hot pluggable or burst mode modules, where it is important for the transmitter to set up the optical power within a given period of time.

At first glance, a closed loop APC scheme seems to overcome all the shortcomings of open loop schemes. Indeed, the circuit can maintain the monitoring photodetector current constant, and to the extent that the optical power is linearly related to the photodetector current, power too will remain constant. For example, unlike the previous methods that did not provide any protection against aging, a feedback-based power control scheme does keep the power constant even as the laser diode ages and its efficiency drops.

However, this method too has its own shortcomings. The basic problem comes from the fact that the relationship between optical power and the photodetector monitor current is not constant over temperature and power. At different temperatures and biasing currents, the ratio of front facet to back facet power does not remain completely constant. Moreover, the coupling efficiency of the back facet power to the monitoring photodetector changes. All these variations result in a tolerance in the amount of photodetector monitoring current for a given amount of optical power.

This tolerance is known as *tracking error* and is usually given in units of dB. The typical value of tracking error for commercially available diode laser assemblies is around ±1.5 dB. This means that even when the circuit keeps the photodetector monitoring current constant, the laser power can vary as much as ±1.5 dB. Another way to state this is that for a constant optical power, the photodetector current can change as much as ±30%, which is roughly equivalent to ±1.5 dB. In some applications, this amount of uncertainty in optical power setting is acceptable. In that case, a closed loop APC scheme would be an acceptable and economical solution, because lasers do not need individual characterizations or tuning anymore. A module just needs to be set at the desired power point, and the feedback loop takes care of the rest.

However, if more accuracy in power setting is needed, tracking error must be compensated, and this again calls for lookup table implementations. But unlike an open loop circuit where the lookup table compensates the bias variations over temperature, in this case the lookup table compensates the tracking error in the power monitor photodetector. Moreover, in order to characterize the tracking error over temperature, either each laser must be characterized individually over temperature beforehand or else each module must be tuned at several temperature points, just like the case of open loop power control. In this case, the advantages of a closed loop APC scheme become less obvious. In such cases the main advantage of a closed loop scheme is protection against power variations as a result of aging. The above discussion shows that controlling the optical power of an optical transmitter especially with accuracies better than approximately ±1.5 dB is a non-trivial problem.

8.4.3 Thermal runaway

In a closed loop scheme, the circuit adjusts the bias current to maintain a constant optical power. As the temperature rises, the threshold current and quantum efficiency of the laser degrade, which means more bias current is needed to maintain the same level of optical power. However, passing more current through the laser diode causes extra heating at the junction, which causes more drop in efficiency. If the junction temperature starts to get too high (say, beyond ~90°C) this process may result in *thermal runaway*: as a result of positive feedback the bias quickly increases until the maximum current capacity of the driving circuit is reached, at which point the feedback loop can no longer be sustained.

The risk of thermal runaway in closed loop APC schemes requires additional safety features to prevent it. For example, a current limiter may be added to the bias circuit which limits the bias current to a certain safe value regardless of the optical power. If the feedback loop attempts to increase the bias current beyond this value, an alarm signal could be activated informing the supervisory systems of a potential thermal runaway situation that needs attention.

Prevention of thermal runaway is just one of the safety features that can be added to an optical transmitter. In practice, several such safety features may be needed in a particular application. For example, most practical laser drivers protect the laser against accidental shortening of either cathode or anode of the laser diode to either supply or ground voltages. Without such protection, an accidental shortening may cause a large amount of current flow through the laser and thus a very high level of optical power, potentially causing damage to the laser [13–15]. It can also pose a risk (in terms of eye safety) to operators that may be working with the system.

8.5 Modulation circuit schemes

Maintaining a constant average power is only one requirement that the transmitter must meet. Another critical parameter in an optical transmitter is the modulation depth or extinction ratio. As discussed in Chapter 1, any form of information transfer requires modulation of some form of energy, and in optical transmitters the light must be modulated in order to transmit information. Therefore, in addition to biasing the laser to generate light, an optical transmitter must also modulate that light in accordance with the data stream that needs to be transmitted.

The simplest form of modulation is direct modulation. Diode lasers can easily be modulated by modulating the flow of current through them, and indeed direct modulation schemes are adequate for many applications with data rates below 10 Gbps. The methods used to modulate the current depend on the application and data rate. At very low data rates, signal integrity issues are not crucial and therefore not much care is needed, the laser diode can be treated like a normal load, with roughly 1–2 V of forward bias. The required modulation current depends on the laser type, and in terms of magnitude it is in the order of the bias current. However, at higher data rates, signal integrity issues start to become important, and special attention must be paid to ensure proper operation.

8.5.1 Basic driver circuit

The basic building block used in creating high-speed modulation is a differential driver [16–19]. In a direct modulation scheme, the differential driver is directly connected to the laser diode. In an external modulator, the differential driver drives a separate modulator. Because of the importance of the differential driver, we will start our discussion with the analysis of a direct modulated design based on a differential driver, as illustrated in Fig. 8.6.

The design is based on a current source whose current can be switched by a differential pair between two branches. The differential pair is driven by a differential signal representing the data that needs to be transmitted. A logic "one" causes one side of the differential pair to conduct, while a logic "zero" causes the other side to conduct. The laser diode is connected to the collector of one of the transistors.

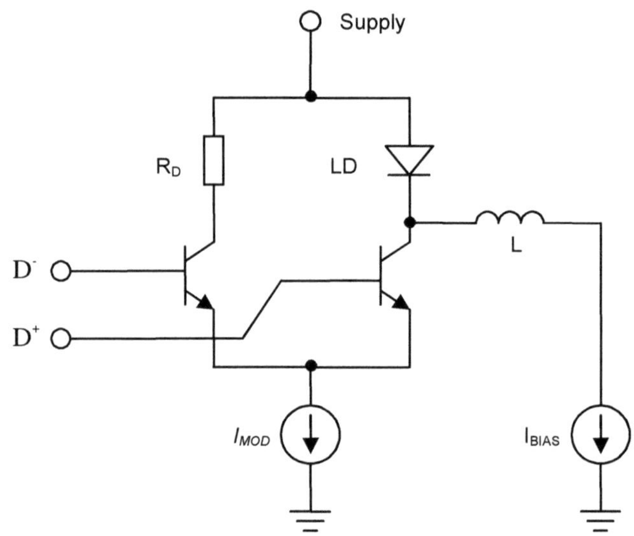

Fig.8.6. Differential structure for high-speed modulation of diode lasers

The other transistor is connected to a dummy load, with characteristic impedance similar to that of the laser diode. Because the current from the current source is switched between the two sides, the current source sets the value of the modulation current. The differential impedance of a forward-biased diode is in the range of few Ohms, and therefore the dummy load is usually in the range of 10–20 Ω.

In addition to the differential modulator, Fig. 8.6 shows a dc path for biasing the diode laser. The bias current is provided by the bias circuit which we discussed in the previous section, either through an open loop or a closed loop scheme. For simplicity, we have shown the bias circuit with a current source. The current source is needed to ensure that the laser always stays above threshold, even during a logic 0 state. The reason is that it takes much longer for the laser to turn on from a completely off state compared to when it is biased slightly above its threshold. Moreover, the spectral characteristic of a diode laser at or below threshold is not suitable for data transmission at high speeds.

The dc path must be isolated from the modulation or the ac path by an element that shows low resistance at dc frequencies but high impedance at modulation frequencies. This is to ensure that the ac current from the differential driver goes through the laser diode and not through the biasing circuit. This is usually done through an inductor or a ferrite bead, both of which have a very low dc resistance.

8.5.2 Transmission line effects

In Figure 8.6 we have implicitly assumed that the load (laser diode) is connected directly to the output transistor of the differential driver. However, in reality the load is usually located at some physical distance from the driver. A typical example is the case of an optical subassembly connected to the driving circuit a few centimeters away. In low data rates, the length of the electrical connections can be neglected, and Fig. 8.6 remains valid. At higher data rates, where the rise and fall times of the signal become comparable to the time it takes for the signal to travel from the driver circuit to the laser, transmission effects and parasitic effects must be taken into account. Indeed in such cases simply connecting the load to the driver (as suggested by Fig. 8.6) with an arbitrary electrical connection will not work. Instead, transmission lines with controlled impedance must be used. Figure 8.7 illustrates a scheme similar to that of Fig. 8.6 but this time explicitly showing the transmission line used in connecting the driver and the load.

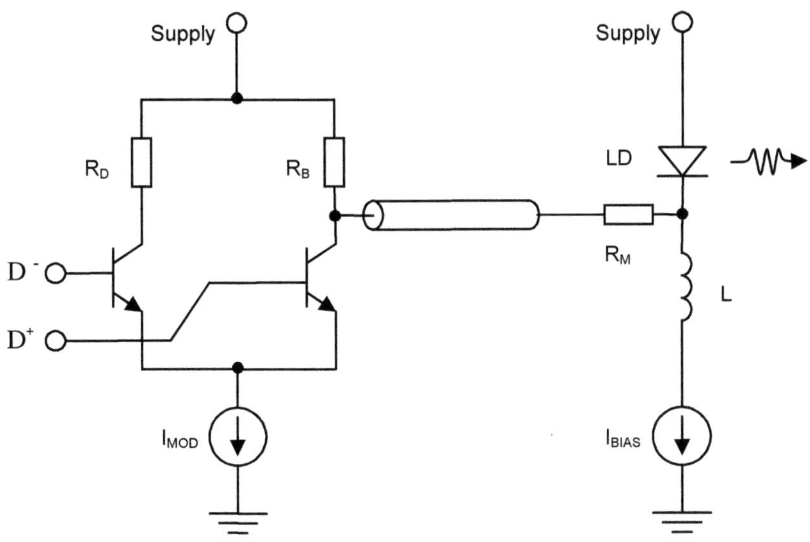

Fig. 8.7. Differential driver with transmission line

In addition to showing the transmission lines, Fig. 8.7 shows some extra details compared to Fig. 8.6. Specifically, in addition to the load, there is a *back termination* resistance shown as R_B at the output of the driving transistor and a matching resistor R_M close to the laser. The reason for adding these resistors becomes clear once we remember from Chapter 3 that for best signal integrity results, a transmission line must be both source and load terminated. The typical impedance of microstrip or stripline transmission lines made in printed circuit boards is in the range of 20–50 Ω. On the other hand, the collector of a transistor acts like a high-impedance current source while the resistance of a forward-biased laser diode is 5–10 Ω. Therefore, without R_B and R_M there will be impedance mismatch both at source and at load. The back termination resistance R_B lowers the high impedance of the current source to the Thevenin equivalent of a voltage source with output impedance of R_B. Thus, the value of R_B should be selected close to the characteristic impedance of the transmission line. On the other hand, generally the impedance of the forward-biased laser diode is only a few Ohms and therefore lower than the impedance of the transmission line. Thus, in order to match the impedance of the load to the characteristic impedance of the transmission line a matching resistor in the form of a series resistor R_M must be added to the load. Therefore, the interface between the driver and the load is both source and load terminated, and this improves the signal integrity. We should also note that impedance matching schemes can become more elaborate by adding frequency-dependent elements. We will discuss some of these schemes in more detail later.

The downside of adding these extra terminations is that they consume extra power. The back termination R_B consumes some of the modulation current, so part of the current that would otherwise go through the load is now going through R_B. The matching resistor R_M acts as a voltage divider, so part of the voltage that otherwise would appear entirely across the load will now be dropped across R_M. To minimize the extra power consumption, usually it is desirable to maximize R_B and to minimize R_M. Typically, R_B is about 30–50 Ω and R_M is about 5–10 Ω.

8.5.3 Differential coupling

The scheme shown in Fig. 8.7 is a generic scheme that can be used in various applications. However, sometimes particular requirements may push this scheme to its limits, at which point we may need to make further modifications in order to address those requirements. One such case is when it is crucial to improve the rise and fall times of the optical output. This is especially important at higher data rates, where slow rise/fall times will cause optical eye closure. In such cases it is desirable to deliver the maximum possible rise and fall times to the load.

Figure 8.8 illustrates a commonly used solution for this problem. This circuit builds on the ideas of Fig. 8.7, with the difference that now the currents from both transistors are delivered to the load. Because of this, usually when reference is made to differential driving schemes, a scheme similar to the one shown in Fig. 8.8 is meant, where both the driving circuit and the way the load is being driven are differential. This is achieved by a differential transmission line that provides the interface between the circuit and the load.

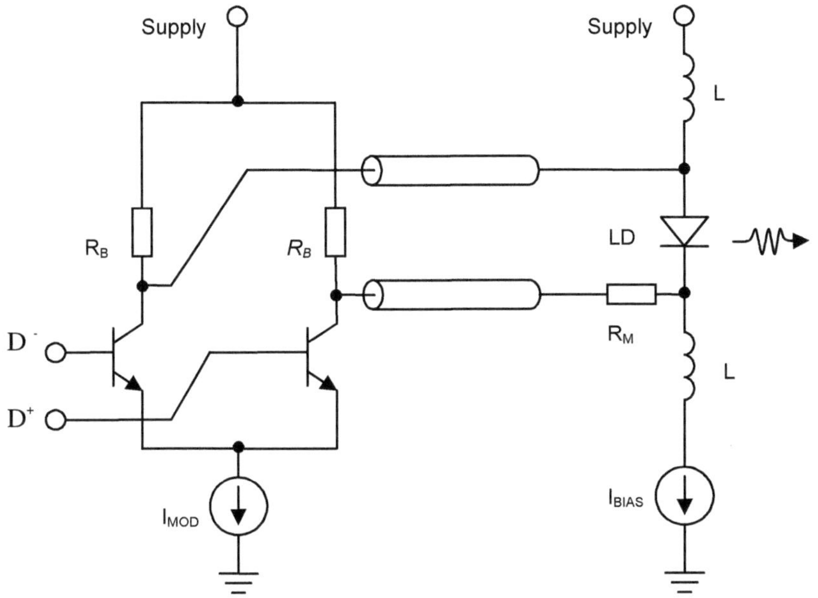

Fig. 8.8. Differential driver with differential transmission line connections

Note that now each transmission line has its own back termination resistor. In this circuit, the rise and fall times can easily be tuned by changing R_B. In general, the optimal value of R_B is equal to the characteristic impedance of the transmission line. Reducing the value of R_B further causes a damping of waveform as seen by the load, and results in increasing rise and fall times. On the other hand, increasing the value of R_B above the optimal value causes a reduction of rise and fall times. However, this usually increases the overshoot too, and excessive overshoot limits the upper limit for R_B.

Note that like before, the bias path is isolated from the modulation path through high-impedance inductors. This circuit is also advantageous in terms of reduction of noise effects on other nearby circuits. Because the currents flowing in the two transmission lines are of opposite phase, the electromagnetic radiation from the switching modulation current is reduced, causing less noise induction into other circuits.

8.5.4 High current drive circuits: ac coupling

One of the important parameters in a laser driver circuit is the maximum bias and modulation current that can be delivered to the load. This is obviously an important factor in high-power applications, because in a laser diode optical power is proportional to current.

To appreciate this, let us consider the dc voltage drop on the diode laser and R_M in Fig. 8.8. The anode of the laser diode is connected with an inductor to the supply voltage, and therefore its dc voltage is almost equal to that of the supply voltage. Many modern electronic circuits operate at relatively low voltages, and a typical value for supply is 3.3 V. At high currents, we may have as much as 2 V drop on the diode. However, there is some additional voltage drop on the series resistor R_M. For example, if we assume a value of 10 Ω for R_M, a modulation current of 50 mA results in a 0.5 V drop across R_M. Assuming a 3.3 V supply, 2 V drop on the diode laser, and 0.5 V drop on R_M, the dc voltage at the collector of the output transistor is pushed down to 0.8 V. Obviously it is hard to have the output transistor operate with a voltage of only 0.8 V. This "headroom" problem is a direct consequence of trying to drive the load at high modulation and bias currents and could result in optical eye distortion and jitter.

A common solution to this problem is to isolate the dc operating point of the circuit from that of the diode laser. This is achieved through ac coupling, an example of which is shown in Fig. 8.9.

In this scheme the output transistor is ac coupled to the diode laser through the capacitor *C*. Moreover, another "pull-up" inductor sets the dc operating point of the collector of the output transistor at supply voltage, thus removing any headroom limitations. This pull-up inductor will not affect the back termination matching provided by R_B because the impedance of this inductor is high at modulation frequencies. The disadvantage of the ac coupling scheme, other than extra complexity, is extra current consumption.

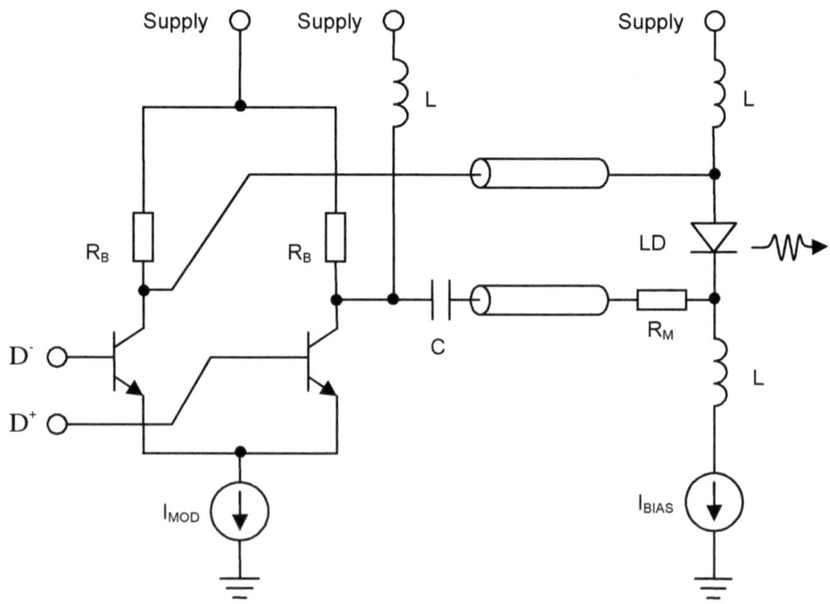

Fig. 8.9. ac coupling scheme to improve the headroom problem at high operating currents

Note that in all the dc coupling schemes that we discussed before, the bias current source I_{BIAS} generates the minimum amount of current that the laser is permitted to operate at, and the modulation current is just added to this minimum bias current. However, in an ac coupling scheme, the bias point corresponds to the mid-point of operation, and the modulation circuit must subtract from or add to this mid-point. This difference is made clear in Fig. 8.10 with reference to the LI curve of a laser. From this figure it can be seen that in an ac coupling scheme the bias point must be higher by $I_{MOD}/2$ compared to a similar dc-coupled circuit. This difference increases at high temperatures, because the drop in slope efficiency of the laser increases the required I_{MOD}.

Another consideration is that the addition of the ac coupling capacitor introduces a low-frequency cut-off point, which means the value of the capacitor must be large enough to ensue that the lowest frequency components of the signal can reach the laser. This holds true not only in the output stage, but also anywhere else where the signal is ac coupled from one stage to the next. The general rule is that the low-frequency cut-off is given by

$$f_{3dB} = \frac{1}{2\pi RC} \tag{8.5}$$

where C is the coupling capacitor and R is the impedance of the circuit that the capacitor is connected to.

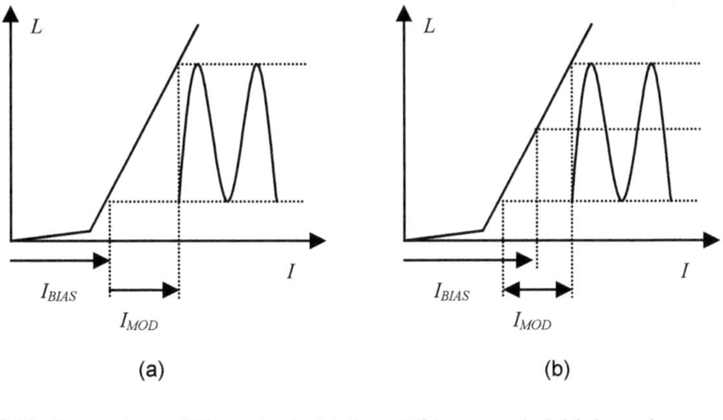

Fig. 8.10. Comparison of bias point in (**a**) dc and (**b**) ac-coupled driving schemes

In Fig. 8.9, the capacitor is mainly charged or discharged through the back termination resistors, and therefore we can estimate the low-frequency cut-off point by substituting R_B in Eq. (8.5). In practice, this is not such an important issue for high data rates (i.e., 100 Mbps and higher) or for applications where the low-frequency contents of the traffic are intentionally removed through coding (such as the

8B10B coding). However, if the traffic has low-frequency contents, then choosing the right value for C becomes more important, especially at lower data rates.

8.6 Modulation control, open loop vs. closed loop schemes

In the circuits we discussed above we implicitly assumed that the magnitude of modulation current (set by the I_{MOD} current source at the base of the differential transistors) is constant. However, in reality the magnitude of modulation current must compensate for any possible changes in the LI curve of the laser. The problem here is essentially the same as what we considered for bias circuits. The LI characteristic of a laser is not constant, especially with temperature. Just as bias control circuits are needed to maintain the optical power constant in the face of these variations, modulation control circuits are needed to maintain a constant optical modulation in spite of changes in laser characteristics.

Like bias control circuits, modulation control circuits can also be divided into open and closed loop schemes. We start with open loop schemes and then continue with considering some popular closed loop schemes.

8.6.1 Open loop modulation control

The approaches used in open loop modulation control are very similar to those used in open loop bias control. The simplest method is to use a temperature-sensitive element such as thermistor to change the value of modulation current with temperature.

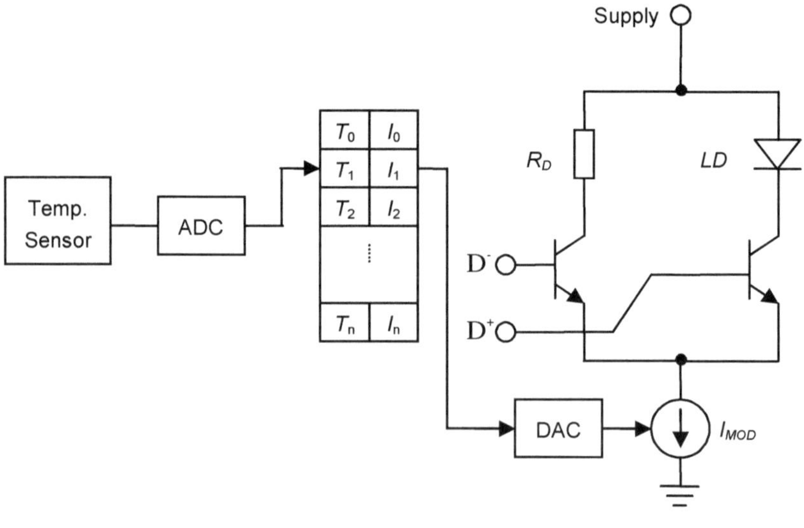

Fig. 8.11. Look-up table scheme for modulation control

However, this approach is very crude and suffers from the same shortcomings that we discussed before. A more robust approach is to use a look-up table, as shown in Fig. 8.11. This scheme is very similar to the look-up table scheme we discussed earlier with respect to power control. Therefore, the concerns we mentioned with the power control apply here too. Specifically, the temperature granularity of the look-up table and the resolution of the DAC must be considered carefully.

In digital transmitters, the extinction ratio (ER) is defined as the ratio of the optical one to the optical zero state. For the circuit of Fig. 8.11, the value of modulation and therefore ER is determined by the current row in the look-up table. Because ER is essentially a ratio, it becomes increasingly difficult to control higher ERs, because a small change in the 0 state results in a large change in ER. Therefore, the resolution of the DAC and also the temperature granularity must be chosen such that the jumps in ER from one value of the DAC to the next will not cause ER to fall out of specification.

From a link budget perspective, it is better to operate a transmitter at higher ERs, because a low ER means the optical power in the 0 state is being wasted. On the other hand, operating a diode laser at a very high ER is not desirable in the case of directly modulated lasers. The reason is that the coherence properties of laser diminish if it is operated too close to threshold. Moreover, once the ER is higher than approximately 10 dB the optical power in the 0 state is small and increasing ER further does not improve the link budget by much. As a result, the most common values of target ER for optical transmitters are around 8–12 dBs.

Although open loop modulation control schemes are conceptually simple, they are hard to implement accurately, mostly because of the difficulty of producing accurate look-up tables for individual laser devices. As a result, it is desirable to have closed loop schemes in which a feedback loop maintains the ER automatically once it is set. This is the topic of the next two sections.

8.6.2 Closed loop modulation control: Pilot tone

ER variations are a result of changes in the slope efficiency of the laser diode. Therefore, if a method can be implemented to measure the changes in slope efficiency, it can be used to compensate any changes in the ER. As noted in the beginning of this chapter, a monitoring photodetector is typically integrated with the laser to provide a feedback signal proportional to the optical power. Therefore, closed loop modulation control schemes must utilize this photodetector current.

One way to achieve this is to inject an extra low-frequency "pilot tone" on the bias current and detect the resulting low frequency tone on the optical output by monitoring the photodetector current [20]. The fact that the tone is low frequency ensures that it is well within the bandwidth of the monitoring photodetector, and therefore it can be recovered properly. Because the amplitude of the pilot tone is constant, any changes in the amplitude of the resulting optical tone can be attributed directly to a change in the slope efficiency of the laser, and therefore it can be used to modify the modulation amplitude. This scheme is illustrated in Fig. 8.12. For a constant value of pilot tone current, I_{TONE}, the slope efficiency is proportional to the amplitude of the resulting optical tone:

$$\frac{\eta_2}{\eta_1} = \frac{P_{TONE2}}{P_{TONE1}} \tag{8.6}$$

where the indices refer to the values of slope efficiency and optical tone at two different temperatures. Therefore, the circuit can "measure" the slope efficiency by measuring the amplitude of the tone in the photodetector signal.

Although this approach has some advantages over open loop control schemes, it has its own problems. Specifically, the amplitude of the pilot tone must be small compared to the data amplitude, because it shows up as extra noise on the optical one and zero levels. To prevent optical eye closure, this amplitude cannot exceed a few percentage of the modulation amplitude, and because the modulation amplitude itself is variable, it is even harder to control the amplitude of the pilot tone accurately. Moreover, an implicit assumption in this approach is that the laser LI curve is linear. Therefore, at high powers where nonlinear saturation effects start to show up this approach is not very effective. As a result of these problems, closed loop modulation control schemes based on pilot tone are hard to implement and rarely used.

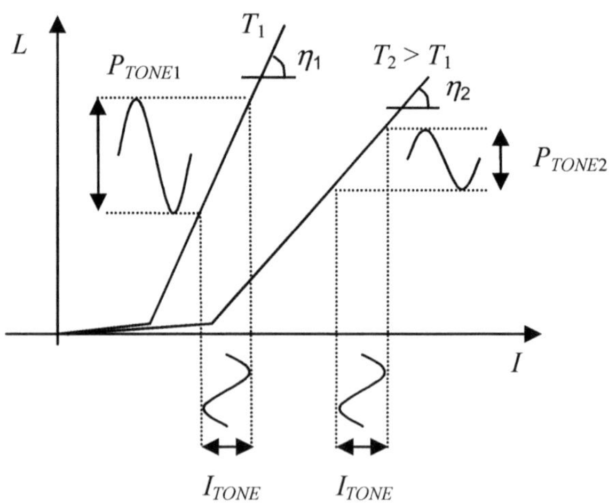

Fig. 8.12. Detection of changes in slope efficiency by an extra pilot tone

8.6.3 Closed loop modulation control: high bandwidth control

The most natural way to control the ER is to directly sense the optical zero and one levels and try to keep their ratio constant. This approach requires a relatively high bandwidth photodetector, enough to clearly convert the optical one and zero levels back to a clear electrical signal that can be used. Because the optical zero level is close to zero, the corresponding photodetector current can be very small, and it would be a challenge to recover a clean true zero level from a small, high-speed, noisy photodetector current. Moreover, if the bandwidth of the photodetector is insufficient, the photocurrent may not have enough time to reach to the steady state one or zero level within short bit sequences, resulting in pattern dependence in the feedback loop.

If these problems are carefully addressed, a reasonable closed loop modulation scheme can be realized. But even in this case the scheme is only as good as the feedback signal, i.e., the photodetector current. As we noted in the section on average power control, the photodetector current does not necessarily have a linear relationship with the optical power. Therefore, like average power control, accurate closed loop control of modulation is a design challenge.[1]

8.7 External modulators and spectral stabilization

Direct modulation schemes are a convenient way of modulating light well into gigabits per second speeds. In fact, short-reach 10 Gbps transmitters are usually implemented with direct modulation. However, for modulating speeds beyond 10 Gbps, and also for long-reach links at lower speeds, direct modulation is not very effective. The reason is that in a diode laser current modulation always causes frequency modulation or chirping, which in turn causes spectral broadening of light. This in turn increases the dispersion and reduces the link budget. Thus, direct modulation links at higher frequencies tend to be dispersion limited rather than attenuation limited. External modulation, on the other hand, essentially removes frequency chirping, because the light source is a CW laser dc-biased well above its threshold with very narrow spectral characteristics. Therefore, use of external modulators significantly increases the attainable link distance. A related issue is that with external modulation higher modulation depths or ERs can be achieved, because in direct modulation, a higher modulation depth causes excessive overshoot, turn-on delay, and turn-on jitter. As a result, very high-speed or very long-reach optical transmitters have to be implemented with external modulators [21–24].

Figure 8.13 shows the typical schematic of an external modulator optical transmitter. The light is generated by a single-mode diode laser which is biased at a dc level through a bias control circuit. This bias control circuit is similar to the bias control circuits we discussed before. The CW light generated by the laser diode now goes through an external modulator which is driven by a modulation driver. The modulation driver is similar to the modulator circuits we discussed be-

[1] Part of the reason for difficulty of ER control is that ER itself is a quantity hard to measure (and even define) beyond a limited accuracy. See Chapter 11 for further discussion of ER measurements.

fore, with the difference that external modulators are typically optimized for higher speed operation.

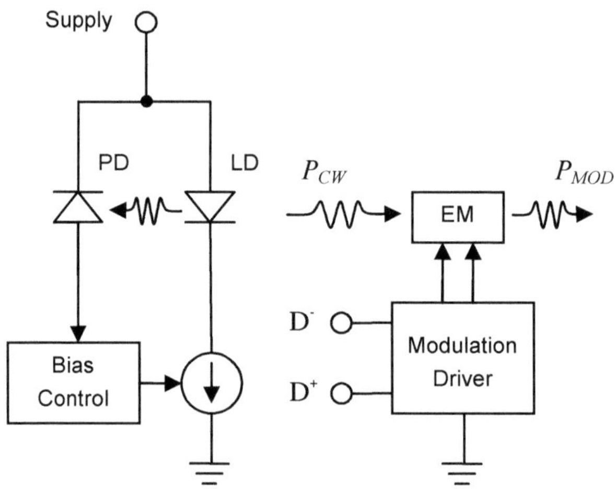

Fig. 8.13. Schematic of an external driver optical transmitter

Thus, for instance, the output transistors must have a very high cut-off frequency. Moreover, the physical structure of the transmitter must support the operating high speeds, i.e., the transmission lines connecting the driver to the modulator must have a flat response beyond the frequencies involved. This requires high-quality, low loss substrates, minimization of parasitic effects, and accurate impedance control. We will discuss some of these high-frequency techniques later in this chapter in more detail.

Optical subassemblies based on external modulators often include additional control circuits. For example, because of the narrow spectral characteristics of external modulators, they are usually used in WDM transmitters. In a WDM system, each channel must have a specific predetermined center wavelength. Moreover, the spectral width of each channel determines how closely the channels can be spaced. Therefore, it is usually necessary to stabilize the center wavelength of WDM transmitters, otherwise temperature changes can result in changes of center frequency of the diode laser and thus cause channel crosstalk and overlap. The main mechanism behind changes in the center wavelength of a laser diode is temperature change. For the same reason, temperature control is the common way to stabilize the center wavelength of a diode laser. Temperature control is usually realized by a *thermoelectric cooler* (TEC).

In a TEC, the direction of current determines the direction of flow of heat. A temperature control circuit can be realized by a temperature sensor and a feedback loop that compares the sensed temperature with the desired set temperature. The error signal can then be used to drive the TEC. If the ambient temperature is the same as the set temperature, the error signal would be zero and the TEC will not be activated. However, as the temperature moves away from the set temperature, the TEC requires more current, either to function as a cooler or as heater. Because it is easier to generate heat rather than cooling an area by removing heat, the set temperature is usually chosen at a point higher than the normal operating temperature, but below the maximum expected operating temperature. As a result, the TEC works as a heater most of the time, and only occasionally works as a cooler. Moreover, when it does have to operate as cooler, it does not have to work that hard, because the set temperature is slightly less than the maximum operating temperature.

Because the temperature sensor, the TEC, and the external modulator must all be thermally coupled, they usually have to be packaged together in the same optical subassembly. Moreover, the thermal control loop must be rather accurate. For example, a typical single-mode diode laser has a wavelength drift of ~0.1 nm/°C, which corresponds to ~25 GHz/°C. Therefore, in a WDM channel spacing application of 50 GHz, a 10% frequency accuracy requires a tolerance of ±5 GHz for center frequency, corresponding to a thermal stability better than ±0.2°C.

8.8 Burst mode transmitters

Within the past decades, the evolution of optical transmitters has mainly been focused on increasing bandwidth, and the proliferation of external modulators and wavelength-stabilized transmitters can be understood in this context. Another expanding area that has affected the industry is passive optical networks (PON). From the perspective of physical layer, what distinguishes PON transceivers is their burst mode operation. As we mentioned in Chapter 2, PON systems consist of a central office transmitter known generically as the optical line terminal or OLT that sends CW traffic downstream to all the receivers that are passively coupled to it. In the upstream direction, the individual users (known generically as optical network terminals or ONUs) send their information to the central office in a burst mode fashion, because the outputs from each of them must be time division multiplexed with the outputs from the others. Thus, at each point in time only one transmitter must be active when it is its turn to transmit, and it must shut down quickly when its turn has passed to allow the next transmitter to send its data. Thus, the burst mode nature of traffic in the upstream direction requires burst mode transmitters for the ONU and burst mode receivers for the OLT. Design of burst mode transmitters and receivers involves special design requirements not encountered in other CW transceivers. Burst mode receivers are discussed in the next chapter. In this section we discuss burst mode transmitters [25–28].

Figure 8.14 shows the critical parameters involved in the output signal from a burst mode laser driver. The laser driver is activated by a burst-enable signal. A critical parameter is the transmitter's response time to the enable signal. The time it takes for the output to reach its steady state, t_{ON}, is the turn-on time. Likewise, the time it takes for the output to fall below a power level that can be considered dark, t_{OFF}, is the turn-off time. Having short turn-on and turn-off times is important because it reduces the overhead by allowing packets from different ONUs to be packed together more tightly. The length of a packet, t_{PKT}, and the time separation between two subsequent packets, t_{DARK}, are also important factors. Ideally, a high-performance burst mode transmitter must be able to support a very short packet length and a very short dark time between each packet to allow for more flexibility in the network. For modern bust mode transmitters, the turn-off and turn-on times are in the order of a few nanoseconds, while the minimum packet length and the minimum dark time between two packets are in the order of tens of nanoseconds.

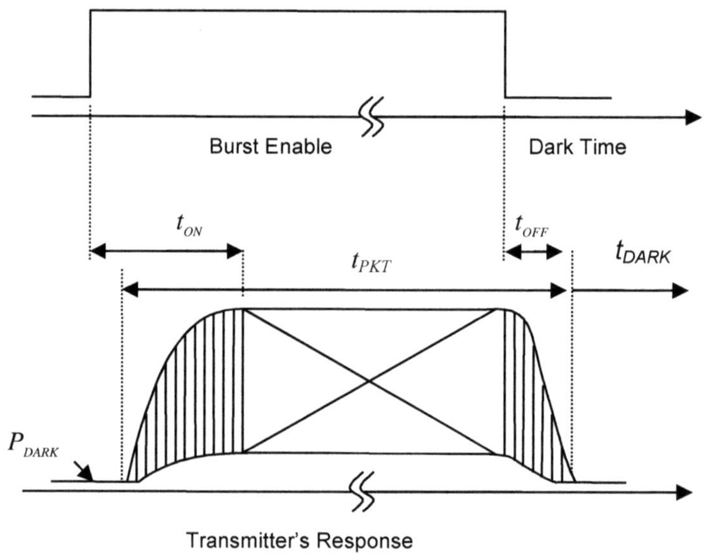

Fig. 8.14. Critical parameters for a burst mode laser driver

It is also important to minimize the dark power level between packets, designated as P_{DARK} in Fig. 8.14. Because usually many ONUs are connected to a PON network, and only one should transmit at any given time, P_{DARK} from other ONUs add up and act as background noise for the transmitting ONU. As a result, it is important for the bias to the laser diode to completely shut down during the dark time; otherwise having a nonzero bias even below threshold could be problematic.

It can be seen that the challenge in a burst mode laser driver is to turn on the diode laser and stabilize its power in a very short time. Turning the laser off is simpler because it involves simply cutting the bias current to the laser. Turning the laser on quickly is even more difficult to achieve if some sort of closed loop power control scheme is also needed. A fast turn-on time can be achieved by modulating the bias to the diode laser in a manner similar to a modulation driver. Therefore, in a burst mode transmitter, a differential driver can be used for both the bias and the modulation currents, as schematically shown in Fig. 8.15.

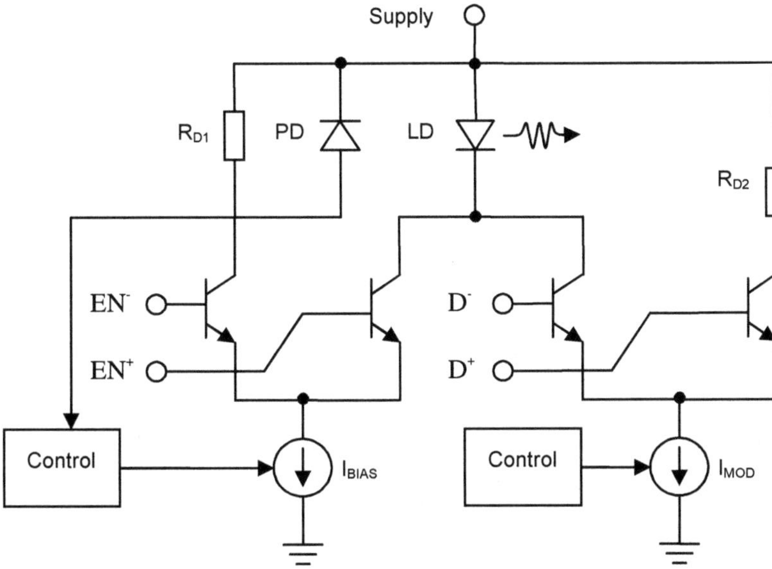

Fig. 8.15. Schematic diagram of a burst mode laser driver

The advantage of a differential driver is that it functions on the basis of current stirring. An essentially dc current (either I_{BIAS} or I_{MOD}) is stirred between the laser diode and a dummy load. In terms of magnitude, both currents are controlled through separate control circuits. Note also that an average power control scheme can be implemented using the feedback current from a photodetector. However, a simple feedback scheme will no longer work here, because the photodetector current turns on and off following the burst nature of traffic. Therefore, the feedback scheme must only use the photodetector current during the on time. Moreover, it is very hard for a traditional feedback loop to stabilize within a few nanoseconds, as expected from a burst mode laser driver.

As a result, modern burst mode drivers use digital power control feedback loops. In a digital loop, like a regular analog loop, the value of the photodetector current is compared to a reference value and the difference error signal is used to adjust the current, with the goal of reducing the error signal to zero. However, in a digital loop this value is stored digitally and used in the next packet. In this manner, the feedback loop is not interrupted at the end of a packet and does not need to initialize from an unknown state every time a new packet arrives. Instead, it simply uses the information it had stored in the previous packet, which would be very close to the steady state for the current packet. Use of digital loops limits the transient response of the loop to only the first few packets after the circuits start to work. After that, the power control scheme can work smoothly by essentially ignoring the photodetector current during the dark times.

8.9 Analog transmitters

In fiber optic communications baseband digital modulation formats are by far more common and more widely used than analog formats. However, analog transmission is still important for areas like video transmission and analog microwave feeds. Even if it were not for these applications, discussion of analog transmitters is still important in its own merit, because in fact all transmitters are fundamentally analog. Even in a digital transmitter, the optical output gradually rises from zero to one state, and it takes time for the output to settle to a steady state. The difference between an analog and a digital system is that in a digital system the receiver samples the signal in regular intervals and as long as the signal is above (or below) a certain threshold level a logic one (or zero) is assumed. Thus, imperfections outside the sampling window as well as a certain amount of noise within the sampling window, as long as the noise is not so much that it results in changing the logical level of the signal, can be ignored. The price to pay for this level of noise immunity is that less information can be transferred for a given amount of bandwidth. On the other hand, in an analog link the received signal is expected to be an accurate representation of the intended signal all the time and for all amplitudes. Therefore, an analog link is more susceptible to any imperfection that may affect the signal quality.

One of the most important characteristics of an analog transmitter is linearity. Suppose the transfer function of the transmitter, $L(I)$, is a nonlinear function of the drive current, I. L(I) can be expanded around some operating bias point, I_0, as

$$L(I) = A + B(I - I_0) + C(I - I_0)^2 + \cdots \tag{8.7}$$

Now suppose the drive signal I is multiplexed on a carrier with a frequency f_0. Then the optical output signal has frequency components at f_0, $2f_0$, etc. The situation becomes more complex if the signal is comprised of more frequencies, say, f_1 and f_2. In that case the output will have components at $f_1 + f_2$ and $f_1 - f_2$, assuming terms beyond second order can be neglected. If higher order terms cannot be neglected, more frequency components will appear in the output. Obviously, the existence of these terms can cause interference and crosstalk between the channels. The amplitude of these components depends on the magnitude of the nonlinear term(s) in Eq. (8.7). Thus, the modulation amplitude is limited by the degree of nonlinearity in the transmitter.

The LI characteristic of a typical diode laser above threshold can be approximated as

$$L = \frac{\eta}{1 + sI}\left(I - I_{TH}\right) \tag{8.8}$$

where I is the drive current, I_{TH} is the threshold current, and s is a parameter modeling saturation. For sI <<1 the transfer function described by Eq. (8.9) is linear. Indeed diode lasers offer a relatively large linear operating region, as long as the operating point I_0 is chosen sufficiently away from the threshold current and from the lasers saturation. On the other hand, externally modulated transmitters are not as linear. For example the output power of a Mach–Zehnder (MZ) modulator is given by

$$L = L_T\left[1+\cos\left(\frac{\pi V}{V_\pi}\right)\right] \tag{8.9}$$

where L_T is the maximum transmitted power, V is the drive voltage, and V_π is the input voltage for which the transmission is zero. Thus, in order to get the best analog performance, the modulator must be operated at the point where it is most linear, which corresponds to the drive voltage of V=0.5V_π. Moreover, the modulation amplitude must be small compared to V_π.

It should be noted that merely operating a diode laser or a modulator in its linear region is not sufficient, because expressions such as Eqs. (8.8) and (8.9) are static in nature and do not include dynamic effects. This can be easily appreciated once we realize that both Eqs. (8.8) and (8.9) do not have any time dependence, implying the output can follow the input instantaneously, in other words, an infinite bandwidth. However, dynamic effects always limit the effective bandwidth of a modulator. For example, as we discussed in Chapter 4, lasers tend to have a flat frequency response up to a point. After that the response shows a peak at the relaxation oscillation frequency of the laser and then it rolls off at approximately 40 dB/decade. The similarity with a second-order system is not merely accidental, because in a laser, like a second-order system, the interplay of two energy storage mechanisms (carriers and photons) with different time constants determines the frequency response. To maintain signal fidelity, the frequency of the input signal must stay well below the resonance frequency to avoid amplitude and phase distortion.

The inherent frequency characteristic of the modulating device is not the only parameter determining the response of a linear transmitter. The electronic driving circuits also have their own frequency responses. Therefore, at high frequencies parasitic effects such as those of the optical subassemblies must be taken into account.

To illustrate these points, consider Fig. 8.16 which schematically shows the circuit elements that constitute a typical laser driver from a frequency response point of view. The signal source is represented by V_s and the output impedance Z_s. The signal must be transferred to the laser diode via a transmission line with a characteristic impedance of Z_T. The laser diode, itself represented by an impedance Z_D, acts as a load for the transmission line. In general, the impedances of source, transmission line, and the diode laser are not equal. Therefore, connecting them without any matching will result in wave reflections from both ends of the transmission line. For example, the reflection coefficient at the load side is given by

$$\Gamma = \frac{Z_D - Z_T}{Z_D + Z_T} \tag{8.10}$$

A similar expression can be written for the back reflection from the source side. In general, we need to have impedance matching circuits to minimize the reflections.

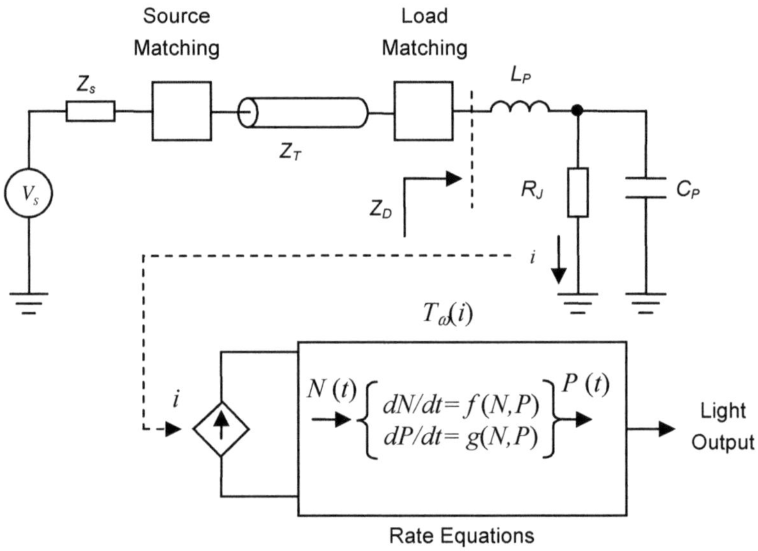

Fig. 8.16. Frequency response block diagram of a typical optical transmitter

We have already discussed such impedance matching circuits when we were discussing digital transmitters, in the form of a back termination resistor at the laser driver and a series resistor at the diode laser side. In the context of our current discussion, a simple resistor can be considered a *broadband* matching circuit, because the behavior of a resistor is (relatively) frequency independent. This is a natural choice for digital transmitters, because as we know digital signals are generally *baseband*, i.e., they have frequency components starting from close to dc. Such broadband matching networks work for analog transmitters too. However, in analog circuits the signals are usually *narrowband*, as they are centered around a carrier frequency. In such cases the matching networks can be narrowband too, i.e., their transfer function can be a function of frequency. The advantage of narrowband matching is that in principle better matching can be achieved, although over a limited range of frequencies.

The impedance of the laser, Z_D, is modeled by a series inductance L_P and a parallel combination of the junction capacitance C_J and the forward-biased resistance of the junction diode R_J. The series inductance represents optical assembly leads as well as the wire bonding to the laser chip, whereas C_J and R_J are functions of the chip structure and biasing condition. The laser driver circuit, transmission line, matching networks, and laser diode junction each have a frequency response and a transfer function. The combined effects of these transfer functions model the deliverance of the signal from the driving source to the diode laser junction. Once the RF signal reaches the junction, it must be converted to light. The dynamics of this conversion are governed by the carrier and photon rate equations, which finally yield to the light output.

Generally speaking, these various blocks each act as a low-pass filter, with at least one dominant pole. In the case of a diode laser, the small signal behavior can be approximated by a second order system, with a resonant frequency. If the RF frequency gets too close to this pole, it will result in phase and amplitude distortion. Thus, for linear transmitters, the RF frequency in general must be kept below the frequency of the dominant pole of all the blocks in the signal path, including the relaxation oscillation of the laser.

8.10 High-frequency design practices

In this section we review major high-frequency techniques used in the design of high-speed circuits. We have already discussed some of these techniques before, when we mentioned differential drive schemes and source/load termination. Here we will consider a few other points that need to be considered in high-speed circuits [29–31].

8.10.1 Power plane

One of the main factors affecting the performance of high-frequency circuits is the quality of power planes. In a printed circuit board (PCB), power planes distribute the supply voltage to different parts of the circuit. Generally, high-performance PCBs are multi-layered, with several layers dedicated specifically to power planes. Figure 8.17 shows an example of a cross section of a six-layer PCB.

The top and the bottom sides of the PCB include the footprints for standard electronic components. These layers are also used for simple and short-distance routing between the components. Layers 2 and 5, which lie immediately after the component layers, are dedicated as ground planes. It is crucial to include complete, uninterrupted ground planes in the design to ensure that all the ground points in the circuit are at the same voltage both at dc and RF frequencies. Small, noncontiguous planes may result in a ground voltage difference between various points. Because of the inductance of small planes, this difference can be higher at RF frequencies, causing voltage differences and ground loops.

Also shown in Fig. 8.17 is a supply plane that distributes the supply voltage to various components. For the same reasons mentioned above, it is usually a good practice to dedicate a solid, continuous plane to the supply voltage. If there is more than one supply in a circuit, each could have its own dedicated plane. Alternatively, the same plane can be divided between them as long as each has a continuous uninterrupted plane geometry.

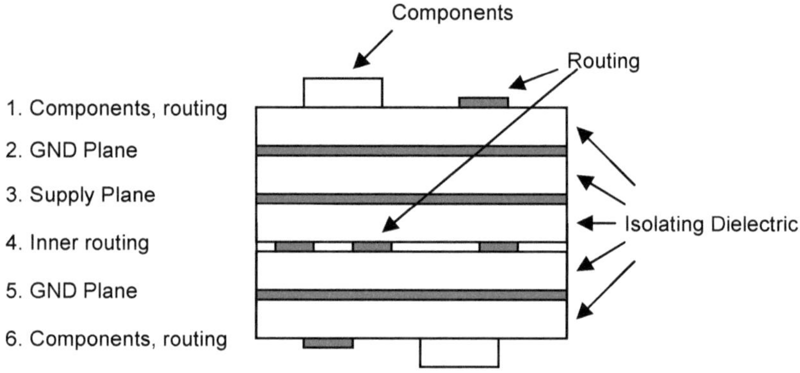

Fig. 8.17. Cross section of a PCB showing the power planes and inner routing layer

The supply and the ground planes are chosen to be next to each other. This structure constitutes a parallel plate capacitor, which increases high-frequency decoupling between the two planes. This decoupling is important because it provides a bypass path for high frequency noise signals, preventing them from developing across the power planes. In effect, this structure ensures that the supply plane has low ac impedance and is thus close to an ideal power plane.

Another important role of the power planes is to provide a reference plane for transmission lines. For example, if in Fig. 8.17 two components on layer 1 need to be connected with a microstrip transmission line, the ground plane on layer 2 can naturally provide the needed reference plane. Likewise, layer 5 can provide the reference plane for any microstrip structure on layer 6.

In some cases the density of the components or geometrical and topological concerns require that transmission lines be implemented in the inner layers. The cross section of Fig. 8.17 can support buried stripline transmission lines on layer 4. In this case, layer 3 and layer 5 provide the reference planes. Note that in this case one of the power planes is a ground plane while the other one is a supply plane. But this is not a problem, because from an ac point of view the supply plane has the same potential as the ground plane, as long as it is sufficiently bypassed.

8.10.2 Circuit layout

As the operating frequencies increase, the standard lumped circuit models that we use for electrical elements start to break. The most obvious case is that of a simple electrical interconnection. In a lumped circuit model we assume that the voltage along an electrical interconnect is the same at any given point in time. In reality, as the operating frequency increases, the corresponding wavelength decreases, and once the length of the interconnects becomes comparable to the wavelength (say, more than 0.1 of the operating wavelength), transmission line effects must be taken into account. In such cases, a controlled impedance transmission line must be used, and source and/or load impedance matching techniques should be utilized. This ties in with the availability of a continuous reference plane, as any discontinuity in the reference plane results in a disconnect for the return path of the involved signals. A corollary of the above statements is that electrical interconnects must be kept as short as possible to minimize current loops and impedance mismatchings.

Parasitic effects also play an increasingly important role at higher frequencies. This means that simple components such as resistors and capacitors may not behave as simple lumped elements. A resistor may show a shunt capacitance and a capacitor may show a series inductance. These parasitic effects cause high-frequency poles and zeros, and should be taken into account if the operating frequency starts to approach those poles and zeros. From the point of view of layout, it is generally a good idea to place the components closer to each other and use smaller land patterns. In this respect, use of surface mount components can significantly improve the high-speed behavior of a circuit. These components do not have leads and therefore their parasitic series inductance is much less compared to their leaded counterparts. This applies both to passive components and active ICs.

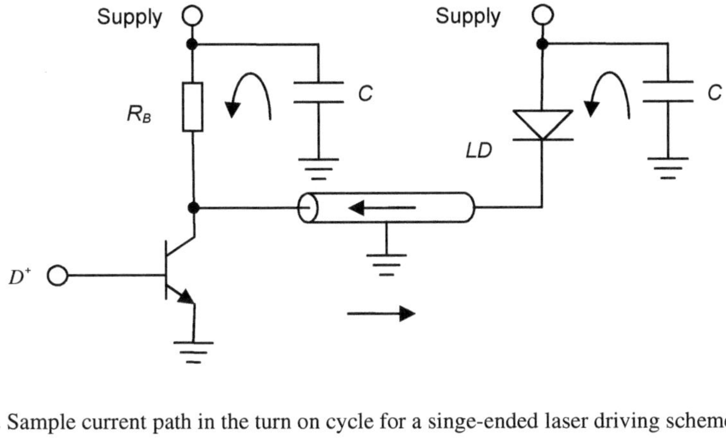

Fig. 8.18. Sample current path in the turn on cycle for a singe-ended laser driving scheme

A continuous reference plane under the layout provides a low-impedance path for ground connections, and so a via can be dropped to provide the ground connection wherever needed. However, the ground plane causes an extra stray capacitance under the connecting pins of the elements. This extra stray capacitance may have an undesirable effect when it comes to the inputs of high-frequency amplifier, as it may reduce the useful bandwidth and degrade the edge rates. So it may be necessary to retreat the ground plane slightly away from under such points to reduce the stray capacitance and improve the frequency response.

Another important practice is the inclusion of bypass capacitors close to the supply point of active devices. These bypass capacitors provide a low-impedance path for high-frequency signals and isolate noise sources from the rest of the circuit. This is especially important for devices that require high current switching, such as laser drivers, or other digital components. Figure 8.18 shows the function of these bypass capacitors during a turn on cycle for a single-ended laser driver circuit.

For instance, when the transistor switch closes, it draws a pulse of current. This pulse is provided through a current division between R_B and the characteristic impedance of the transmission line. The current pulse travels through the transmission line until it reached the diode laser. Both R_B and diode laser are connected to the supply voltage plane. If it were not for the two bypass caps, the current had to be provided by the voltage source. Even with a low-impedance supply plane, the current pulse would have caused a voltage glitch because of the large rate of change of current, approximately given by

$$\Delta V = L\frac{\Delta I}{\Delta t} \tag{8.11}$$

where L is the inductance of the power plane, ΔI is the modulation amplitude, and Δt is the signal rise time.

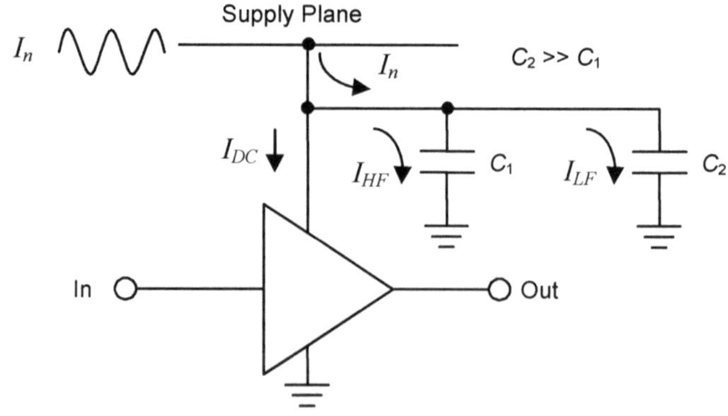

Fig. 8.19. The role of bypass capacitors in isolating noise from a sensitive device

For instance, a modulation current of 50 mA with a rise time of 100 ps results in a voltage glitch of 50 mV over and inductance of 100 pH. This voltage glitch can appear as supply noise for other components connected to the supply plane. However, the bypass capacitors provide a charge reservoir for the current pulse and reduce the magnitude of glitch for the rest of the circuit.

The bypass capacitors can also help reduce the effects of the noise already existing in power planes on a sensitive device. This is a scenario perhaps more relevant to analog parts. Figure 8.19 illustrates an example of an amplifier tied to a noisy power plane. Without the bypass capacitor, the noise currents can enter the amplifier through the supply connection. But the bypass capacitors provide a low-impedance path for noise currents, preventing them from disturbing the sensitive nodes within the amplifier.

In practice, it is common to use more than one capacitor for bypassing critical points. This is because each capacitor has a resonant frequency, above which its impedance starts to rise. Using more than one capacitor ensures that the whole frequency range of interest is covered. As can be seen from Fig. 8.19, the high- and low-frequency components of the noise current are bypassed by C_1 and C_2 respectively, where $C_2 >> C_1$.

In many circuits, both analog and digital circuits exist. An example of a sensitive analog circuit in fiber optic systems is the receiver, which has to convert a very weak signal received from a PIN or APD detector to a strong digital or analog signal. Digital circuits may include circuits like laser drivers and microprocessors. It is a good practice to increase the isolation between these parts in order to reduce crosstalk noise. Oftentimes the easiest way to realize this separation is to use separate power planes. For example, each circuit can be implemented in a separate section of the PCB, with its own power and ground planes. If the power planes have to be dc connected, a low resistance but high-impedance element (such as an inductor or a ferrite bead) can be used to allow dc currents but isolate high-frequency noise.

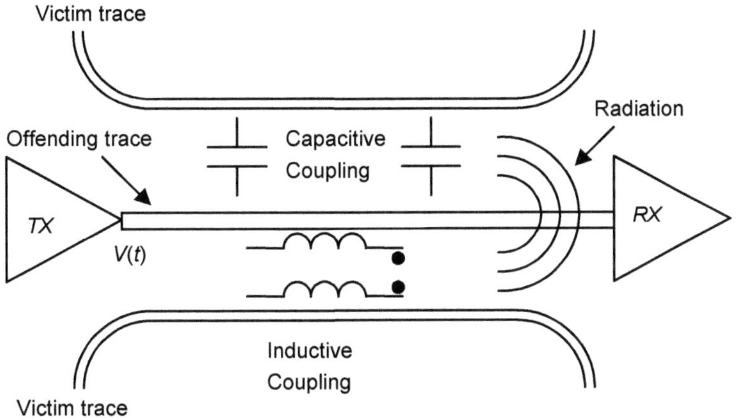

Fig. 8.20. Noise transfer due to capacitive, inductive, and radiation coupling

Crosstalk noise can also result from energy coupling between an "offending" element and a "victim" element. An example of an offending element can be a long trace carrying a high-frequency signal. Such a trace can act as an antenna and can radiate electromagnetic energy which can be picked up by other elements in the circuit. There can also be noise energy transfer due to capacitive or inductive couplings.

Figure 8.20 is an illustration of different mechanisms for transfer of energy from an offending trace to adjacent points in the circuit. To minimize noise coupling, therefore, traces carrying high currents or fast switching signals must be kept separate and at a distance from other points in the circuit, especially sensitive high-impedance points and traces such as amplifier inputs.

To minimize radiation, the area of the loops carrying high currents (consisting of the trace itself and the return path) should be minimized. Having a continuous reference plane for such high current traces or using differential signaling schemes can achieve this goal. Using differential signaling for the victim traces can also reduce the noise reception, because noise signals appear as a common mode signal which will be rejected by a differential receiver.

8.11 Summary

In this chapter we discussed some of the main issues involved in the design of fiber optic transmitters. First, we introduced the concept of a transmitter optical subassembly (TOSA). Almost all practical transmitters use a TOSA as their light source. TOSAs perform a wide variety of functions and make otherwise delicate optical sources practically usable.

To operate a TOSA, we need a driver circuit. In many cases the laser is modulated directly. Direct modulation transmitters are in general simpler to design and operate. We started from the simplest modulation schemes and gradually introduced various topics such as power and modulation control as well as particular circuits used in these transmitters. We discussed issues such as thermal runaway, transmission line effects, and ac coupling.

Next we turned our attention to other classes of optical transmitters. Specifically, we discussed external modulators, burst mode transmitters, and analog transmitters. Our treatment of these topics, although brief, can serve as a general framework for many different classes of optical transmitters used in fiber optic applications.

Finally, we turned our attention to high-frequency design practices involved in these transmitters. Power plane management, layout techniques, and noise immunity were among some of the topics we covered.

We will return again to some other aspects of optical transmitter design in Chapter 10, where we discuss reliability. Reliability issues, including thermal, mechanical, optical, and electrical concerns, are an integral part of any design. Therefore, topics covered in Chapter 10 can be considered a complementary follow-up for the subjects discussed in this and the next chapter.

References

[1] H. Stange, "Optical subassemblies," in *Handbook of Fiber Optic Data Communication*, Edited by C. DeCusatis, 2nd Ed., Academic Press, New York, 2002

[2] D. Kim, J. Shim, Y. C. Keh, and M. Park, "Design and fabrication of a transmitter optical subassembly (TOSA) in 10-Gb/s small-form-factor pluggable (XFP) transceiver," IEEE Journal of Selected Topics in Quantum Electronics, Vol. 12, pp. 776–782, 2006

[3] M. S. Cohen et al., "Low-cost fabrication of optical subassemblies," IEEE Transactions on Components, Packaging, and Manufacturing Technology, Part B: Advanced Packaging, Vol. 20, pp. 256–263, 1997

[4] T. Shih et al., "High-performance and low-cost 10-Gb/s bidirectional optical subassembly modules," Journal of Lightwave Technology, Vol. 25, pp. 3488–3494, 2007

[5] L. Coldren and S. W Corzine, *Diode Lasers and Photonic Integrated Circuits*, John Wiley & Sons, New York, 1995

[6] Y. Yoshida et al., "Analysis of characteristic temperature for InGaAsP BH laser with p-n-p-n blocking layers using two-dimensional device simulator ," IEEE Journal of Quantum Electronics, Vol. 34, pp. 1257–1262, 1998

[7] T. Higashi, T. Yamamoto, S. Ogita, and M. Kobayashi, "Experimental analysis of characteristic temperature in quantum-well semiconductor lasers," IEEE Journal of Selected Topics in Quantum Electronics, Vol. 3, pp. 513–521, 1997

[8] D. M. Gvozdic and A. Schlachetzki, "Influence of temperature and optical confinement on threshold current of an InGaAs/InP quantum wire laser," IEEE Journal of Selected Topics in Quantum Electronics, Vol. 9, pp. 732–735, 2003

[9] R. Sobiestianskas et al., "Experimental study on the intrinsic response, optical and electrical parameters of 1.55-μm DFB BH laser diodes during aging tests," IEEE Transactions on Device and materials, Vol. 5, pp. 659–664, 2005

[10] J. W. Tomm et al., "Monitoring of aging properties of AlGaAs high-power laser arrays," Journal of Applied Physics, Vol. 81, pp. 2059–2063, 1997

[11] L. Day-Uei et al., "A 3.8-Gb/s CMOS laser driver with automatic power control using thermistors," IEEE International Symposium on Circuits and Systems (ISCAS), pp. 2546–2549, 2007

[12] Application note, "Selecting and using thermistors for temperature control," ILX Lightwave. Available from www.ilx.com

[13] W. R. Smith, "Mathematical modeling of thermal runaway in semiconductor laser operation," Journal of Applied Physics, Vol. 87, pp. 8276–8285, 2000

[14] R. Schatza and C. G. Bethea, "Steady state model for facet heating leading to thermal runaway in semiconductor lasers," Journal of Applied Physics, Vol. 76, pp. 2509–2521, 1994

[15] P. G. Eliseev, "Optical strength of semiconductor laser materials," Progress in Quantum Electronics, Vol. 20, pp. 1–82, 1996

[16] S. Galal and B. Razavi, "10-Gb/s limiting amplifier and laser/modulator driver in 0.18-μm CMOS technology," IEEE Journal of Solid-State Circuits, Vol. 38, pp. 2138–2146, 2003

[17] R. Schmid et al., "SiGe driver circuit with high output with high output amplitude operating up to 23-Gb/s," Journal of Solid-State Circuits, Vol. 34, pp. 886–891, 1999

[18] J. W. Fattaruso and B. Sheahan, "A 3-V 4.25-Gb/s laser driver with 0.4-V output voltage compliance," IEEE Journal of Solid-State Circuits, Vol. 41, pp. 1930–1937, 2006

[19] H. M. Rein et al., "A versatile Si-bipolar driver circuit with high output voltage swing for external and direct laser modulation in 10-Gb/s optical fiber links," Journal of Solid-State Circuits, Vol. 29, pp. 1014–1021, 1994

[20] HFDN-18, application note, "The MAX3865 laser driver with automatic modulation control," Maxim, 2008. Avilable from www.maxim-ic.com

[21] L. Chen et al., "All-optical mm-wave generation by using direct-modulation DFB laser and external modulator," Microwave and Optical Technology Letters, Vol. 49, pp. 1265–1267, 2007

[22] J. X. Ma et al., "Optical mm-wave generation by using external modulator based on optical carrier suppression," Optics Communications, Vol. 268, pp. 51–57, 2006

[23] S. Hisatake et al., "Generation of flat power-envelope terahertz-wide modulation sidebands from a continuous-wave laser based on an external electro-optic phase modulator," Optics Letters, Vol. 30, pp. 777–779, 2005

[24] H. Yang et al., "Measurement for waveform and chirping of optical pulses generated by directly modulated DFB laser and external EA modulator," Optics and Laser Technology, Vol. 37, pp. 55–60, 2005

[25] S. H. Park et al., "Burst-mode optical transmitter with DC-coupled burst-enable signal for 2.5-Gb/s GPON system," Microelectronics Journal, Vol. 39, pp. 112–116, 2008

[26] Y. H. Oh et al., "A CMOS burst-mode optical transmitter for 1.25 Gb/s Ethernet PON applications," IEEE Transactions on Circuits and Systems II-Express Briefs, Vol. 52, pp. 780–783, 2005

[27] D. Verhulst et al., "A fast and intelligent automatic power control for a GPON burst-mode optical transmitter," IEEE Photonics Technology Letters, Vol. 17, pp. 2439–2441, 2005

[28] J. Bauwelinck et al., "DC-coupled burst-mode transmitter for 1.25 Gbit/s upstream PON," Electronics Letters, Vol. 40, pp. 501–502, 2004

[29] B. Young, *Digital Signal Integrity*, Prentice Hall, Englewood Cliffs, NJ, 2001

[30] H. W. Ott, *Noise Reduction Techniques in Electronic Systems*, 2nd Ed., Wiley, New York, 1988

[31] M. I. Montrose, *EMC and the Printed Circuit Board Design: Theory and Layout Made Simple*, IEEE Press, Piseataway, NJ, 1998

Chapter 9
Optical Receiver Design

9.1 Introduction

In this chapter we consider issues related to the design of optical receivers. As signals travel in a fiber, they are attenuated and distorted, and it is the function of the receiver circuit at the other side of the fiber to generate a clean electrical signal from this weak, distorted optical signal.

An optical receiver consists of an optical detector, usually a PIN or APD diode, which converts the optical signal to an electrical signal. However, the signal generated by a detector is generally too weak to be usable, and so the detector is followed by several stages of further amplification and signal conditioning. For digital receivers these typically include a transimpedance amplifier, a limiting amplifier, and a clock and data recovery unit. For analog receivers, the amplification may be combined with proper filtering and frequency domain signal conditioning. We will discuss some of the general requirements of these circuits in this chapter.

Traditionally, optical receivers have been working in continuous (cw) mode. However, with the advent of fiber-to-home and PON networks, burst mode receivers have become increasingly important. Design of these receivers poses particular challenges, and we will discuss some of these challenges in this chapter as well.

We should also note that an optical receiver, like an optical transmitter, handles serial data at the nominal link data rate, and thus it should be a high-speed circuit possessing the whole bandwidth of the optical signal. As a result, many of the high-speed techniques we have discussed previously apply here too. However, because we have covered these techniques in the previous chapters, we will not cover them here in any detail, and just make references to the previous chapters wherever needed.

9.2 Receiver optical subassembly (ROSA)

As we noted before, the first stage of an optical receiver circuit consists of an optical detector. The detector is usually part of a receiver optical subassembly, or ROSA. The role of a ROSA is very much similar to that of a TOSA in an optical transmitter. Thus, we do not go over the details here as they are in most part similar to what we covered in Chapter 8. In fiber optic receivers, the detector is usually a PIN or APD device. As discussed in Chapter 6, a PIN or APD is essentially a reverse-biased diode with very small dark current. From a circuit point of view,

M. Azadeh, *Fiber Optics Engineering*, Optical Networks,
DOI 10.1007/978-1-4419-0304-4_9, © Springer Science+Business Media, LLC 2009

the photodetector acts like a light-controlled current source. As light hits the junction, more current flows, and it is the changes in this current that the circuit must detect and amplify to reconstruct the original signal.

The current generated by a PIN or APD detector is generally very weak and could be in the μA range or even less. Obviously, such a weak signal is not directly usable, and so it must be amplified. Moreover, as we mentioned in the previous chapters, most high data rate circuits are based on differential signaling. It is therefore necessary to convert the current output of the detector to a differential signal, so that next stages can handle it more easily.

9.2.1 Transimpedance amplifier (TIA)

To achieve these goals a ROSA often includes a pre-amplifier, also known as a *transimpedance amplifier* (TIA), immediately after the detector. Figure 9.1 shows a schematic diagram of a detector and a TIA.

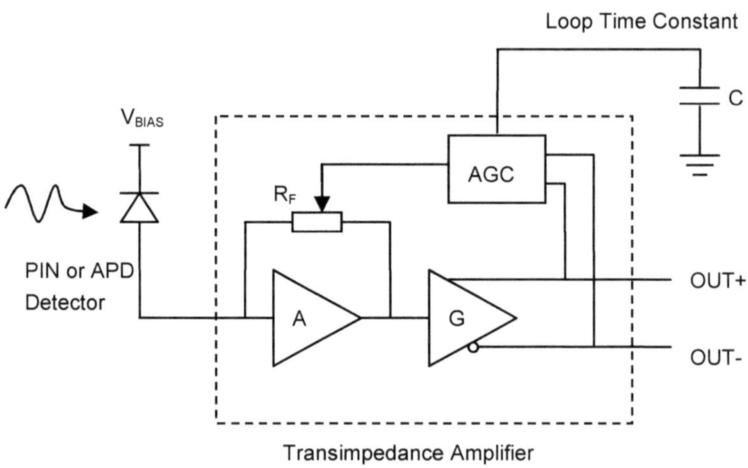

Fig. 9.1. Schematic diagram of a detector and a transimpedance amplifier

As can be seen from the figure, the detector is reverse biased at a given voltage V_{BIAS}. The detector current flows into a high-gain amplifier whose closed loop gain is determined by a feedback resistor, R_F. The output of this first-stage amplifier is then converted into a differential signal which constitutes the output of the transimpedance amplifier.

A common problem for all receivers is the wide *dynamic range* of received signals, because depending on the distance and the power of the transmitter, the power level at the receiver can vary greatly [1–4]. Here dynamic range refers to the ratio of the maximum to the minimum optical power that the receiver will encounter under normal operating conditions and is usually expressed in units of dBs. For instance, a dynamic range of 30 dB (which is not unusual) means that the lowest optical power that the receiver should be capable of recovering is 1000 times smaller than the highest power that it should be able to work with. Such a wide dynamic range makes it difficult to keep all the stages in their linear mode.

For instance, suppose at maximum optical power, the maximum peak–peak value of the TIA output is 2 V. A dynamic range of 30 dB means that the same output can fall to only 2 mV for the weakest optical signal that this receiver must work with. Naturally, a 2 mV signal is not large enough for subsequent stages in the circuit. Wider dynamic ranges only exacerbate this ratio. Note that simply increasing the gain does not solve the problem, because a larger gain will help the low power signals but will cause an increase in the maximum peak-to-peak value of output when the input power is high. In short, one (or any number of) linear amplifier stage(s) cannot change the dynamic range of a signal.

To improve the dynamic range performance of the receiver, usually an *automatic gain control* (AGC) circuit is used. The AGC circuit is implemented via a feedback loop that samples the output of the TIA and controls the feedback resistor in the first stage. Using a nonlinear block such as an AGC circuit will greatly improve the situation, because then the gain will be kept high when it has to be high, i.e., at low powers. But at higher levels of received power when the gain does not need to be too high, the gain is lowered. This concept is illustrated in Fig. 9.2.

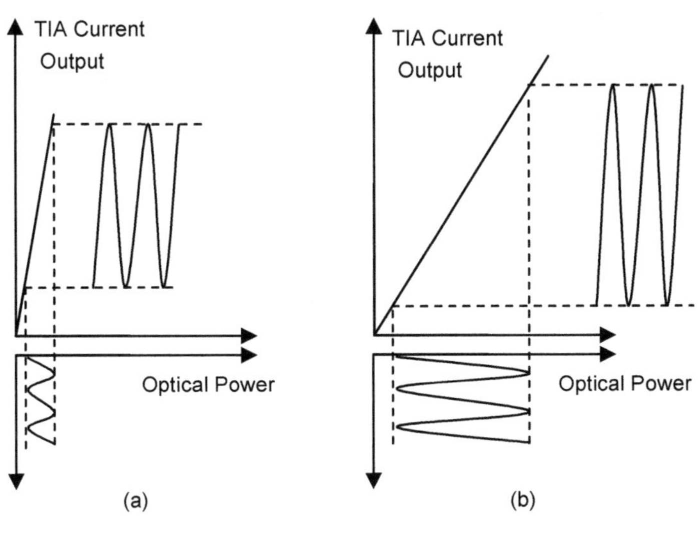

Fig. 9.2. Gain compression in an AGC amplifier. (**a**) Gain is higher when the input power is low and (**b**) gain is lower when the input power is high

Figure 9.2a represents the case where the received optical power is low. The AGC circuit in this case increases the gain of the TIA. In Fig. 9.2b the received optical power is higher. As a result, the AGC has lowered the gain of the TIA, and although the resulting output is still higher compared to case (a), the ratio of the two output is much less than the ratio of the two inputs. In other words, the AGC has *compressed* the gain. As a result of this gain compression, the dynamic range requirement for the subsequent stages in the circuit is relaxed.

The AGC circuit is a feedback loop, and like any other feedback loop, has a time constant [5]. This time constant has a direct bearing on the pattern dependency of the receiver. A fast time constant reduces the immunity of the receiver to long sequences of consecutive identical digits (CIDs), because the AGC circuit cannot distinguish between long CIDs and actual power changes. A slow time constant, on the other hand, reduces the ability of the receiver to track the changes in the level of the received signal.

In most applications the received power does not change significantly in short periods of time, and therefore a slow time constant does not impact the tracking capability of the receiver. In such cases the slowest possible time constant must be used to minimize pattern dependency, which could become especially annoying at lower data rates and longer CIDs. The exception to this rule of thumb is burst mode applications, where the receiver must respond to quick dynamic changes in power between packets coming from transmitters at different distances. We will discuss these receivers separately later in this chapter.

9.2.2 Detector/TIA wire bonding in optical subassemblies

It is well known that in order to maximize the signal-to-noise ratio (SNR) of a communication system, it is crucial to improve the SNR at the first stage when the signal is weakest. In other words, any noise added to a signal at the first stage will be amplified by subsequent stages, and thus it will be hard (if not impossible) to remove. For fiber optic receivers, the implication is that the current output from the detector must be amplified as soon as possible. In other words, it must not be routed for long distances, because any degradation due to transmission line effects or coupled noise from other parts of the circuit can seriously impact the overall performance of the receiver. This is a concern especially at higher data rates. As a result, for data rates above several hundred Mbps the TIA and the detector must be mounted in the same optical subassembly.

As a result, virtually all TIA circuits available are in the form of integrated circuits in die form, intended to be wire bonded to the detector [6,7]. Direct wire bonding reduces the distance that the signal must travel in order to get to the TIA. Moreover, the signal does not have to go through the pins of standard packaged ICs, as those pins inevitably introduce more parasitic effects. Figure 9.3 shows the top view of the wire-bonding schematic for a simple photodetector and TIA in a ROSA. To have an idea of the dimensions in such structures, it is useful to keep in mind that the length of a typical TIA die is usually less than 1mm.

The photodetector is placed at the focal point of a lens which focuses the light at the center of the detector. Detector's cathode and anode are wire-bonded to the TIA. The TIA provides the bias voltage for the detector's cathode and receives the photocurrent from the anode. The outputs of the TIA are wire-bonded to the transmission lines or other electrical interconnections that carry the signal out of the optical subassembly. As can be seen from Fig. 9.3, the distance that the signal must travel from the detector to the input of the TIA is very short. It is sometimes possible to extend the effective bandwidth of the interface by slightly increasing the length of the wire bond and creating a peak in the frequency domain transfer function from the detector to the TIA. However, such a peaking will introduce additional phase shifts and it should generally be avoided unless the set-up is carefully studied and tested.

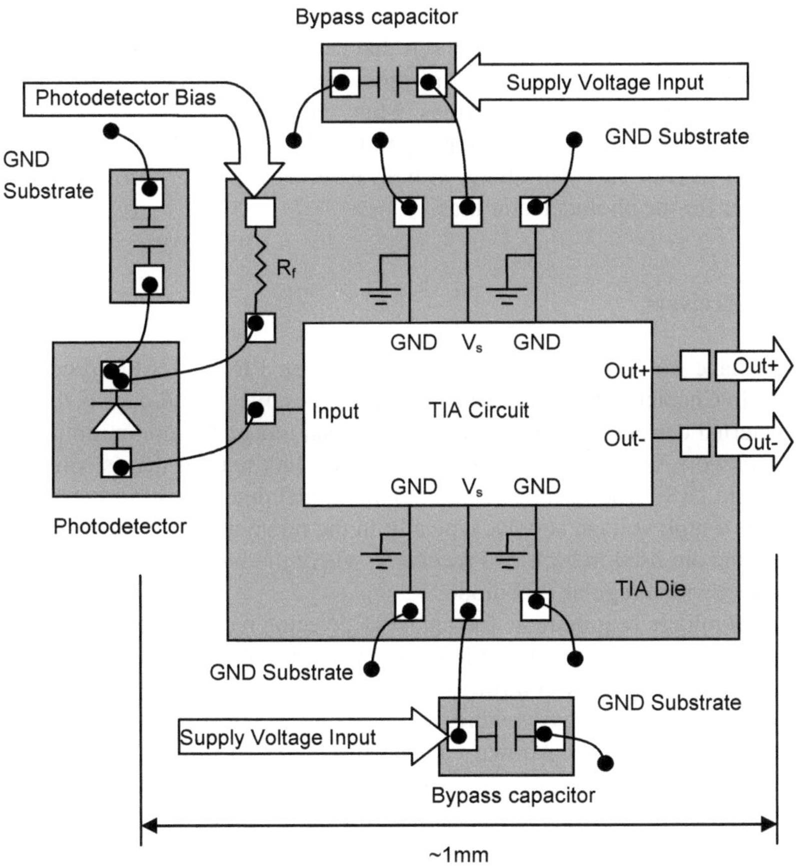

Fig. 9.3. Typical wire bonding diagram for a photodetector–TIA in a ROSA

Another critical parameter is the size of the detector. Larger detectors are easier to work with, as it would be easier to align them with the focal point of the light. Also, the focal point does not need to be very small, and therefore cheaper optical components can be used. The drawback of a large detector, however, is higher junction capacitance. This could impose a problem at higher data rates. For example, a typical PIN or APD detector with an active area diameter of 50 μm may have an approximate junction capacitance of 0.25–0.5 pF. This junction capacitance is proportional to the area of the detector, which varies with square of the diameter. On the other hand, the capacitance of the detector and the input resistance of the TIA determine the bandwidth of the interface between the detector and the TIA. Thus, increasing the size of the photodetector will increase the capacitance of the junction and cause a reduction in bandwidth.

It is also important to provide sufficient bypassing capacitors close to the TIA and the photodetector. This bypassing is needed to isolate the noise from other parts of the circuit from coupling into the TIA. Supply and ground bondings must be kept short to minimize the ground loops that can cause ac drop. Moreover, usually several supply and ground pads are provided to reduce the effective inductance for power and ground connections. Many integrated TIAs also provide an additional filtering scheme for the supply point of the photodetector in the form of an integrated resistor (R_f) that along with an external capacitor, provides a low-pass RC filter for the photodetector bias.

9.2.3 APD receivers

So far we have not explicitly distinguished between PIN and APD diodes. As mentioned in Chapter 6, the advantage of APD diodes over PIN diodes is that they provide internal current gain. This inherent gain can reduce the minimum power a receiver can work with by a few decibels, an extra link margin that is sometimes very desirable [3,8]. The disadvantage is that an APD detector is more expensive and requires a high voltage supply, typically in the range of 30–70 V. As a result, APD detectors are used in high-end receivers, where the link budget requirements cannot be met with a regular PIN diode.

The high-voltage requirement for an APD detector requires a separate high-voltage step-up converter circuit. Moreover, for optimal performance, the APD bias voltage must be closely controlled over operating temperatures. As discussed in Chapter 6, at any given temperature, the best SNR is achieved when the APD is biased slightly below its breakdown voltage. If the APD is biased higher, its current gain will increase; however; because of excessive noise the performance of the receiver will suffer. On the other hand, if the APD is biased lower, the current gain will decrease, and consequently the APD gain is not fully utilized. This leaves a relatively narrow range of voltages, say 2 V, in which a given APD delivers the best performance.

However, there are two additional factors that complicate the task of an APD bias circuit. First, each individual APD has a breakdown voltage that is unique to it and may slightly differ from the breakdown voltage of a similar APD. Therefore to maximize the utilization of an APD, the breakdown voltage of each device may have to be measured individually and the bias circuit for that device must be tuned accordingly.

The second complication comes from temperature. The breakdown voltage of an APD detector increases with temperature. Therefore, the bias circuit must change the APD bias voltage as a function of temperature. In most cases the breakdown voltage varies linearly with temperature. This means the bias circuit must increase the voltage linearly with temperature too.

We faced the same problem of temperature dependence in the previous chapter where we were discussing temperature compensation of diode laser bias and modulation currents. As we mentioned there, it is possible to address such requirements by adding a simple temperature sensor to the electronic circuits. However, as a result of the nonlinear characteristics of such temperature-sensitive devices, it is very difficult to achieve the required accuracy in the control of the desired parameter. An alternative approach, based on look-up tables, is generally the preferred method. Figure 9.4 shows the schematic of this approach in a typical APD receiver.

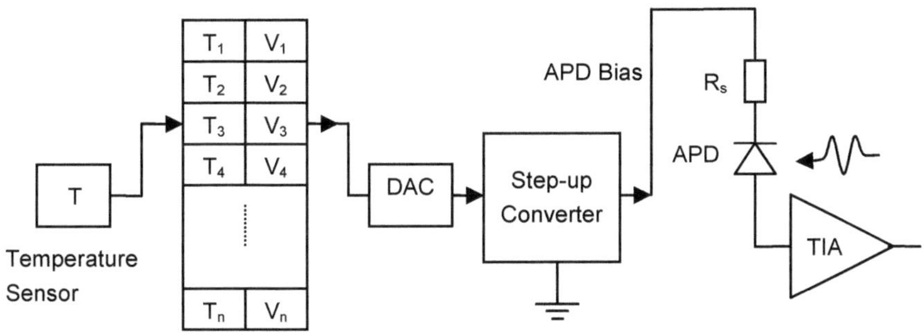

Fig. 9.4. A look-up table implementation for an APD bias circuit

In a look-up table approach, a temperature sensor measures the current temperature. This temperature is then used as an index into a two-dimensional table, which is pre-programmed with the desired value for the APD voltage corresponding to that temperature. The look-up table is implemented digitally, for instance, in a microcontroller or a digital potentiometer. Therefore, the digital value corresponding to the desired temperature has to be converted to an analog value through a digital-to-analog converter (DAC). The result is then amplified by a step-up converter to a high-voltage level needed for the APD bias. In this manner, any desired voltage vs. temperature profile can be accommodated. As noted before, usually the breakdown voltage increases almost linearly with temperature. A typical slope is around 0.1 V/°C.

Figure 9.4 also shows a resistor R_s in series with the APD detector. This resistor is a very useful tool for enhancing the performance of an APD receiver. Its most important role is improvement in the dynamic range of the receiver. We know that the gain of the APD reduces if the APD voltage is decreased. On the other hand, the highest gain is needed when the optical power is small, while the same high gain can saturate the TIA at high optical power. Adding this series resistor provides a feedback mechanism whereby the APD voltage – and therefore its gain – is reduced at high input powers. This is because a higher input power increases the photocurrent, which in turn causes a voltage drop on R_s. As a result of this gain reduction, the maximum optical power that the receiver can work with will increase. This gain reduction is also a useful feature for safety. Because without this feature, if for any reason the receiver is exposed to a high level of optical power, excessive current combined with high voltage can cause excessive power dissipation and damage to the ADP.

9.3 Limiting amplifier

In digital receivers, the next stage after a TIA is a limiting amplifier (LIA). As shown in Fig. 9.5, the function of the limiting amplifier is to provide additional gain and signal conditioning. The output of the TIA is still essentially an analog signal. By comparing this analog signal with a threshold level, and because of its high gain, the LIA acts as a comparator and provides a full-swing digital signal that can be used by the subsequent blocks in the rest of the circuit.

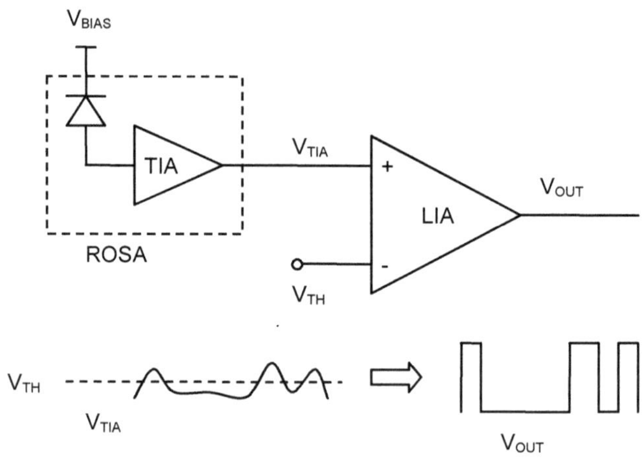

Fig. 9.5. Conceptual illustration of the limiting amplifier's function

As can be seen from Fig. 9.5, choosing the right threshold level is a crucial factor in the performance of the receiver. Typically, the threshold level must be chosen in the mid-point of the TIA's output swing to minimize the probability of error. If we think of the output of the TIA in the form of an eye pattern, the threshold must be chosen in the middle of the vertical eye opening, where the combined effect of the noise from the one and zero level signals is minimum.

Note that the threshold voltage is a dynamic value that needs to be generated based on the current state of the signal. This is because the common-mode voltage of the signal coming out of the TIA is not constant, but can be a function of a variety of parameters, including the received power level, supply voltage, and temperature. In a simple circuit such as the one shown in Fig. 9.5, one way to generate the threshold is by using an integrating circuit that acts as an average detector. Such a scheme will work as long as the signal is balanced in terms of average number of zeros and ones. For an unbalanced signal, more sophisticated schemes may be necessary to detect the average of the signal.

The scheme shown in Fig. 9.5 is based on a single-ended interface between the TIA and the LIA. However, in practical circuits almost always a differential coupling between the TIA and LIA is used, a feature needed to minimize the effects of common-mode noise. Also in a differential scheme the effective amplitude of the signal that is received by the LIA will be doubled. Figure 9.6 shows a practical LIA circuit that is differentially coupled to the TIA.

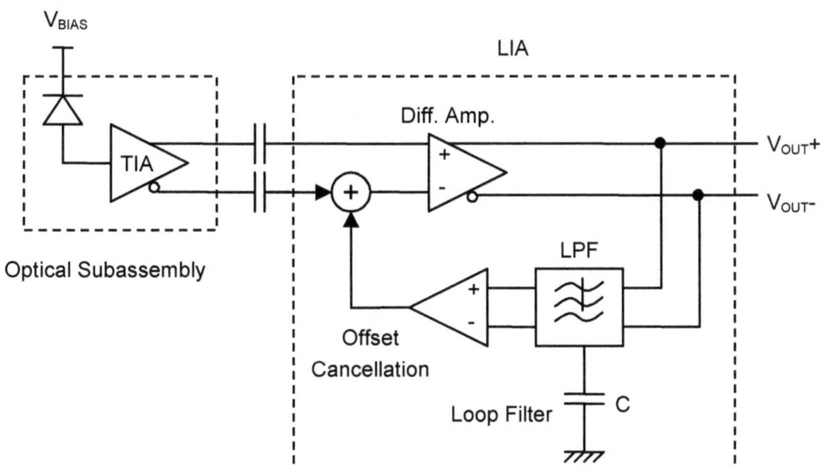

Fig. 9.6. Schematic of a differential limiting amplifier circuit with automatic offset adjust

First, notice that the TIA and LIA are ac-coupled. This isolates the dc offsets between the two blocks, and therefore parts with different dc levels can be connected together without disturbing each other's dc biasing point. However, adding these coupling capacitors results in a low-frequency cut-off point in the path of the signal [9,10]. Larger values of coupling capacitors yield a lower cut-off point, thus allowing operation at lower data rates. If these capacitors are not chosen large enough, pattern dependency may be observed. On the other hand, the larger the value of the capacitor is, the lower its resonance frequency will be. A capacitor does not behave like an ideal capacitor above its resonance frequency because of the parasitic inductance of the package. Therefore, large values of coupling capacitors may cause problems at higher data rates. Generally speaking, a value of 0.1 μF is typical for data rates above a Gbps. Lower data rates may require larger capacitors.

Figure 9.6 also includes a schematic representation of the threshold adjust circuit, which is now based on a dc offset cancellation loop. This loop samples the differential voltage at the output of the LIA. The differential signal then goes through a low-pass filter (LPF), at which point it is converted to an offset voltage that is added to the input voltage of the LIA. As usual, the time constant of the loop is controlled by a capacitor *C*.

The operation of the dc cancellation loop can be explained as follows. The LIA can be considered as a high-gain differential amplifier which receives a low-amplitude differential signal from the TIA. If the common-mode voltages of the two differential signals at the input of the amplifier are the same, the signal is balanced, and therefore the output of the amplifier will also be balanced. This situation is shown in Fig. 9.7a. In this case the average voltages for the two outputs of the high-gain amplifier are the same, the output of the LPF will be zero, and no threshold adjustment is applied to the LIA's input. This represents the equilibrium operating point of the loop.

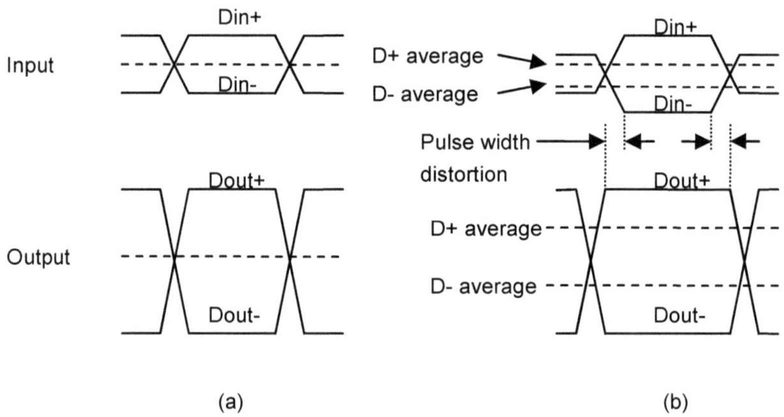

Fig. 9.7. Operation of the dc offset cancellation loop. (**a**) Balanced inputs and outputs, (**b**) dc offset in inputs, causing pulse distortion and dc offset at output

On the other hand, consider Fig. 9.7b which shows the inputs and outputs of a high gain amplifier with no dc-cancellation loop, and where the differential inputs have a dc offset between them. Because of the high gain of the amplifier, the outputs will be swinging between the same high and low points as before. However, the width of the zero and one bits will change, resulting in an offset between the average voltage of the two output signals. As can be seen, a dc offset at the input of the differential amplifier has caused pulse width distortion and dc offset at its output.

The addition of a dc cancellation loop will solve this problem by injecting an additional dc offset to the LIA input, forcing the dc offset at the output to go to zero. In this way, a feedback loop dynamically maintains the average voltage of the signals as they are applied to the limiting amplifier, effectively optimizing the threshold level continuously. This also underlines the importance of choosing the right time constant for the feedback loop. A long time constant results in a slow loop response. This would be useful for slow data rates or for data patterns with long identical sequences. But such a long time constant could cause problems when it is necessary for the loop to adjust itself to new conditions in a speedy manner.

9.4 Clock and data recovery

The limiting amplifier provides a full-swing digital signal. However, any jitter present in the signal will be transferred directly through the TIA and the LIA. In other words, although these circuits will "clean" the signal in terms of its amplitude, they do not perform any conditioning in the signal timing. If a signal has accumulated jitter, that jitter will still be present in the output of the LIA. Moreover, to recover the original signal correctly, the receiver must have information about the timing of transitions. Therefore, the receiver must somehow have clock information about the signal so that it can sample the signal at the right times, i.e., when each bit has reached its correct logic level. The most common way of obtaining this information is to recover the clock from the received signal itself.

This is done through a clock and data recovery (CDR) circuit [11–13]. Normally, a CDR consists of a phase lock loop (PLL) along with a flip–flop that synchronizes the transitions in the bit stream with the transitions in the clock signal generated by the PLL. Figure 9.8 illustrates the block diagram of a typical CDR circuit.

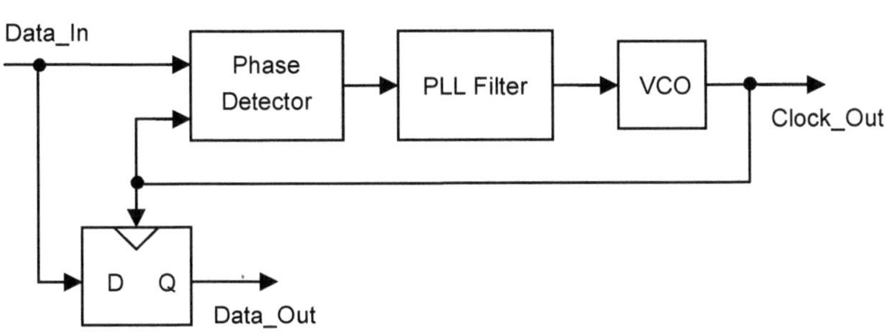

Fig. 9.8. Clock and data recovery (CDR) circuit

The phase detector compares the phase of the incoming signal with the clock and produces an analog voltage proportional to the phase difference. The analog voltage then goes through a loop filter. Typically the loop filter is in the form of a proportional-integrator (PI) controller. The output voltage of the filter controls the frequency of a voltage control oscillator (VCO). In this scheme, the output of the phase detector is the error signal, and the feedback loop attempts to minimize this error signal. If the loop filter has a pole at the origin (i.e., it includes an integrator), any non-zero error signal will be integrated over time and will cause the loop to move. As a result, at steady state, the error signal must be zero. This corresponds to the case where the PLL is locked, and the output of the VCO is at the same frequency as the data. Consequently, the output of the VCO can be used as the recovered clock. At the same time, this recovered clock can be used to retime the serial input data, for instance by having the input data go through a D flip–flop which is triggered on the rising edge of the recovered clock.

In this way, the CDR can be used, not only to extract the clock from the received serial stream, but also to "clean up" the jitter in the received data. This is shown in Fig. 9.9.

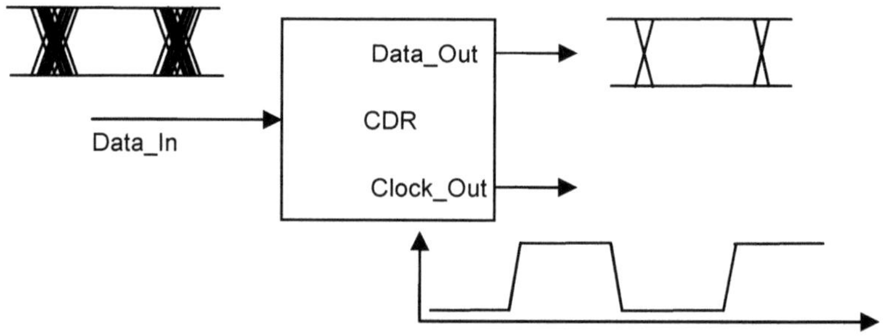

Fig. 9.9. Clock and data recovery from a single stream of data

We should note that a CDR's behavior, including its response to jitter, is greatly influenced by the loop filter. For instance, if the jitter present at the input signal exceeds a certain limit, the CDR may not be able to clean the signal anymore, a property characterized by the CDR's jitter tolerance. We will discuss issues regarding the characterization of a CDR's performance later in this chapter.

9.5 Performance of optical receivers

So far we have discussed the circuits that make up an optical receiver. In this section we will discuss the ways that the performance of optical receivers can be characterized and measured.

9.5.1 Signal-to-noise ratio (SNR) and bit error rate (BER)

The performance of a digital receiver is mainly determined by the signal-to-noise ratio at the decision point in the receiver. Figure 9.10 shows the eye diagram of a digital signal that can have two values, x_1 corresponding to a logical 1 and x_0 corresponding to a logical 0. The threshold level x_{th} is set between x_1 and x_0. Once a threshold level is selected, any value above the threshold is interpreted as a 1 and any value below the threshold is interpreted as a 0.

In reality, because of the inherent noise in the detection process, the 0 and 1 levels are not single-valued, but instead are characterized by probability distributions.

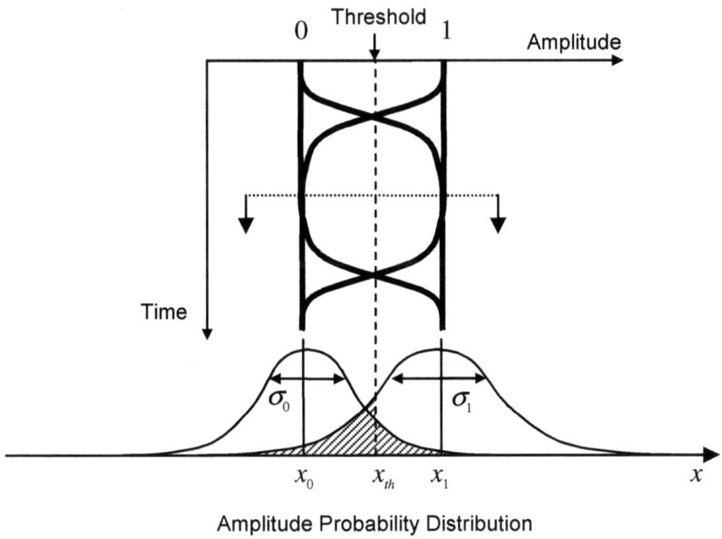

Fig. 9.10. Noise probability distribution in a digital signal

The exact form of these distributions can be very complex [14–21]. However, a Gaussian probability distribution is a simple and often acceptable approximation that is widely used. In that case, the two distributions shown in Fig. 9.10 are characterized by their mean values, x_1 and x_0, and by their standard deviations, σ_1 and σ_0.

We are interested in correlating the bit error rate at the detection point to the signal-to-noise ratio of the detected signal. To do this, we need to find the probability that a particular bit is detected in error, i.e., a 1 bit is detected as 0 or a 0 bit is detected as 1. Assuming the decision threshold is set at point x_{th}, the probability that a transmitted 1 is erroneously detected as a zero is given by

$$P_{ERR\_1}(x_{th}) = \frac{1}{\sqrt{2\pi}\sigma_1} \int_{-\infty}^{x_{th}} \exp\left(-\frac{(x-x_1)^2}{2\sigma_1^2}\right) dx \tag{9.1}$$

Likewise, the probability that a transmitted 0 is erroneously detected as a 1 is given by

$$P_{ERR\_0}(x_{th}) = \frac{1}{\sqrt{2\pi}\sigma_0} \int_{x_{th}}^{\infty} \exp\left(-\frac{(x-x_0)^2}{2\sigma_0^2}\right) dx \tag{9.2}$$

In balanced data, the probability that a transmitted bit is 1 is equal to the probability that it is 0, and they are both equal to 0.5. Therefore, the overall probability of error, or the bit error rate BER, as a function of the threshold level, is given by

$$BER(x_{th}) = 0.5 \times P_{ERR\_1}(x_{th}) + 0.5 \times P_{ERR\_0}(x_{th}) = \\ \frac{1}{2\sqrt{2\pi}\sigma_1} \int_{-\infty}^{x_{th}} \exp\left(-\frac{(x-x_1)^2}{2\sigma_1^2}\right) dx + \frac{1}{2\sqrt{2\pi}\sigma_0} \int_{x_{th}}^{\infty} \exp\left(-\frac{(x-x_0)^2}{2\sigma_0^2}\right) dx \tag{9.3}$$

The crosshatched area in Fig. 9.10 is a graphical representation of BER for a given threshold level. Equation (9.3) can be simplified greatly in terms of the Q factor, which represents the signal-to-noise ratio at the optimal threshold level [21]:

$$Q = \frac{x_1 - x_{th}}{\sigma_1} = \frac{x_{th} - x_0}{\sigma_0} \tag{9.4}$$

Equation (9.3) can then be written as

$$BER(Q) = \frac{1}{2} \operatorname{erfc}\left(\frac{Q}{\sqrt{2}}\right) \tag{9.5}$$

where erfc(.) is the complementary error function defined as

$$\operatorname{erfc}(x) = \frac{2}{\sqrt{\pi}} \int_{x}^{\infty} \exp(-\tau^2)\, d\tau \approx \frac{1}{x\sqrt{\pi}} \exp(-x^2) \tag{9.6}$$

Although the erfc function is tabulated [22], the approximation given by Eq. (9.6) is excellent for $x > 3$ [21].

The Q factor is a convenient representation of the SNR of the signal and is widely used for characterization of signal quality. The Q factor can be interpreted more clearly if we eliminate x_{th} from Eq. (9.4) to obtain

$$Q = \frac{x_1 - x_0}{\sigma_1 + \sigma_0} \tag{9.7}$$

which is indeed the ratio of the amplitude of the signal $(x_1 - x_0)$ to the sum of the rms noise power at the 0 and 1 levels.

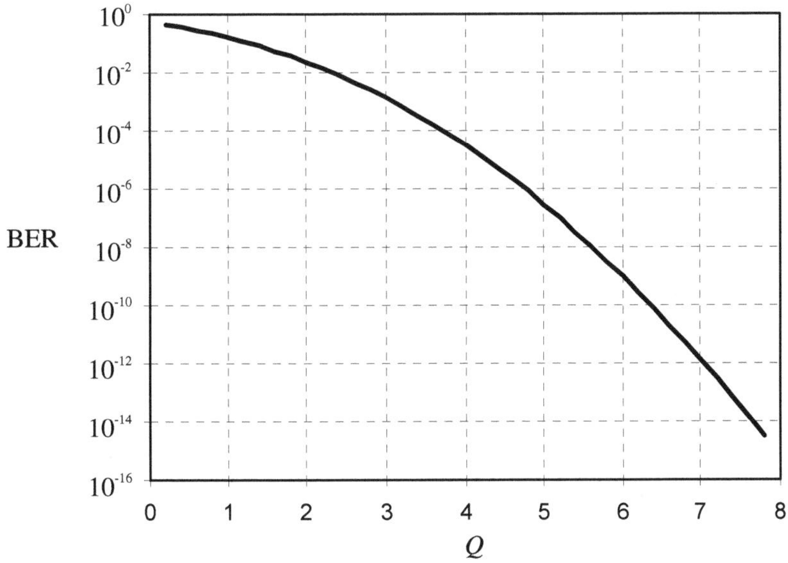

Fig. 9.11. Bit error rate as a function of Q factor

The exponential dependence of BER on Q is evident from Fig. 9.11, which shows a plot of BER as function of Q. For instance, a BER of 10^{-10} corresponds to $Q \approx 6$. To improve BER to 10^{-12}, Q must be increased to ~7.

9.5.2 Sensitivity

The SNR and BER are directly related to the *sensitivity* of a receiver. Sensitivity refers to the lowest optical power under which a receiver can operate and maintain a BER above a target value. A receiver with better sensitivity is capable of working with a weaker optical signal. Because sensitivity is related to optical power, it is measured in units of decibel-milliwatt. Thus, if the sensitivity of a receiver is −30 dBm, it can work with signals with an optical power as low as −30 dBm or

1 μW. The target value of BER at which sensitivity is defined depends on the application. Generally a BER of 10^{-12} is adequate for most applications, although in some cases lower values like 10^{-10} are acceptable too.

Sensitivity is also a function of the extinction ratio (ER) of the signal. Because information is carried by the modulation of the signal, for a constant average power, a lower ER causes a lower modulation amplitude and thus link penalty. In other words, to compensate the effect of decreased vertical eye opening the initial launch power must increase. This effect, known as *ER power penalty*, or simply ER penalty, can be estimated easily using Eq. (3.31) Chapter 3. Consider two optical signals with the same average power, but with different linear extinction ratios ER_1 and ER_2 ($ER_1 > ER_2$). The power penalty for the reduction of extinction ratio from ER_1 to ER_2 is the ratio of the eye amplitude in the two cases, expressed in decibel-milliwatt. This ratio can be obtained from Eq. (3.31):

$$ER_{Penalty} = 10\log_{10}\left[\frac{(1-ER_1)(1+ER_2)}{(1+ER_1)(1-ER_2)}\right] \tag{9.8}$$

For example, if the extinction ratio of a transmitter drops from 11 to 9 dB, the link suffers a power penalty of 0.4 dB. We can also consider the case of very high extinction ratio (ER→∞) as a reference and calculate the penalty of any other ER with respect to that reference. Setting $ER_1 \to \infty$ in Eq. (9.8), the ER penalty for any other ER can be calculated:

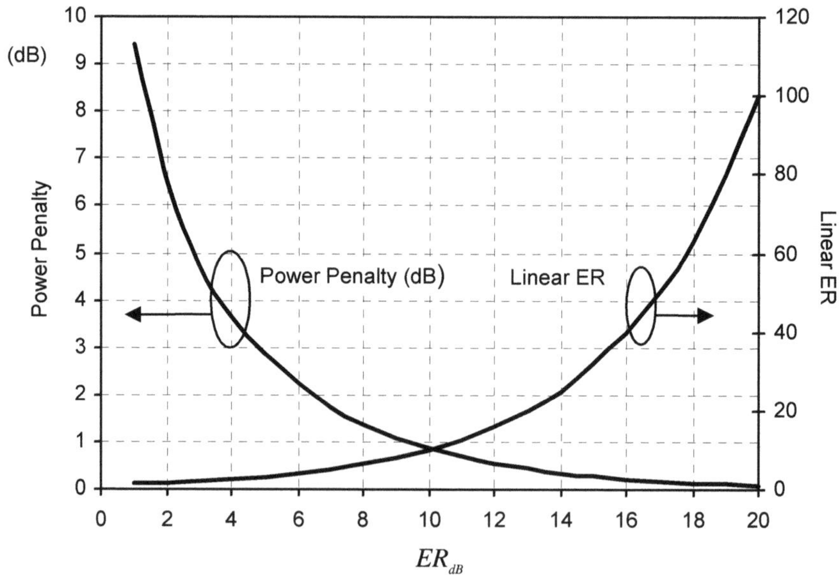

Fig. 9.12. Extinction ratio power penalty as a function of logarithmic ER

$$ER_{Penalty} = 10\log_{10}\left(\frac{1+ER}{1-ER}\right) \quad (9.9)$$

Figure 9.12 shows a plot of ER penalty as a function of logarithmic extinction ratio. For reference, the linear ER is also shown in this figure. As expected, for any given average power, lowering ER results in a power penalty. For instance, lowering ER from 15 to 10 dB results in a penalty of around 0.6 dB. However, lowering ER another 5 dB (from 10 to 5 dB) results in 2 dB of penalty. The nonlinear nature of this curve implies that extinction ratios less than 10 dB are exceedingly "expensive," because much of the optical power is dc light and does not contribute to the transmission of information.

As noted above, for sensitivity to be clearly defined, the target BER and the signal's ER must be clearly specified. It is therefore not uncommon to characterize a receiver more extensively over a range of parameters through characteristic curves. One of the most common ways to do this is through waterfall curves. A waterfall curve is a plot of BER as a function of received power. Figure 9.13 illustrates representative waterfall curves for an optical receiver.

For instance, the receiver characterized in Fig. 9.13 has a sensitivity of −31 dBm at BER=10^{-12} at ER=12 dB. It can also be seen that a 1 dB decrease in optical power results in an increase of more than one order of magnitude in BER. Thus, the waterfall curve provides a means of relating the BER tolerance of a system to the link budget. Also, it can be seen that increasing the ER will improve the BER for a given power, although at ERs above 10 dB the return starts to diminish.

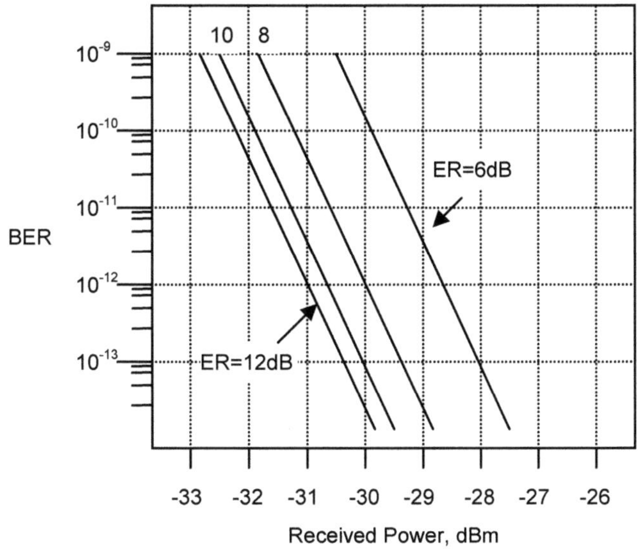

Fig. 9.13. Waterfall curve representation of a receiver's performance

We should note that Fig. 9.13 is only a representative example of waterfall curves. Theoretically, a waterfall curve is a consequence of the exponential dependence of BER on SNR. In reality, however, waterfall curves may show more complex behaviors. For instance, the slope of the line may change significantly at different powers. Such anomalies highlight deviations of the actual noise statistics from the simple Gaussian distribution which we assumed in the previous section. Therefore, because waterfall curves can easily be obtained by measurement, they provide an excellent tool for analysis of a receiver and the nature of noise statistics at the detection point.

9.5.3 Overload

A receiver can work properly within a limited range of optical power. As noted in the previous section, sensitivity specifies the lower side of this limit. From a theoretical point of view, sensitivity represents a point where SNR falls below the level that corresponds to the target BER. On the other hand, as the received power increases, eventually we get to a point where the receiver begins to saturate. For instance, the current output from the photodetector may get so large that the transimpedance amplifier can no longer handle the signal. Saturation of the input stage of the amplifier will then lead to excessive distortion and will cause the receiver to not function properly. The maximum optical power that a receiver can receive while still working properly is known as the *overload*. Like sensitivity, overload is expressed in units of decibel-milliwatt.

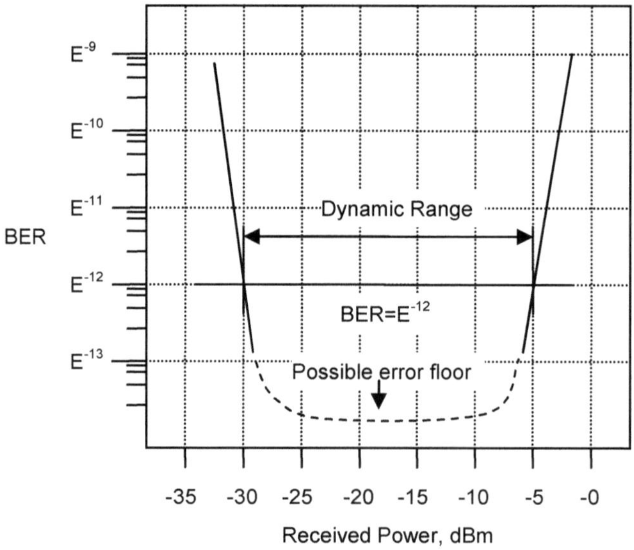

Fig. 9.14. Representative behavior of a receiver from sensitivity to overload

Waterfall curves provide a useful tool to characterize overload too. Figure 9.14 shows a sample waterfall curve over a larger range of optical power and exemplifies the behavior of a receiver from sensitivity to overload. Let us assume that we are dealing with an application where the maximum acceptable BER is 10^{-12}. In this example, the sensitivity of the receiver is −30 dBm, because if the power falls below this level, the BER will exceed 10^{-12}. On the other hand, for optical powers higher than −5 dBm the BER will exceed the acceptable threshold level of 10^{-12} again. Therefore, the overload point in this example is −5 dBm.

An interesting question is what happens to the BER for power levels between sensitivity and overload. From a theoretical point of view we may expect the BER to fall to exceedingly low values. In reality, however, an *error floor* may exist that limits the lowest achievable BER. This may be due to a variety of phenomena that could result in deviations from an ideal Gaussian noise distribution. However, it is hard to verify the existence of such error floors, because by definition they represent very low BERs. As an example, a BER of 10^{-16} at a bit rate of 1 Gbps is equivalent to one error every 115 days. Therefore, it is impractical to obtain experimental data points for such low BERs.

Another parameter related to sensitivity and overload is *dynamic range*. As shown in Fig. 9.14, dynamic range represents the ratio between the maximum and minimum powers that the receiver can work with. Dynamic range (in units of decibel) can be obtained by simply subtracting the sensitivity from the overload (in decibel-milliwatt). In the example of Fig. 9.14, the dynamic range is 25 dB.

9.6 Characterization of clock and data recovery circuits

As discussed earlier, an optical receiver typically requires a clock and data recovery (CDR) circuit to extract the clock signal from the received serial data. Moreover, the extracted clock can be used to retime the serial data itself, thus reducing the amount of jitter that is present in the data.

Intuitively, we expect that there should be a limit to the amount of jitter that can be present in the data stream for CDR to be able to operate properly. The two important concepts involved in the characterization of a CDR are its *jitter tolerance* and *jitter transfer* functions. We shall discuss these two concepts next.

9.6.1 Jitter transfer

Jitter transfer refers to the relationship between the jitter applied at the input of a digital circuit to the resulting jitter at the output of the circuit [23,24]. It is assumed that applied jitter has a sinusoidal time relationship. Figure 9.15 illustrates the concept of jitter transfer function.

Let us assume that a sinusoidal signal is used to modulate the exact transition time (or phase) of a signal. When the sinusoidal is zero, the transition takes place exactly at the time it is supposed to occur. This corresponds to no jitter. As the value of the sinusoidal increases, the transitions start to take place at an earlier time. The magnitude of jitter can be measured in terms of the ratio of the time deviation to the total length of a bit, or a Unit Interval (UI). Thus, a jitter of 0.1 UI means the time deviation of an actual transition with respect to when it should have occurred is 0.1 of the length of the bit. Note that using UI as the unit of time makes jitter description independent of the data rate. Therefore, the amplitude of the sinusoidal jitter function can be expressed in units of UI. Using a sinusoidal jitter function, then, we can add a precisely defined amount of jitter with precisely defined frequency to the digital signal. This signal can be applied to a digital circuit, such as a CDR. The output of the circuit can then be analyzed for the amount of jitter present in it.

The ratio of the amplitude of jitter in the output signal to the amplitude of the jitter in the input signal at a given frequency is the jitter transfer function at that frequency. Usually this ratio is expressed in units of decibels, and the jitter transfer function is thus defined by

$$T_J(\omega) = 10\log\left[\frac{J_{OUT}(\omega)}{J_{IN}(\omega)}\right] \tag{9.10}$$

where $T_J(\omega)$ is the jitter transfer function, and $J_{IN}(\omega)$ and $J_{OUT}(\omega)$ are the magnitudes of jitter for the input and output signals.

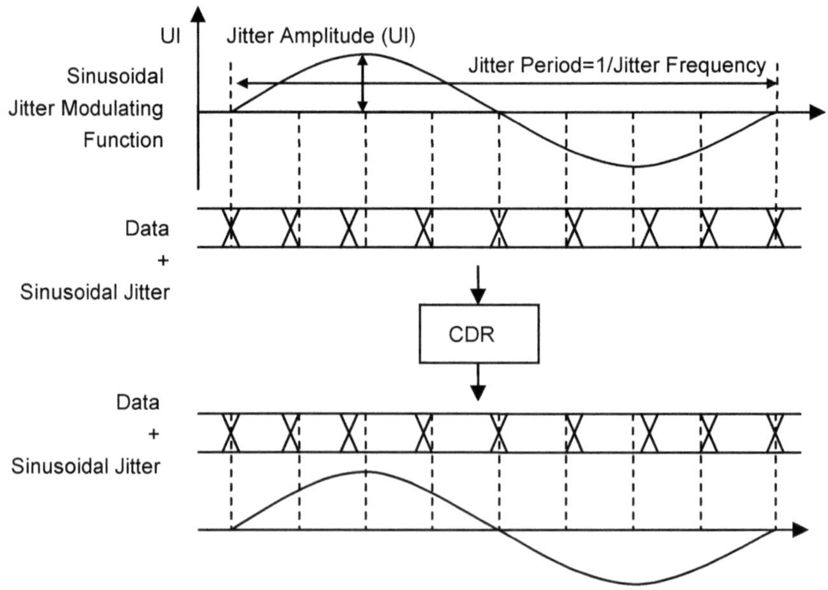

Fig. 9.15. Illustration of jitter transfer function

Therefore, by scanning the frequency over a specified range, we can obtain the jitter transfer function over that frequency range.

Figure 9.16 shows a representative jitter transfer function for a CDR. The jitter transfer function starts at 0 dB at low frequencies. Thus, for very low frequencies, the magnitude of the jitter at the output of the CDR is equal to the magnitude of jitter at input. This behavior is expected, because a very low-frequency jitter represents a phase shift which varies slowly with time, and the feedback loop in the PLL tracks the phase shift. But as the frequency of jitter increases, the amplitude of jitter at output starts to drop. This is where the PLL feedback loop maintains the phase of the output signal, in spite of the instantaneous changes of phase in the input. Note that ideally, a CDR should not show any jitter peaking in its jitter transfer function. In other words, the CDR itself should not add jitter to the signal. Moreover, note that the transfer function shown in Fig. 9.13 represents a first-order transfer function with a dominant pole at the transition frequency f_t.

Also shown in Fig. 9.16 is a *jitter transfer mask*. In this example, the jitter mask shown is from ITU-T's specification for STM-16 signals [25]. The jitter mask imposes a maximum allowable magnitude for the jitter transfer function, which means the transfer function of a complainant CDR must fall below this mask. The mask starts at the 0.1 dB point, which means that a compliant CDR is allowed to add a small amount of jitter to the input at low frequencies. However, for frequencies higher than 2 MHz, the CDR is expected to attenuate the jitter at a rate of 20 dB/decade. In other words, the CDR should filter out high-frequency jitter from the signal.

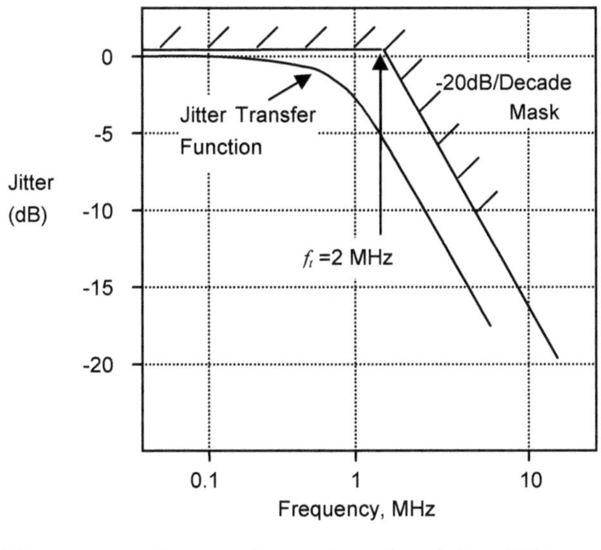

Fig. 9.16. Representative jitter transfer function and mask for a CDR

9.6.2 Jitter tolerance

Closely related to jitter transfer is the concept of jitter tolerance [26–28]. As the name suggests, jitter tolerance is a measure of the ability of a digital circuit, in our case a CDR, to work error-free at the presence of input jitter, and it is defined as follows. At a given jitter frequency, the amplitude of jitter at the input of the CDR is increased while the bit error rate at the data output of the CDR is being measured. This can be done by sampling the data with the CDR's extracted clock. If the CDR is working properly, the recovered clock must follow the phase of the signal and therefore no errors should be observed. The amplitude of jitter is then increased until errors are observed. This amplitude, again in terms of Unit Intervals (UIs), is the jitter tolerance of the CDR at that jitter frequency. Scanning the jitter frequency yields the jitter tolerance over the entire jitter frequency of interest. Figure 9.17 shows the typical shape of jitter tolerance for a CDR.

Also shown, as an example, is a *jitter tolerance mask* for an STM-16 signal, as defined by ITU-T [25]. A jitter tolerance mask outlines the minimum jitter tolerance expected in a particular application. The mask starts at 1.5 UI at low frequencies and goes down to 0.15 UI at high frequencies. The important point to notice about the jitter tolerance of the CDR is that it is complimentary to the jitter transfer. For frequencies below f_T, the CDR tracks the jitter, and therefore it can tolerate large amounts of jitter. On the other hand, for high-frequency jitter, the CDR's ability to track the input jitter diminishes, resulting in a lower jitter tolerance.

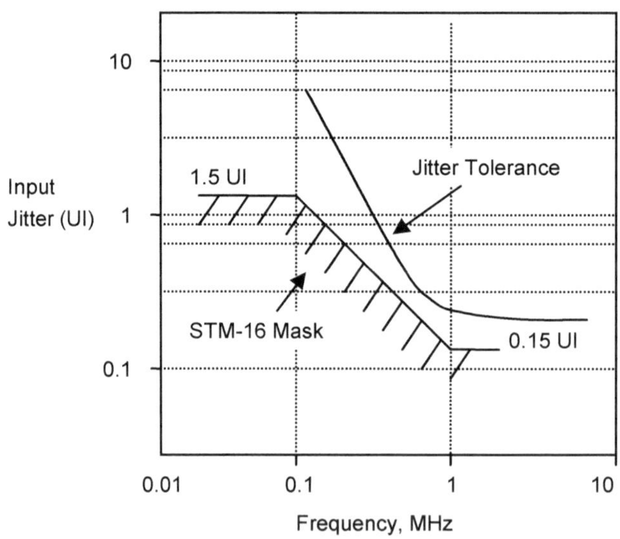

Fig. 9.17. Representative jitter tolerance and mask for a CDR

9.7 Burst mode receivers

The receivers we have been discussing so far can be categorized as continuous mode or CW because the received optical power remains relatively constant. Thus, it is easy for the receiver feedback loops to catch up and adjust with any long-term change in power. However, there is a class of applications where the received power can change in a very short period of time. This is the case in passive optical networks (PON), where a single receiver has to work with multiple optical transmitters that in general can be at different physical distances. The outputs of these transmitters are multiplexed through a time division multiplexing scheme, resulting in a bursty traffic. As a result, a PON receiver has to be able to work with consecutive packets with widely different optical power. Such burst mode receivers (BMRs) are challenging to design, because they need to adjust themselves to a new received power level in a very short period of time. In this section we review the concepts involved in these BMRs [29–34].

9.7.1 Dynamic range challenges in burst mode traffic

Many of the challenges in a burst mode receiver originate from the nature and format of the data stream. To better appreciate various aspects of the received optical signal in a burst application, it is useful to view the signal in a logarithmic scale, as shown in Fig. 9.18.

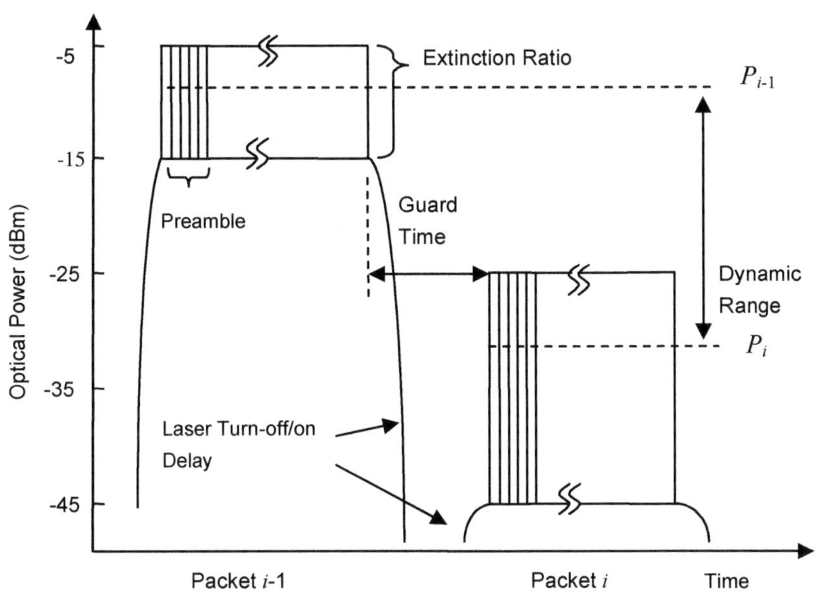

Fig. 9.18. The parameters involved in a burst mode stream at the receiving point

This figure shows two packets that are sent by two different transmitters. The first packet has originated from a transmitter at a closer distance, thus, its average power is higher. The second packet has traveled for a longer distance and therefore its average power is lower. The two packets are separated by a *guard time*, to ensure packet collision is avoided. The guard time is also needed to ensure the transmitter sending packet i has enough time to shutdown completely, while the transmitter sending packet i+1 has enough time to turn on and reach steady state. The ratio of the average power for the two consecutive packets is the *instantaneous dynamic range* that the receiver has to accommodate:

$$DR_{dB} = 10 \log\left(\frac{P_{i-1}}{P_i}\right) \tag{9.11}$$

It is important to distinguish the instantaneous dynamic range from the usual dynamic range encountered in standard receivers, which in this context can be called steady-state dynamic range. The reason can be seen from Fig. 9.18: a BMR has a very short time to adjust to the new optical power level in an incoming packet. As a result, the sensitivity of a BMR generally suffers. This *burst mode sensitivity penalty* is a major challenge in the design of BMRs [35–37].

Moreover, note that the extinction ratio (ER) of the optical signal is not infinite. Therefore, a situation can be encountered where the 0 state in the previous packet has more optical power than the 1 state in the next packet, as is the case in Fig. 9.18. In this figure, the ER of the first packet is 10 dB and DR_{dB} is 20 dB. In practical PON applications, the maximum value of DR_{dB} can be in the range of 15–20 dBs, and the guard time can be as short as a few tens of nanoseconds.

Each packet starts with a preamble field consisting of (usually) a sequence of alternating 1s and 0s. The preamble does not carry any information; however, it is used by the BMR to extract the decision threshold for the rest of the packet. The preamble is followed by some extra overhead bits and the actual payload data. Because the preamble can also be very short (in the range of a few tens of nanoseconds) the challenge in a BMR is to extract an accurate threshold level in a very short time.

9.7.2 Design approaches for threshold extraction

From the previous discussion it follows that the main difference between a standard receiver and a BMR is that in the latter case, the receiver must extract the optimal threshold for each packet separately, and it has a very short period of time to do this. This threshold is the dc level represented by P_{i-1} and P_i in Fig. 9.16.

In general, there are two approaches to threshold level recovery, feed-forward and feedback. Figure 9.19 provides a comparison of these two methods.

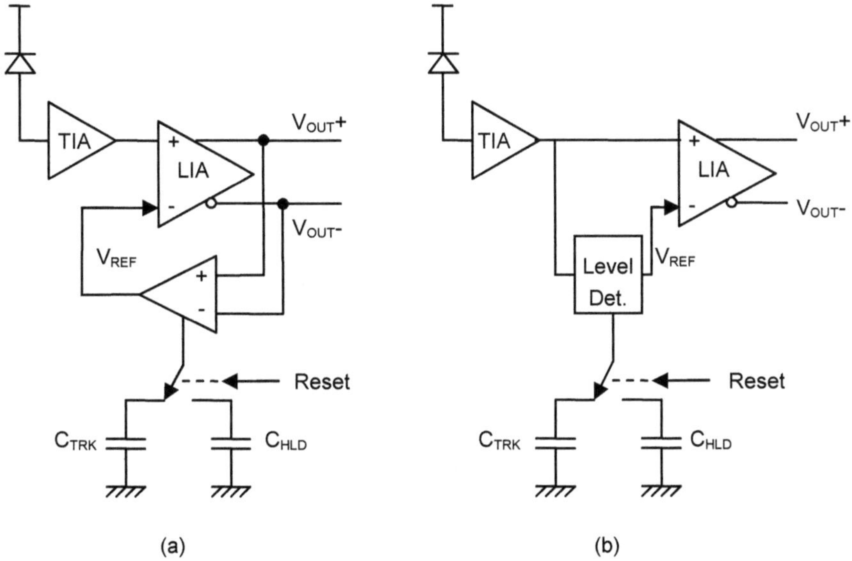

Fig. 9.19. Threshold extraction in a burst mode receiver based on (**a**) feedback, and (**b**) feed-forward

In a threshold extraction scheme based on feedback, an amplifier samples the outputs of the LIA and generates a dc reference voltage based on the difference between the common mode voltages of the two outputs. This is shown in Fig. 9.19a. The feedback loop controls the reference voltage continuously. Thus, once a new packet arrives, a new value for the reference voltage is generated which is based on the average power in the new packet. In many ways, this scheme is similar to the scheme used in standard receivers, for example, as shown in Fig. 9.6.

However, in many BMR systems, an additional "reset" input is necessary. This is because oftentimes a single time constant cannot accommodate all the timing requirements involved. During the guard time, the circuit must have a quick time response so that it can adjust the threshold level fast enough, especially when it tries to respond to a high instantaneous dynamic range situation. On the other hand, during a given packet, the time response must be much slower so that long sequences of identical digits do not disturb the threshold level. A dual time constant approach can resolve this issue. As shown in Fig. 9.19, these conflicting requirements can be met if the circuit can select between a shorter "tracking" time constant and a longer "hold" time constant by means of an external reset signal. Of course, providing such a reset signal requires prior knowledge of the timing of the packets. This knowledge cannot be recovered from the data stream at the physical level. Instead, it must be provided by the higher layers within the receiver system. In PON systems, this is done through a process of ranging.

On the other hand, Fig. 9.19b shows a BMR based on a feed-forward scheme. Here no feedback loop exists. Instead, the threshold level is extracted directly from the signal coming out of the TIA by a level detector circuit. The actual implementation of a level detector may use different approaches. For instance, it can be based on a linear scheme such as an average detector, or a nonlinear scheme such a peak and bottom detector. Regardless of the actual implementation, here too an external reset signal is usually needed to distinguish between the conflicting timing requirements within individual packets and during the transition from one packet to the next.

There are tradeoffs to be considered between a feedback and a feed-forward scheme. The most important challenge in a feedback scheme is the response time of the feedback loop. A BMR is expected to use up a small portion of the preamble to extract the correct power level for a packet. If the timing requirements are relaxed, it is possible to implement feedback loops fast enough to fulfill the design specifications. The advantage of a feedback loop is that the threshold is acquired more accurately. Also, large dynamic range variations between packets can be accommodated. However, if the timing requirements are strict, a feedback loop may not be able to reach steady-state point fast enough and, like any other closed loop system, stability concerns arise. For instance, in GPON applications, the preamble can be as short as a few tens of nanoseconds. In such cases, a feed-forward scheme provides an alternative solution. However, the drawback is that the extracted threshold may not be as accurate, and thus burst mode sensitivity penalty may increase.

9.7.3 Burst mode TIAs

So far we have not explicitly discussed the implications of burst mode traffic on TIA operation. In practice, TIAs also need to be modified to accommodate burst mode traffic [38].

In a BMR, the primary factor that is affected in a TIA is the AGC loop. As noted before, the AGC loop increases the dynamic range of the TIA and it does so through a feedback loop with a given time constant. In normal applications, the time constant can be made arbitrarily long. However, in a BMR, a long time constant means that the TIA will not respond to the quick changes between packets fast enough. As an example, a high-power packet can reduce the gain of the TIA through the AGC loop. However, if this high-power packet is followed by a low-power packet, the TIA needs time to increase its gain. As a result, the initial part of the low-power packet experiences a lower gain, causing potential errors.

To mitigate this problem, several approaches can be taken. One approach is to disable or remove the AGC loop altogether, and therefore remove the time constant limitation. The advantage of this approach is that it is easy to implement, and many commonly available TIAs can be used. The disadvantage is that once the AGC loop in a TIA is disabled the dynamic range of the TIA suffers, resulting in signal distortion, sensitivity degradation, or overload deterioration.

A more elaborate approach is to have two time constants, a fast time constant during transition from one packet to the next, and a slow (or infinite) time constant during each individual packet. This approach provides a much wider dynamic range, because it effectively adapts the gain of the TIA for each individual packet in the data stream. The disadvantage of such an approach is that it is more complex and requires additional circuitry and control connections between the TIA and the rest of the receiver circuitry. As noted earlier in this chapter, a photodetector and the TIA are usually interconnected by wire bondings and enclosed in a ROSA. Therefore, accommodation of complex interconnections or additional signals between the inside and outside of the ROSA is not very easy.

9.8 Summary

In this chapter we discussed the basic concepts involved in an optical receiver. In a fiber optic link, the receiver is responsible for converting the optical signal that has traveled through the fiber from optical domain back to electrical domain. As discussed in Chapter 5, optical signals suffer both in amplitude and in timing as they travel through the fiber. Thus, the receiver must be able to recover the signal in the face of both amplitude and time (phase) degradation.

A fiber optic receiver circuit uses a photodetector at the front end. In most cases this is a PIN diode. However, in high-end receivers an APD detector may also be used. An APD device provides inherent current gain at the very first stage, and therefore allows for the detection of weaker optical signals. The disadvantage of an APD detector is added cost and complexity, and the requirement for a high-voltage source.

Regardless of the type of detector, the current output of the detector is too weak for most practical purposes. Therefore, the detector is almost always followed by a transimpedance amplifier (TIA) that provides additional gain. Moreover, to accommodate a wider dynamic range, the TIA usually provides a means of automatic gain control, whereby the gain is reduced at higher signal levels.

To minimize the parasitic effects and noise coupling, the photodetector and the TIA are usually integrated into a receiver optical subassembly (ROSA) where they are connected together with very short wire bondings. Moreover, the ROSA is usually hermetically sealed to improve reliability against environmental factors such as contamination and humidity.

In digital receivers, the next stage after the TIA is to compare the signal with a threshold level to extract a clean, digital signal. This is the responsibility of a limiting amplifier (LIA), which must both extract the threshold level and provide the comparison. Usually, the threshold level is extracted from the signal through a dc-offset cancellation loop. In these cases, the time constant of the loop is an important parameter, because it determines the response of the LIA to long sequences of identical digits.

A limiting amplifier removes the amplitude noise and provides a clean digital signal. However, the signal may still include phase noise or jitter. Moreover, in order to interpret the signal correctly, the system needs to extract the clock information from the signal. These two functions are achieved by a clock and data recovery (CDR) circuit, which oftentimes follows the LIA. A CDR is typically based on a PLL loop, which locks the frequency of a voltage-controlled oscillator to that of the signal to generate a clock signal. The extracted clock can then be used as a gating signal for the serial data, synchronizing the transitions in the data stream with the clock and effectively removing the jitter from the serial data.

The quality of the signal at the decision point is characterized by SNR. SNR in turn, determines the BER and the sensitivity of the receiver. Sensitivity refers to the minimum optical power at which the receiver can operate below a given BER. More generally, waterfall curves are used to characterize sensitivity over a wider range of BERs. On the other hand, the performance of CDRs is characterized by jitter transfer function and jitter tolerance. Practical applications determine specific values for all these parameters.

A subclass of digital receivers, known as burst mode receivers (BMRs), must work with packet-based data streams where the average power can change quickly from one packet to the next. These receivers are an important part of passive optical networks, and together with burst mode transmitters constitute the basis of PON links.

References

[1] L. L. Wang, "Ultra-wide dynamic range receiver for noise loaded WDM transmission systems," Optics Express, Vol. 16, pp. 20382–20387, 2008

[2] R. Vetury et al., "High sensitivity and wide-dynamic-range optical receiver for 40 Gbit/s optical communication networks," Electronics Letters, Vol. 39, pp. 91–92, 2003

[3] H. Matsuda et al., "High-sensitivity and wide-dynamic-range 10Gbit/s APD/preamplifier optical receiver module," Electronics Letters, Vol. 38, pp. 650–651, 2002

[4] M. S. Park, C. H. Lee, and C. S. Shim, "Optical receiver design with high sensitivity and high dynamic range using feedback and lossy noise-matching network," Optical and Quantum Electronics, Vol. 27, pp. 527–534, 1995

[5] J. M. Khoury, "On the design of constant settling time AGC circuits," IEEE Transactions on Circuits and Systems-II, Analog and Digital Signal Processing, Vol. 45, pp. 283–294, 1998

[6] T. Kurosaki et al., "Low-cost 10-Gb/s optical receiver module using a novel plastic package and a passive alignment technique," IEEE Journal of Lightwave Technology, Vol. 23, pp. 4257–4264, 2005

[7] HFAN-3.2.0, "Improving noise rejection of a PIN-TIA ROSA," Application note from Maxim Integrated Products, 2008. Available from www.maxim-ic.com.

[8] J. M. Baek et al., "High sensitive 10-Gb/s APD optical receivers in low-cost TO-can-type packages," IEEE Photonics Technology Letters, Vol. 17, pp. 181–183, 2005

[9] M. Stephen, L. Luo, J. Wilson, and P. Franzon, "Buried bump and AC coupled interconnection technology," IEEE Transactions on Advanced Packaging, Vol. 27, pp. 121–125, 2004

[10] HFAN-1.1, "Choosing AC-Coupling capacitors," Application Note from Maxim Integrated Products, 2008. Available from www.maxim-ic.com

[11] T. Kok-Siang et al., "Design of high-speed clock and data recovery circuits," Analog Integrated Circuits and Signal Processing, Vol. 52, pp. 15–23, 2007

[12] F. Centurelli and G. Scotti, "A high-speed low-voltage phase detector for clock recovery from NRZ data," IEEE Transactions on Circuits and Systems I, Regular Papers, Vol. 54, pp. 1626–1635, 2007

[13] G. T. Kanellos et al., "Clock and data recovery circuit for 10-Gb/s asynchronous optical packets," IEEE Photonics Technology Letters, Vol. 15, pp. 1666–1668, 2003

[14] M. Ahmed and M. Yamada, "Effect of intensity noise of semiconductor lasers on the digital modulation characteristics and the bit error rate of optical communication systems," Journal of Applied Physics, Vol. 104, Article Number 013104, 2008

[15] K.C. Jong, H.W. Tsao, and S.L. Lee, "Q-factor monitoring of optical signal-to-noise ratio degradation in optical DPSK transmission," Electronics Letters, Vol. 44, pp. 761–762, 2008

[16] J. D. Downie, "Relationship of Q penalty to eye-closure penalty for NRZ and RZ signals with signal-dependent noise," Journal of Lightwave Technology, Vol. 23, pp. 2031–2038, 2005

[17] E. W. Laedke et al., "Improvement of optical fiber systems performance by optimization of receiver filter bandwidth and use of numerical methods to evaluate Q-factor," Electronics Letters, Vol. 35, pp. 2131–2133, 1999

[18] D. R. Smith and I. Garrett, "Simplified approach to digital optical receiver design," Optical and Quantum Electronics, Vol. 10, pp. 211–221, 1978

[19] HFAN-3.0.0, "Accurately estimating optical receiver sensitivity" Application Note from Maxim Integrated Products, available from www.maxim-ic.com

[20] HFAN-9.0.2, "Optical signal-to-noise ratio and the Q-factor in fiber-optic communication systems," Application note from Maxim Integrated Products, available from www.maxim-ic.com

[21] N. S. Bergano, F. W. Kerfoot, and C. R. Davidsion, "Margin measurements in optical amplifier system," IEEE Photonics Technology Letters, Vol. 5, pp. 304–306, 1993

[22] M. R. Spiegel, *Mathematical Handbook*, Schaum's Outline Series, McGraw-Hill, New York, 1992

[23] Y. D. Choi, D. K. Jeong, and W. C. Kim, "Jitter transfer analysis of tracked oversampling techniques for multigigabit clock and data recovery," IEEE Transactions on Circuits and Systems II-Analog and Digital Signal Processing, Vol. 50, pp. 775–783, 2003

[24] M. J. E. Lee et al., "Jitter transfer characteristics of delay-locked loops – theories and design techniques," IEEE Journal of Solid State Electronics, Vol. 38, pp. 614–621, 2003

[25] G.783, Transmission Systems and Media, Digital Systems and Networks, ITU-T, 2006

[26] C. F. Liang, S. C. Hwu, and S. L. Liu, "A Jitter-tolerance-enhanced CDR using a GDCO-based phase detector," IEEE Journal of Solid State Circuits, Vol. 43, pp. 1217–1226, 2008

[27] M. Hayashi et al., "Analysis on jitter tolerance of optical 3R regenerator," IEEE Photonics Technology Letters, Vol. 15, pp. 1609–1611, 2003

[28] S. Y. Sun, "An analog PLL-based clock and data recovery circuit with high input jitter tolerance," IEEE Journal of Solid-State Circuits, Vol. 24, pp. 325–330, 1989

[29] K. Hara, et al., "1.25/10.3 Gbit/s dual-rate burst-mode receiver," Electronics Letters, Vol. 44, pp. 869–870, 2008

[30] E. Hugues-Salas et al., "Fast edge-detection burst-mode 2.5 Gbit/s receiver for gigabit passive optical networks," Journal of Optical Networking, Vol. 6, pp. 482–489, 2007

[31] X. Z. Qiu et al., "Development of GPON upstream physical-media-dependent prototypes," Journal of Lightwave Technology, Vol. 22, No. 11, pp. 2498–2508, Nov. 2004

[32] J. M. Baek et al, "Low-cost and high-performance APD burst-mode receiver employing commercial TIA for 1.25-Gb/s EPON," IEEE Photonics Technology Letters, Vol. 17, No. 10, pp. 2170–2172, Oct. 2005

[33] Q. Le et al., "A burst-mode receiver for 1.25-Gb/s Ethernet PON with AGC and internally created reset signal," IEEE Journal of Solid State Circuits, Vol. 39, No. 12, pp. 2379–2388, Dec 2004

[34] H. Wang and R. Lin, "The parameter optimization of EPON physical layer and the performance analysis for burst mode receiver," Proceedings of SPIE, Vol. 4908, pp. 105–114, 2002

[35] A. J. Phillips, "Power penalty for burst mode reception in the presence of interchannel crosstalk," IET optoelectronics, Vol. 1, pp. 127–134, 2007

[36] P. Ossieur et al., "Sensitivity penalty calculation for burst-mode receivers using avalanche photodiodes," Journal of Lightwave Technology, Vol. 21, No. 11, pp. 2565–2575, Nov. 2003

[37] P. Ossieur et al., "Influence of random DC offsets on burst-mode receiver sensitivity," Journal of Lightwave Technology, Vol. 24, No. 3, pp. 1543–1550 March 2006

[38] S. Nishihara et al., "10.3 Gbit/s burst-mode PIN-TIA module with high sensitivity, wide dynamic range and quick response," Electronics Letters, Vol. 44, pp. 222–223, 2008

Chapter 10
Reliability

10.1 Introduction

Reliability is a critical concern in all engineering systems. Reliability issues present significant challenges to the designers, because usually it is hard to predict them in advance. As a result, oftentimes problems associated with reliability show up later in the design phase. Moreover, the inherent statistical nature of failures makes them harder to understand and deal with. From an economic perspective, reliability issues are often very expensive, because they expose problems late in the design phase or even when the systems are deployed and are under normal operation. Such failures may require costly and time-consuming maintenance, debugging, or recalls.

All systems have a finite life, and therefore eventually fail. However, there are a wide variety of factors that may hasten the failure. Some of the more common ones that are relevant to most systems include temperature and mechanical stress. On the other hand, there could be specific design flaws that increase the risk of failure under certain conditions. While environmental stress factors are generally outside the control of the designers and therefore uncontrollable, the design itself is controllable, and therefore design flaws (at least in principle) can be minimized by proper design practices and testing.

In most general terms, we can divide reliability methodologies into three categories of design, test, and quantification. On the design level, the designer's goal is to reasonably predict and accommodate the risk factors that the system may encounter during its normal operation within its life time. Ideally, the design phase is the best time to address reliability issues, because it is easiest to implement any necessary changes at this phase before the design has solidified and proceeded to later phases in its life cycle.

Regardless of how much care has gone into the design of a product or system, it is always likely that some bugs escape the original design. To discover these bugs, the design must go under a slew of tests aimed specifically at exposing any potential reliability risk factors. These tests should simulate and even amplify the real stress factors that the product is likely to encounter during its normal life. The existence of reliability standards is a great help in this phase, as these standards attempt to classify risk factors in specific categories and define reasonable levels of stress in each category. Thus, a strong design methodology must include adequate reliability testing as part of the design cycle itself.

Reliability testing can act as a bridge to the third reliability category, i.e., quantification. Quantification methods try to address the problem of predicting the probability of a failure for an existing device, module, or system, regardless of the design practices and test methodologies that have gone into it. These methods are

M. Azadeh, *Fiber Optics Engineering*, Optical Networks,
DOI 10.1007/978-1-4419-0304-4_10,

of necessity statistical in nature and are based on statistical and probabilistic assumptions and models.

In this chapter our main concern is design practices aimed at improving reliability of fiber optic modules and systems. Because a successful design must always include adequate reliability testing, we will also discuss various tests specifically aimed at exposing reliability risks. The last section of this chapter discusses reliability quantification and statistical modeling of reliability risks.

We should also note that in reality reliability is also a reflection of a whole range of quality procedures associated with all aspects of the manufacturing process, including supply chain management, quality control, documentation, vendor qualification, and production process control, to name a few. A robust design is a crucial and necessary, but by no means sufficient, part of this overall quality picture. However, for the most part such issues fall outside the scope of this book. Our goal here is to focus only on design considerations and practices that can improve reliability, given other required quality management elements are in place.

10.2 Reliability, design flow, and design practices

Reliability is commonly defined as the probability of a device functioning properly over the expected life time of the device under normal working conditions. This definition shows that the concept of reliability includes of several elements. First and foremost, reliability is an inherently probabilistic concept. This means that a "reliable" device may fail at any moment, while it is possible for an "unreliable" device to function for a long period of time. Thus, reliability concepts start to converge with reality only for large populations of devices, or operations over long periods of time. The second element of the definition has to do with the "proper" function of the device. Generally, this requires the existence of a set of specifications that clearly define the proper function of a device. Examples of these specifications include manufacturer data sheets and industry standards. Thus, without clearly defined specifications, reliability remains a vague concept at best. The third element in the definition is life time of the device. This is important, because no device lasts forever. Therefore, reliability can only make sense in the context of a limited device life time. Finally, the definition includes the concept of normal working conditions. Even the most reliable device fails under sufficient stress. Therefore, inherent in the concept of reliability are the normal working conditions under which the device or the system under question is intended to work.

Approaching design from a reliability perspective, therefore, requires paying attention to these elements and their ramification. Reliability in engineering systems is a vast and emerging field, especially in the light of disasters that have cost dearly in terms of human life, environmental impact, or economic effects [1–7].

10.2.1 Design flow

Let us start by an overview of a general design flow for a typical engineering product. Typically, a design starts with a set of target specifications over a range of operating conditions, and the goal is to come up with a design that meets or exceeds these specifications. As noted in this chapter's introduction, a designer attempts to address all the concerns in the initial design phase. However, almost always there are various hidden problems that escape that first initial design phase. To expose these problems various tests must be performed. The best place to start is to test the target specifications over the range of intended working conditions. Moreover, because of the random nature of many failures, it is best to conduct these tests on a number of samples. If and once new failures are found, the design is modified to address the failures and the risks associated with them. Once all the failures are addressed, the product can proceed to the next phase, for instance, a pilot run. Figure 10.1 illustrates a flow chart for this process.

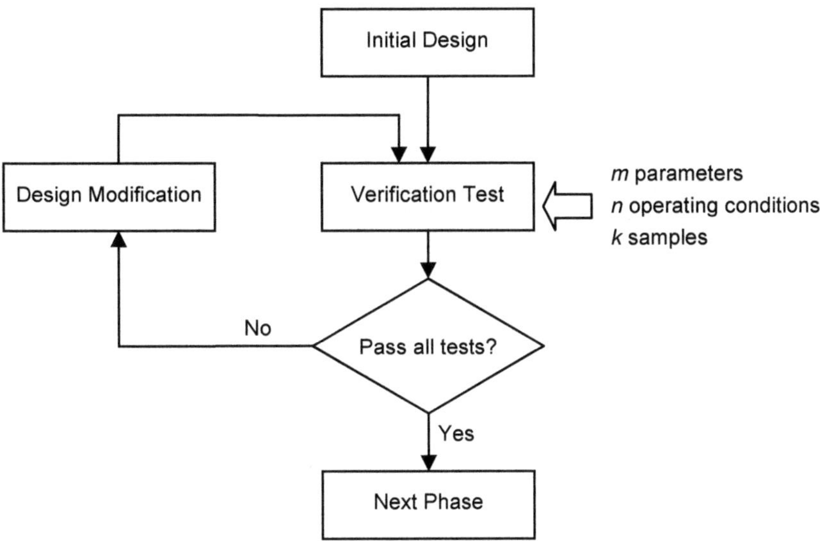

Fig. 10.1. Typical design and verification cycle

The flowchart depicted in Fig. 10.1 highlights some of the problems with this approach. Let us assume as suggested in the figure that we want to test k samples for m parameters, over n operating conditions. The number of operating conditions defines an n-dimensional space, and at each point in this space, $k \times m$ measurements must be made. This clearly represents a huge parameter space which exponentially expands with number of operating conditions. Covering all this space thoroughly requires a significant amount of time and testing, which in many cases may simply be impractical. Moreover, the design may have to go through many iterations before all the problems are worked out. Obviously this poses a problem in terms of resources, cost, and time to market. Therefore, although the process of Fig. 10.1 can potentially resolve most or all the hidden bugs in a design, we need to refine it to make it more practical.

10.2.2 Modular approach

One logical approach that can simplify this process is to divide the design into several different blocks, where the operation of each block is relatively self-contained and each block interacts with the rest of the system through a few well-defined "interfaces." In fact, this is the appropriate approach for all complex engineering systems: starting from the bottom, the individual blocks are defined separately and each are validated independently. Then, more complex blocks are made from these simpler, already verified blocks, and in this manner the whole system is built. The advantage of this modular "bottom-up" approach is that it is more likely that bugs are found earlier in the design where it is easier and cheaper to fix them.

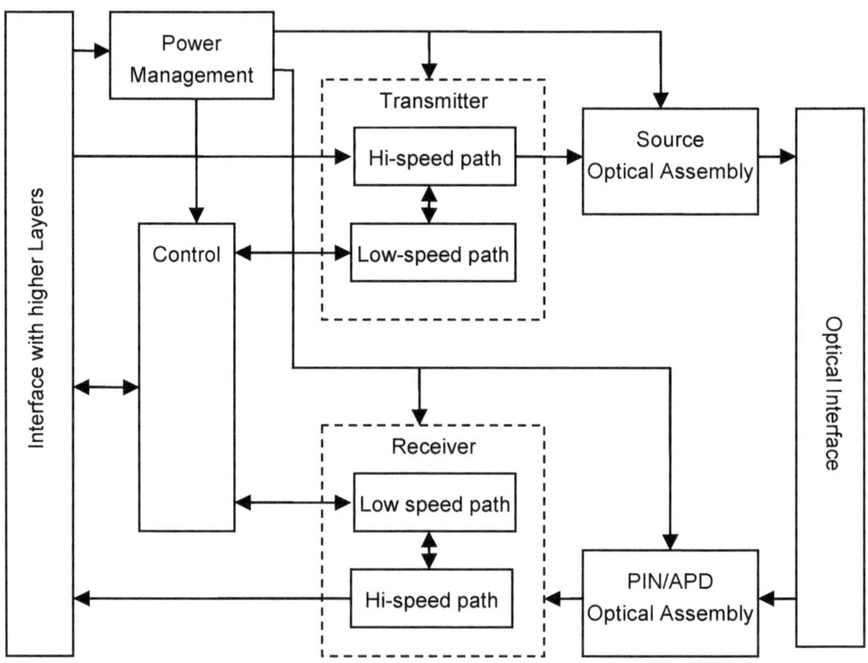

Fig. 10.2. Physical layer building block and functional interconnections

Moreover, even later in the design it would be much easier to isolate a bug to a specific block as the interfaces between various blocks are clearly defined. Fortunately it is not that hard to divide fiber optic systems into relatively independent building blocks. Figure 10.2 illustrates a typical block diagram.

A fiber optic system usually consists of a transmitter and a receiver block. Each of these blocks consists of a high-speed and a low-speed path. The high-speed path refers to the data path, which includes the data carrying signals. The data path operates at the nominal rate or frequency of the optical ink. The low-speed path involves the supporting functions, which include control and monitoring signals.

The control block usually includes a microcontroller or a microprocessor that is responsible for control of the hardware as well as establishing communication with the higher layers.

The optical subassemblies include a light source (i.e., LED or diode laser) on the transmitter path and a PIN diode or avalanche photo-diode (APD) on the receiver path. In most cases these units are enclosed in a hermetically sealed enclosure to minimize environmental effects to the sensitive optical device. Moreover, the receiver optical subassembly typically includes a transimpedance amplifier that has to be in physical proximity of the PIN or APD photodetector to minimize noise. The optical subassemblies interface electrically to the circuit from one side and through optical couplers to the fiber on the other side.

Finally, the power management circuits provide the necessary supply voltage and currents for each block. Most modern circuits operate under low operating voltages, usually 3.3 V and below. An exception is an APD receiver which needs a high voltage (~40–70 V).

Once we separate the physical layer transceiver system into these building blocks along with the interfaces that each block must support, it becomes easier to isolate hidden bugs and therefore improve the system's robustness against reliability risks.

10.2.3 Reliability design practices and risk areas

It is useful to incorporate the definition of reliability with sound design practices such as the modular "bottom-up" approach and summarize the results in the form of general recommendations, as follows:

- The requirements of the design, including its target performance and specifications, its life span, and its intended working conditions, must be clearly defined.
- A complex system is only as reliable as its components. It is therefore critical to use well-tested, well-characterized, high-quality components from approved vendors.
- By "decoupling" the individual blocks, and validating them separately, a single bug in one block does not affect the design of the rest of the blocks. Thus, not all blocks shown in Fig. 10.1 need to go through the same number of iterations.

- Once a certain block passes the test of time and its robustness is proven, it can be reused in other designs with much less potential risk. The use of such proven blocks can significantly reduce the design cycle time and reliability risks.
- Simplifying the design can improve reliability. Because smaller blocks become more generic, off-the-shelf components along with circuit recommendations from manufacturers can be used. To this end, many manufacturers offer useful application notes, guide lines, and design procedures for their circuits. Following these recommendations can significantly reduce the risks of future failures.
- In cases where reliability goals cannot be fulfilled with other means, parallel paths or redundancies provide an alternative. A redundant component, device, or system in critical areas improves reliability.
- It is crucial to gain as much knowledge as possible about any failure in an already existing circuit or system. A complete understanding of the root cause of failure as well as its effects on other parts is crucial before attempting to address the failure.
- Optoelectronic devices, including lasers and detectors, are semiconductors. As such, they are generally very sensitive to temperature. All fiber optic components, modules, and systems must be characterized and tested over their intended operating temperature thoroughly before they can be considered reliable.
- Reliability standards also provide a valuable resource for designer. Standards often embody a large body of knowledge and experience. These standards outline not only design practices, but also test and validation procedures [8–14].

These recommendations summarize some of the useful design practices that once followed properly can improve reliability and reduce the design cycle and cost. However, in order to arrive at more specific design practices, we need to have further refinements to the generic flowchart of Fig. 10.1. One way to achieve this is to divide reliability risk factors into different categories and try to focus on specific issues within each category, as illustrated in Fig. 10.3.

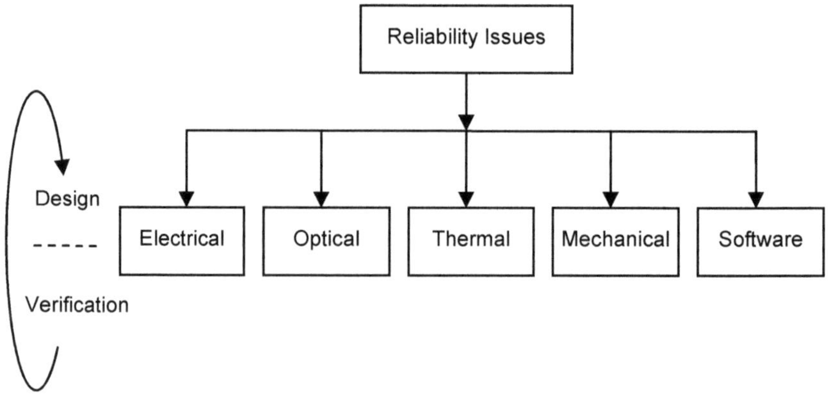

Fig. 10.3. Classification of reliability issues for fiber optic systems

The idea is that instead of trying to verify a design against a broad category called reliability, it is useful to classify various potential reliability issues into separate categories and actively search for, identify, and address risk factors associated with each category. In each category, the issues must be identified and worked out in both design and test stages, and the cycle should be iterated as long as the risk factors remain. In the following sections we will focus on each area separately and discuss techniques that can be utilized to improve the design phase and techniques that can be utilized to define tests with improved probability of exposing design faults.

10.3 Electrical issues

A fiber optic system, as shown in Fig. 10.2, includes a series of electrical circuits to support its functions. These circuits are implemented as printed circuit boards (PCB) and are made of passive and active components. Passive components typically refer to resistors, capacitors, and inductors, while active components refer to elements such as amplifiers and drivers that require a power supply and are usually packaged as ICs.

From an electrical perspective, the circuits are designed to carry out a number of functions. Breaking down these functions into simpler blocks as suggested by Fig. 10.2 would help the designer in evaluating the functions that each block is supposed to support. Because the circuits are intentionally designed to support these functions, the designer can usually check for them in a systematic way. However, for reliability purposes, a number of additional considerations must be made. We discuss some of the issues related to electrical reliability in the following sections [8,10,12,13].

10.3.1 Design margin

One of the best ways to reduce the reliability risks is to add margin to the design. This extra margin can address two areas: margin in the parameter space over which the circuit is supposed to function and margin over the value of each target specification.

To clarify this point, let us consider a practical example. Figure 10.4 illustrates a two-dimensional parameter space consisting of operating temperature and supply voltage as related to a laser driver circuit. As noted before, temperature represents one of the most important stress parameters in optical transceiver. Let us assume the temperature can vary within a range of T_{MIN} and T_{MAX}. For commercial grade parts the temperature range is typically 0–70°C, while for industrial grade parts it can be −40 to 85°C. Because the operating characteristics of all semiconductors (especially diode lasers) vary greatly with temperature, temperature variations must be carefully considered both in the design and in the test stage. The other variable shown in Fig. 10.4 is power supply. In a big mother board the supply voltage is provided by a separate supply unit. In an integrated or pluggable transceiver, the supply is just an input to the circuit that is provided through a connector. In both cases, as far as the laser driver circuit is involved, the supply voltage is an input variable and can vary over a range of V_{MIN} and V_{MAX}. The space *S* defined by V_{MIN}, V_{MAX}, T_{MIN}, and T_{MAX} is a rectangular domain in the (T,V_S) plane.

Now let us consider an output parameter of the transmitter that needs to comply with a given specification. For instance, let us assume that the optical power of the transmitter, *P*, must be within a range given by P_{MIN} and P_{MAX}. In general, the optical power can vary with temperature and supply voltage. Therefore, we can think of *P* as a function of V_s and *T*. Therefore, the design objective can be stated as

$$P_{MIN} \leq P(T, V_S) \leq P_{MAX}, \quad [V_S, T] \in S \tag{10.1}$$

To add margin in the parameter space, we need to increase the area within which we try to meet the specification stated by Eq. (10.1). The area denoted by S′ adds additional margin in both the temperature and the supply voltage dimensions. Extending the domain beyond *S* ensures that any potential risk area lying immediately beyond *S* is caught and guarded against.

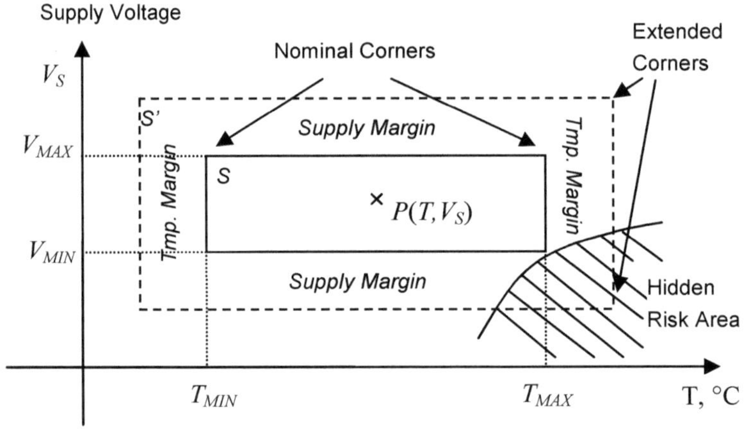

Fig. 10.4. Sample parameter space for an optical transmitter

The dashed area shown in Fig. 10.4 depicts a hypothetical example where a slight increase in temperature or reduction in the supply voltage beyond S causes the optical power to fall outside specifications. Although technically this area is outside the specified operating condition, its existence poses a reliability risk, because it is always possible that the operating condition point moves slightly beyond *S* due to various uncertainties. Alternatively, it is possible for the risk area itself to expand with time and to encroach upon *S*.

The risk will be even higher if the risk area represents a potential *hard failure* mechanism. A hard failure, as opposed to a *soft failure*, represents a case where a small change in the parameter space causes a big, possibly unrecoverable, change in a system output. In our example, this can be expressed as

$$\left(\frac{\partial P}{\partial V_S}\right)^2 + \left(\frac{\partial P}{\partial T}\right)^2 \geq M \tag{10.2}$$

where M is some large limit. The meaning of Eq. (10.2) is that the magnitude of the gradient of P is large, in other words, P varies rapidly in the direction specified by ∇P. An example for an optical transceiver is the thermal runaway condition, where at high temperature the average power control circuitry attempts to maintain the optical power constant by increasing the bias current to compensate for the efficiency degradation of the laser diode. However, increasing the bias current heats up the laser even more, causing a further reduction in efficiency. Therefore, a positive feedback loop is established which will cause a high current failure in a short time.

Another common practice is to add additional margin for the output specification given by Eq. (10.1). For instance, we can modify Eq. (10.1) to

$$P_{MIN} + P_{MAR} \leq P(V_S, T) \leq P_{MAX} - P_{MAR} \tag{10.3}$$

where P_{MAR} is some additional power margin. Obviously, tightening the target specification reduces the risk of the output falling out of nominal specification for various reasons over time, which can happen as a result of process variations, component aging, environmental factors, and the like.

The concept illustrated in Fig. 10.4 applies not only to the design phase, but also to the test phase. During validation tests, the design must be tested at as many points as possible to ensure it meets its targets. In practice, however, it is usually not possible to cover the S (or S') domain completely and thoroughly because of time or resource limitations. As a result, certain crucial, representative points must be selected in such a way as to maximize the probability of risk or fault exposure. Obviously the nominal operating point should be included. In addition, as Fig. 10.4 suggests, either the *nominal* or *extended corner points* are good candidates for design qualification. Use of these extreme points can reduce test time and increase the efficiency of the tests.[1]

Adding extra margin to the target speciation and verifying the design beyond the nominal parameter space are common ways of improving the reliability of a design. However, they come at a cost, because they could involve more complex design, more testing, possibly more expensive components, and reduced yields.

---

[1] The implicit assumption is that the output function (in our example $P(V_S, T)$) varies monotonously in the space S or S'. Usually, this is a reasonable assumption.

Therefore the extra margin in both the parameter space and the target specification is a compromise for which engineering concerns are only one factor.

10.3.2 Printer circuit boards (PCBs)

Electrical circuits are commonly built with printed circuit board (PCB) technology. A PCB provides a mechanical substrate for electronic components. It also provides the necessary electrical connections between circuit components where they need to be connected. Consequently, the PCB is in fact a very critical component in an electrical circuit, a point sometimes overlooked by the designers.

As a first step, following established PCB quality standards helps reliability. Many of these standards are available from IPC [15]. In terms of PCB layout, the manufacturing technology of a PCB is defined by a series of design rules. Following these design rules ensures that the layout does not result in geometrical sizes, patterns, and shapes that are hard to manufacture or have a high risk of failure. Table 10.1 below summarizes typical values for some of the key design rules that state-of-the-art PCB manufacturing processes can achieve. Although actual values vary depending on the individual manufactures, Table 10.1 gives an approximate idea on the limits that can be presently achieved. It might be tempting to push the technology in a particular PCB layout to the very limit allowed by the design rules. However, this is generally not a good idea from a reliability perspective. Very narrow traces, very small via sizes, and high layer counts are convenient from a layout perspective, as they allow for a denser layout and cleaner routing. However, pushing the technology to its very limits brings with it not only additional cost, but also reliability risks. In general, all PCBs must be tested for connectivity and isolation, and these tests are offered by most PCB manufactures.

Table 10.1. Typical values of advanced PCB technology

Parameter	Value
Minimum trace width	4 mils/ 0.1 mm
Minimum trace separation	4 mils/ 0.4 mm
Minimum drill hole size	10 mils/ 0.25 mm
Maximum layer count	40
Maximum aspect ratio of vias	12:1
Minimum board thickness	30 mils/ 0.75 mm
Maximum board thickness	300 mils/ 7.5 mm
Controlled impedance	30–150 Ω
Controlled impedance tolerance	±10%

The risk of using advanced technology is not that PCBs are defective from the beginning. Instead, the risk is having a PCB that passes connectivity and isolation tests at the time of manufacturing, but with marginal physical integrity.

For example, some of the traces may be narrower than the nominal minimum as a result of extra etching, some vias may be partially plated, or some connections may not be physically strong. As a result, a circuit built with such a PCB may pass all the tests, but later on fail due to environmental stress factors. Because of these risks, it is always wise to use the least level of technology possible in the design of a PCB, because usually a lower technology ensures higher long-term reliability.

This also holds for the extra design features used in a PCB. In a state-of-the-art PCB, the designer can use a variety of features, such as blind and partial vias, filled vias, micro vias, high aspect ratio vias, hybrid PCBs consisting of a mixture of flexible and rigid PCBs, or tight tolerance geometries. While all these features facilitate the design and yield more flexibility, their use must be carefully considered and weighted against their reliability risk factors. As before, the general rule of thumb is that the higher and the more specialized the technology, the higher the reliability risks.

10.3.3 Component selection

Selection of components is a critical part of an electrical design. A physical component is always an approximation to the model used in the schematic of an electrical circuit, and therefore the designer must carefully study the limitations of the physical components he or she selects. Often times the best place to start is the datasheet provided by the manufacturers. It is always a good practice to study these datasheets and all the parameters they discuss very carefully.

One effective way of improving reliability is the practice of *derating*. All components can tolerate only a limited range of inputs, and the designer must ensure those ranges are met with a sufficient margin under all circumstances. The easiest way to achieve sufficient derating is to use components with larger *absolute value ratings*, usually available from manufacturer datasheets. The manufacturers also provide application notes and design guidelines about their products. Taking advantage of this information is an easy way of finding about certain reliability risks associated with the use of the relevant components.

Often times the same component is available in a variety of packages and sizes. Broadly speaking, components can be divided into *leaded* and *surface mount*. The performance and small size advantages of surface mount components make their use unavoidable. In fact, many modern components are available only as surface mount. However, the surface mount components themselves are available in a variety of sizes. Passive components are generally available in sizes ranging from a few millimeters to sub-millimeters.

Table 10.2 Standard sizes available for passive components

Series	Length × Width (in.)	Length × Width (mm)
01005	0.016 × 0.008	0.4 × 0.2
0201	0.024 × 0.012	0.6 × 0.3
0402	0.040 × 0.020	1 × 0.5
0603	0.060 × 0.030	1.5 × 0.75
0805	0.080 × 0.050	2 × 1.25
1206	0.120 × 0.060	3 × 1.5

Table 10.2 summarizes the standard available sizes for passive components. Not all values are available in all sizes. Smaller value capacitors are usually available in different package sizes, whereas larger capacitor values are generally not available in very small package sizes. On the other hand, resistors are more readily available in a more variety of physical sizes.

Obviously use of components in smaller size packages is attractive for designers. A higher component density can be achieved for a given board space, and smaller components generally show less parasitic effects. However, from a reliability point of view, use of very small components may be problematic. From an electrical point of view, smaller parts generally have lower voltage, current, and power rating. This reduces the design margin and increases the probability of failures. From a mechanical perspective, smaller components are more fragile, harder to assemble, and harder to replace. Thus, mechanical and thermal stress factors can adversely affect the reliability of these components. Finally, very small components can be more expensive too. As a result, very small component sizes must be avoided if possible.

Active components and integrated circuits are also usually available in different packages and sizes. The above comments are generally true for active components too. Very small size packages should be avoided if possible. Very small packages have less thermal and mechanical stress tolerance and require higher technology PCBs.

10.3.4 Protective circuitry

In many cases, the designer can add extra components or circuits to protect against certain risks or failures. The standard fuses used in a variety of electronic equipment are obvious examples. If for any reason the circuit draws more current than it should, the fuse disconnects the supply voltage to the circuit, thus protecting it from catastrophic failure.

There are other forms of protective circuitry that can be used depending on the application. For example, diodes are widely used for protection against transients or protection against reverse biasing [16–18]. Adding a series resistor to an APD bias circuit results in a negative feedback loop which reduces the bias voltage of the APD at higher currents, thus protecting the APD from damage. Laser driver circuits often have safety circuits that shut down the laser in the case of a short circuit, preventing it from launching high levels of optical power. In all these cases, extra elements or circuits are added in anticipation of a potential failure mode, resulting in designs that are more reliable in the face of uncertain operating conditions.

10.4 Optical issues

Optoelectronic systems are distinguished from other electronic systems in that in one way or the other they involve the generation, propagation, and detection of light. For fiber optic applications, there is also the additional concern of coupling light to and from optical fibers, which includes a host of challenging mechanical issues. In this section we discuss some of the reliability issues related to the optical devices and subassemblies.

10.4.1 Device level reliability

Diode lasers are the most widely used optical devices, and they have been studied extensively from a reliability point of view [19–22]. The main failure mode for laser diodes is internal degradation due to crystal defects. The rate at which these defects are introduced in the crystal is proportional to the operating current and temperature. As a result of these defects, as the laser ages, its threshold current increases while its slope efficiency decreases. At a constant bias current, the optical power from the laser is found to decrease almost exponentially with age, where the time constant of the decrease is proportional to bias current. Another failure mechanism associated with lasers is facet degradation and damage. This is usually a result of intense optical power passing through the facet and an increase in nonradiative recombination at facets. The overall effect of these degradations is a reduction in facet reflectance, causing an increase in laser threshold current.

These failure modes point to the strong correlation of reliability with operating current and temperature. In fact, these two factors are often intertwined: operating the laser at a higher bias current by itself increases the temperature. On the other hand, in circuits with automatic power control, increasing the temperature causes an increase in bias current to maintain optical power.

The important conclusion is that from a reliability point of view it is extremely beneficial to operate lasers (and other optoelectronic devices) at lower temperatures and a lower bias current. Moreover, if possible, it is beneficial to limit the rate of change of temperature that the optical system experiences. Fast changes can cause thermal stress and thermal disequilibrium. As a result, the circuits that compensate such changes can over- or under-compensate the temperature change. A clear example is the average power control loop for a diode laser.

Although environmental conditions under which a device is to operate are for the most part outside the control of the designer, there are still some aspects that can be controlled at the design level. For example, well-designed thermal management can reduce the operating temperature of the devices, and therefore improve reliability. We will discuss thermal issues separately in this chapter.

10.4.2 Optical subassemblies

It is very hard to work with a laser die or a PIN diode directly. Therefore, these devices are usually packaged in the form of hermetically sealed subassemblies such as TO cans or butterfly packages [23,24]. These subassemblies provide a mechanical platform to hold the active optoelectronic device and any additional lenses, coupling optics, or optical filters in place. They also provide electrical interconnects to these devices. These interconnections consist of wire bondings inside the device and standard pins outside the package for interconnection to the rest of the circuits. More complex optical subassemblies may also include additional components such as RF termination or TEC cooler devices.

Obviously, the manufacturing quality of these subassemblies is critical for the reliability. Humidity is a serious risk factor for optoelectronic devices, and therefore package hermeticity is very important as it ensures better resistance against humidity. A leak test with a tracer gas can be a useful test in accessing the hermeticity of a particular package [14]. A damp heat test can also be a useful tool in testing the effects of humidity on the optical subassembly [25].

Another important concern is the risk of failure due to electrostatic discharge (ESD). Because optical subassemblies are often handled by human operators like standard electronic components, they need to have some level of resistance to ESD damage. At the same time, these devices have to be handled in accordance with standard ESD safety protocols, otherwise latent failures can occur which pose reliability risks during the lifetime of the device.

It is also important to characterize optical subassemblies over all relevant operating conditions. As usual, temperature is a critical factor. Parameters such as optical couplings, bonding of die level chips to substrates, gain of feedback signals such as monitoring photodetector currents can all be affected by temperature. Thus, these critical parameters must be characterized carefully over a large enough sample size to ensure that the design is safe over all operating conditions.

10.4.3 Optical fibers and optical coupling

Coupling light into and extracting light from the fiber is the ultimate goal of an optical subassembly. Because of the small core dimension of optical fibers, especially single-mode fibers, this can be a delicate task. Therefore, mechanical stress can impact the performance, as even a minute change in alignment can cause a big change in optical coupling efficiency.

Optical subassemblies are available in two form, pigtail and receptacle. In pigtail type, the assembly must be tolerant to factors such as mechanical pulling, twisting, and side forces. In the receptacle type, factors such as mating loss repeatability, durability, and wiggle are important factors that can affect long-term reliability [26–29]. In any case, it is important to realize that optical subassemblies are delicate parts and they constitute one of the most crucial reliability risk factors in optical systems.

Another factor that is especially important for single-mode lasers is back reflection. It is well known that even a tiny amount of optical feedback can change the behavior of a single-mode laser significantly. As a result, care must be taken to minimize any back reflection from the fiber, because by design the fiber is tightly coupled to the laser. Using an optical isolator reduces the back reflection significantly and the risks of failures due to unknown or uncontrolled amounts of reflection. Using angled optical connectors, which reduce the amount of reflection, coupled into the fiber also helps in the reduction of back reflections.

10.5 Thermal issues

We have already emphasized the role of temperature as a major reliability risk factor. In general, failures increase exponentially with temperature [9]. In fact, as we will see at the end of this chapter, the most common way to accelerate the aging processes in a device and expose failures in a design is to run the device under high temperature conditions.

Generally speaking, the operating temperature of an optoelectronic device or module is an environmental parameter and is outside the control of a designer. Commercially graded parts are expected to operate in the range of 0–70°C, while extended or industrially graded parts should be able to operate in the range of −40°C to 85°C. An exception is thermally stabilized devices such as TEC-cooled WDM lasers, which by design operate at a constant temperature. In spite of this, good thermal management practices can reduce the temperature stress by removing the generated heat from the device. This is usually done by reducing the thermal resistance from the hot device to a cooler environment or by isolating hot spots in a system from thermally sensitive areas.

The rise in temperature for any device with respect to a reference point is proportional to the power dissipation within the device and inversely proportional to the circuit's thermal resistance to a thermal reference point, as shown in Fig. 10.5.

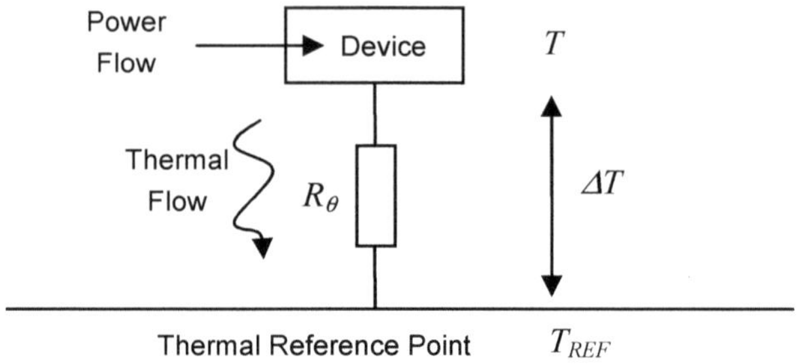

Fig. 10.5. Thermal resistance and temperature difference in a simple model of heat dissipation

Thus, in a manner similar to an electrical resistance, we have

$$T - T_{REF} = \Delta T = R_\theta P \quad (10.4)$$

where T is the device temperature, T_{REF} is the temperature of the reference point, P is the power consumption in watts, and R_θ is the thermal resistance in degree Celsius per watt. The thermal reference point is assumed to have a large thermal capacity so that its temperature will not rise as a result of heat transfer from the device. It can be seen from Eq. (10.4) that reducing the temperature can be handled from two parallel paths: reduction of power consumption and reduction of thermal resistance.

10.5.1 Power reduction

Reduction of power consumption can be addressed in part by choosing inherently lower power parts, whenever possible. IC manufacturers constantly improve their products, and one of the areas that they try to address is reduction of power consumption. Use of newer, lower power technologies can be a help not only in terms of performance, but in terms of the amount of heat that is being generated within the circuit.

Power reduction can usually be achieved through other means too. In general, continuous lowering of power must in itself be a design target and should be addressed on its own merit in each design cycle. Choice of higher resistance pull-down and pull-up resistors, biasing the circuits at lower operating current, elimination of extra functionalities, and partial or optional disabling of circuit blocks when they are not needed can all help.

In laser driver circuits, considerable power savings can be achieved if the laser is biased at the lowest target power possible. In direct modulated lasers, dc coupling (vs. ac coupling) can result in additional power savings. Essentially, the bias current required in an ac coupling scheme is more than the required current in a dc coupling scheme by as much as half the modulation current. Considering the fact that for a constant extinction ratio the required modulation current increases exponentially with temperature, using dc coupling can be very beneficial, because it yields more savings at high temperatures where it is most critical to reduce the power dissipation.

10.5.2 Thermal resistance

The temperature rise in a circuit can also be minimized by minimizing the thermal resistance between "hot" devices and some external "thermal ground." By thermal ground we mean a relatively large body mass with very good conductivity which is able to absorb heat without experiencing a rise in temperature. Examples include the chassis of a system or an external heat sink that is being cooled by forced air circulation.

For hot ICs mounted on a printed circuit board, this goal can be achieved in part by providing thermal relief on the PCB. Most high-power ICs have a metallic exposed pad which acts both as a ground (or sometimes supply) connection and as a path for heat transfer. Connecting this pad to the reference plane on the board through multiple thermal vias can provide a low thermal resistance path for heat dissipation, because oftentimes the reference planes are relatively large and can absorb the heat from hot points and distribute it. Of course, reference planes in a PCB are not ideal thermal grounds, as they do heat up because they have to absorb the heat from all the running components that are placed on them.

Another effective way to reduce the thermal resistance is to add thermal pads which can fill the air gap between the PCB and the chassis. It is well known that air is not a good heat conductor. Specifically, where there is no forced air circulation a thin layer of air which acts as a thermally isolating coating tends to stick to the surfaces, resulting in a substantial increase in the thermal resistance. Removing the air by filling the space with some low thermal resistance material can substantially improve the thermal performance of the system.

The use of heat sinks is also an effective way of lowering the thermal resistance. A heat sink is usually a piece of metal with a large surface to volume ratio. The large surface and the high thermal conductivity of the metal reduce the thermal resistance. However, heat sinks are especially effective when they are exposed to forced air circulation, which may not be an available option in some applications.

At this point we should also make a note on the importance and limitations of forced air circulation. Forced air circulation can greatly enhance the removal of heat from hot surfaces. In the model of Fig. 10.5, the rise in temperature for the heat dissipating part is measured against the temperature of a thermal reference point. If this thermal reference point has a very large heat capacity compared to the heat generation rate in the system, then we can think of it as a "thermal ground." In that case, its temperature is constant, and thus the temperature rise in the device will be minimized. However, this is not always the case. A typical example is the metal chassis in an electronic system. Although the chassis can absorb heat, ultimately its temperature will rise as more and more heat is dumped into it. As a result, the device temperature will also rise more, compared to the ideal case of an infinite heat absorbing chassis. The role of forced air circulation is to lower the thermal resistance from this chassis to a more ideal thermal reference, say, outside air temperature, which supposedly is constant. Thus, from the point of view of the heat dissipating device, forced air circulation makes the thermal reference point more like an ideal thermal ground. This picture holds even in the simple case of a device exposed to air without any extra heat sinking. In this case the air plays the role of thermal reference. Without air circulation, the temperature of the air immediately at the surface of the device will rise quickly. Thus, the air cannot play the role of a good thermal ground. However, with forced air circulation, the hot air film sticking to the surface is continuously changed with fresh cool air. In this case, the air behaves more like an ideal thermal ground.

10.6 Mechanical issues

Another class of reliability risks that need to be considered is those having to do with the mechanical integrity [30,31]. Specifically, these risks are associated with the stress resulting from factors such as mechanical vibration and shock, as well as forces such as pull, push, and torque that will inevitably act on the individual components to various degrees. It is expected that an optical module be able to tolerate reasonable amounts of such stress factors. We will consider some of the main mechanical reliability concerns in the following sections.

10.6.1 Shock and vibration

The most obvious cases of mechanical failures are those resulting from mechanical shock and vibration. When an electronic module or system is accidentally dropped, it is hoped that it survives the mechanical shock. Likewise if the system is subjected to a prolonged period of vibration, for instance during transportation, it is expected that the system retain its mechanical integrity.

In essence, both shock and vibration stress the system by exerting large forces on individual components. The force experienced by a component is proportional to the acceleration and is given by Newton's second law:

$$F = m\frac{dv(t)}{dt} \tag{10.5}$$

where F is the force in Newtons, m is the mass of the component, and $v(t)$ is the speed of the component. Therefore, the larger the mass, and the larger the acceleration (or deceleration), the larger the force. It must be ensured that this force is properly distributed and passed on to a solid frame. For example, consider Fig. 10.6 which depicts several scenarios where the force exerted on a component results in stress accumulation in a particular point, increasing the risk of mechanical failure at that point.

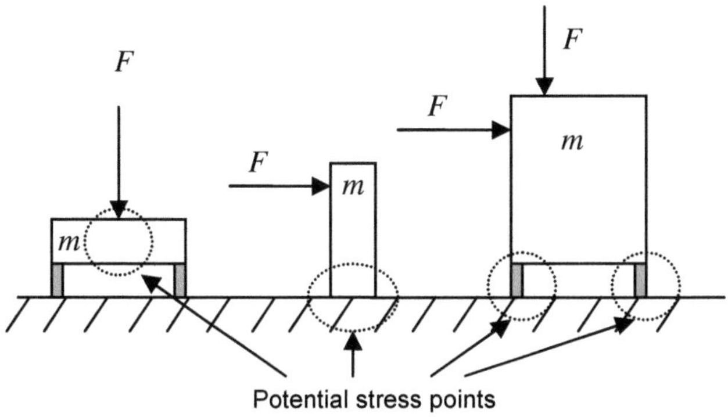

Fig. 10.6. Mechanical stress points resulting from lateral and vertical acceleration

We should mention that the failure risks could be slightly different between shock and vibration. A shock input stimulates the "impulse" response of the system by applying an intense pulse of acceleration. As we shall see in Chapter 12, it is expected that small components withstand accelerations in the order of hundreds of *g*s.[2] Although this may sound excessive, this is the acceleration a part could experience if it falls on a hard surface from even distances as short as a meter. In fact, this is exactly how parts are tested for shock resistance: they are secured to a platform that can fall from a distance that is calibrated for acceleration. Vibration, on the other hand, includes the application of mechanical energy in a specific frequency for a prolonged period of time. In reliability tests, the frequency of vibration is changed to cover all relevant frequencies that a part may encounter.

As is well known from the theory of linear systems, an impulse signal has a very wide range of frequencies, and therefore it can stimulate a wide range of mechanical frequencies. If the shape of the impulse is approximated with a Gaussian function, the profile of the shock pulse in frequency domain is also Gaussian, with its width inversely proportional to the time width of the impulse.[3]

The shock pulse duration used in reliability tests for optoelectronic modules typically varies from 1 to 10 ms depending on the weight of the component [14]. This corresponds to a mechanical frequency range of up to several kilohertz. Keeping the frequency domain picture in mind is useful in studying the effects of vibration on the mechanical integrity of a module as well, because vibration is essentially the application of mechanical energy in a narrow frequency range. In practical reliability tests for optoelectronic testing, usually the vibration frequency

---

[2] One *g* is the acceleration of a free falling body near earth and is equal to 9.8 m/s.

[3] This is the mechanical version of the Fourier uncertainty principle discussed in Chapter 3.

is varied from several hertz to several kilohertz [14]. Therefore, both shock resistance and vibration resistance require a part to be safe against the application of mechanical energy up to several kilohertz. Thus, stability and integrity of components up to such high mechanical frequencies must be considered in the mechanical design of a module.

An important issue that needs to be considered is the existence of mechanical resonance points within this frequency range. Such resonance points may result in accumulation of mechanical energy in a narrow oscillation frequency, causing reliability risks and failures. Examples include long extensions of PCB boards in space without mechanical support, receptacle-based optical subassemblies that may need to float in a module to allow for tight optical coupling with an inserting connector, long wire-bonds to optoelectronic chips, and flexible circuits that may be used to move signals between PCB assemblies.

Mechanical stress could be more demanding on optical subassemblies especially when it comes to coupling efficiencies. A major requirement for optical coupling is to maintain the alignment intact. Take, for example, the coupling of light output from a diode laser to a fiber (Fig. 10.7).

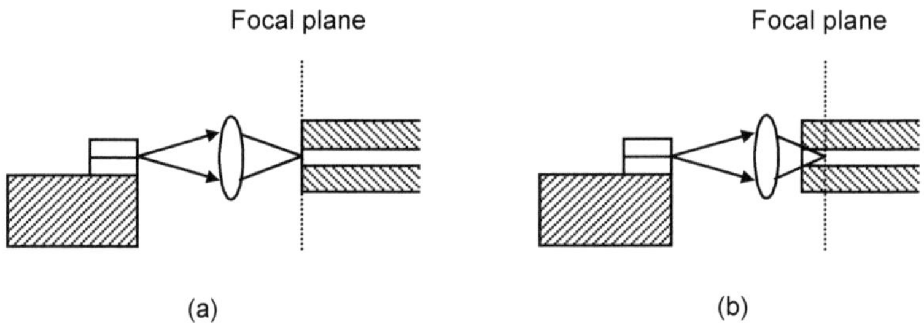

Fig. 10.7. Sensitivity of coupling efficiency to optical alignment. (**a**) focal imagining, (**b**) out of focus imaging

Usually a lens is used to image the output light of the laser into the core of the optical fiber. To maximize the efficiency, the light must be focused tightly to the core of the fiber. However, the drawback in this scenario is that any displacement can greatly change the coupling efficiency. On the other hand, the fiber can be placed in a plane slightly off the focal point of the lens. In this case the coupling efficiency is lower, but the coupling is now not very sensitive to displacements. If the additional coupling loss can be tolerated, this is both an assembly and a reliability advantage.

10.6.2 Thermal-induced mechanical failures

Certain mechanical problems can have thermal roots. This usually happens when two parts that are mechanically coupled tightly have different thermal expansion coefficients. As a result of temperature variations, these parts expand or contract at a different rate, causing mechanical tension to build up. Moreover, fast changing temperatures will stress PCB components and soldering. This is especially true for surface mount parts that are soldered to the PCB only through their contacts. A fast temperature ramp can stress these parts and may result in microcracks. For the same reason, subjecting a part to a rapid temperature ramp can expose potential risks by accelerating the failure mechanism. As such, it can be used as a method for identifying weak points in an assembly.

10.6.3 Mechanical failure of fibers

Glass fiber cables must be treated with more care compared to a regular copper cable. The exposed surface of an optical fiber is susceptible to dust gathering. Failure to clean these surfaces can result in considerable power loss. Moreover, because fiber-to-fiber coupling relies on physical contact between aligned fiber cores, small particles of dust could result in damage in the fiber surfaces. Any air gap between the cores of fiber assemblies can also result in light reflection, which may considerably degrade the performance of sensitive parts such as DFB lasers.

Another source of power loss in a fiber assembly is fiber bending. Any bending in a fiber assembly will cause some loss, and sharper bends with a smaller radius will cause more loss. Therefore assemblies that involve multiple fiber loops must be treated with considerable care as they could result in a significant (and somewhat unpredictable) loss. In particular, the bend radius should not be less than what is specified in manufacturers' datasheets.

Moreover, glass fibers inevitably have some level of flaws in the form of microcracks and crystal defects. These defects grow in time, in the form of both *static* and *dynamic fatigue*. Statistical models are used to predict the growth of these defects as a function of time and stress levels [32]. As these microcracks expand, the loss associated with the fiber can also increase.

10.7 Software issues

Software reliability has received considerable attention in recent years, in part because of the ubiquitous presence of computers and microcontrollers in most electronic products and systems [33–36]. Obviously, software is a key part of communication networks. In fact, as we mentioned in Chapter 2, higher network layers become increasingly more software intensive. All the intelligence behind the control of traffic, routing, and switching is provided by either high-level software that run on dedicated computers or embedded firmware within microcontrollers and other similar devices.

Software also plays an important, albeit less central, role in the physical layer. In modern communication systems most critical blocks are expected to have some level of intelligence, and fiber optic modules transceivers are no exception. A clear example is the digital diagnostics requirements spelled out in the SFF-8472 standard. Essentially, they require an optical transceiver to provide a two-wire serial interface through which a number of critical parameters can be interrogated, and to some extent controlled. These parameters include static module identification formation, as well as real-time data about critical dynamic variables in the link such as the transmitted and received optical power.[4] These requirements can sometimes be fulfilled by special-purpose ICs. But a more flexible solution is to use a microcontroller to implement these functions. In other instances, a microcontroller may supervise the function of common blocks such as laser drivers and receivers and determine the behavior of the system in exceptional cases, such as during a laser thermal runaway situation. Such operations require supervisory software, and this makes software reliability a part of reliability concerns for fiber optic systems.

10.7.1 Software reliability

There are some important differences between software and hardware reliability [37]. Physical devices are subject to wear and tear, and therefore hardware failures eventually increase with time. On the other hand, there is no wear and tear in a piece of software. Unlike hardware failures, software failures result from logical flaws and programming errors that have human origin. Repairing a piece of hardware restores it to its original condition before failure, whereas fixing a bug in a piece of software results in a new piece of software.

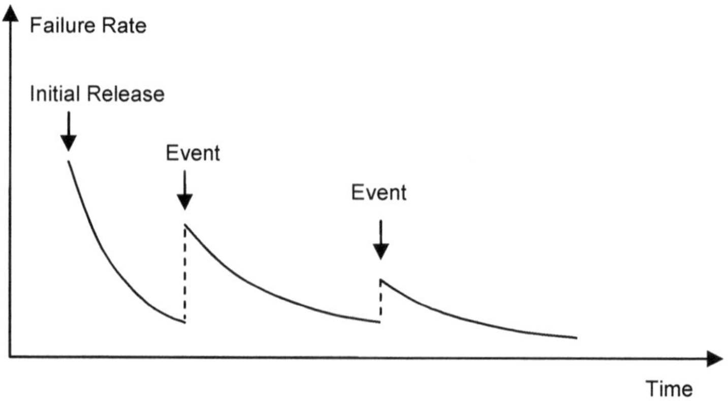

Fig. 10.8. Typical failure rate for a piece of software

[4] See Chapter 12 for more details on the SFF-8472 requirements.

As software complexity increases, it becomes virtually impossible for software designers to take into account all the logical possibilities. As a result, a new piece of software almost always starts its life with some hidden bugs, and only continuous testing can expose these hidden bugs. As the code matures, it is expected that more bugs are found, and therefore the reliability should only increase. However, usually within the life cycle of a piece of software there are critical times where it goes through major changes.

These change events may be internal upgrades to some piece of code within the software, changes in the operating conditions, a variation in the range or format of the inputs the software is handling, hardware changes, etc. Generally, the likelihood of new failure modes increases after such events, and a new cycle of debugging is needed to iron out these new bugs. The result is a somewhat different software failure curve compared to the standard bathtub curve for hardware failures. An example of such a software failure curve is shown in Fig. 10.8.

Although it is hard to make generalizations, this curve highlights the fact that passage of time in itself does not affect software reliability. Instead, the failure risk is influenced by certain kinds of events during the life cycle of a piece of software.

10.7.2 Failure rate reduction

Improving software reliability requires careful, systematic attention in all phases of software lifecycle. The lifecycle of a piece of software usually includes five stages: initial analysis, design, coding, testing, and operation [37]. Clear definition of problems, requirements, specifications, needed functionalities, and interfaces are some of the tasks during the initial analysis.

For the design phase, it is important to define the general architecture of the software in terms of clear building blocks. Here the operation of each block and its interface with other blocks as well as the overall control flow of the software should be clearly defined. The next phase of the design stage includes the implementation of the defined blocks. Each block must be implemented and tested carefully against the expected specifications and over the range of its intended inputs. Documentation is also an important aspect of design, because without proper documentation it becomes increasingly difficult to maintain and debug the software in later stages.

The coding phase is where the software is translated into actual lines of code in a particular language. Well-known programming practices, such as readability, simplicity, documentation, reusing previously tested codes, and structured programming, should be observed.

The next critical phase is testing. This is an especially important phase for reliability purposes. Here the modules in a program are integrated and the overall piece of software is tested against the requirements and specifications. Specifically, the software must be tested over a range of conditions that it may encounter during its operation. Any potential bugs found must be studied and completely understood before being fixed. In some cases, it is also necessary to test the software in the "real" systems and under real operating conditions to ensure compatibility.

Finally, during the operation phase, the software typically requires maintenance, documentation, potential upgrades, etc. In all these phases, design reviews, internal documentation, paying attention to the logic and avoiding sloppy patches, careful analyses of any failures, and design of specific stress tests to ensure software robustness in all working conditions are among the practices that improve the software quality and reliability.

10.8 Reliability quantification

So far we have been discussing design practices to improve reliability. Another aspect of reliability, to which we turn in this section, deals with reliability quantification of an existing device or system. Here we are primarily concerned with two questions. The first question deals with the ways reliability can be modeled and analyzed mathematically. To this end, a variety of probabilistic and statistical models have been proposed and used. The second question is how to estimate the reliability of an actual device. In this respect, we need to use the available reliability models and fit them with experimentally measured parameters from the device we are interested in. By combining the theoretical models with experimental data, we attempt to come up with reasonable estimates regarding the probability of failure for the device. We will deal with each of these questions in the following sections.

10.8.1 Statistical models of reliability: basic concepts

As we have noted before, reliability is an inherently probabilistic concept. Therefore, it is generally impossible to predict the exact failure time of a particular device or system, and we can only talk about the probability of failure for a given part. However, when the population of devices starts to become large, these probability statements take the form of statistical predictions which can be correlated with experimental observations.

For reliability estimations, the basic parameter we are interested in is the time it takes for a given device to fail or t. Because we can never know t exactly, we assume it is a random variable, with a probability density function given by $f(t)$. The exact expression for $f(t)$ depends on the model we choose to use to describe the failure mode(s) we are interested in and is not important for our current discussions. Based on $f(t)$, we can define two additional functions, $F(t)$ or the *failure function* and $R(t)$ or the *reliability function*, as follows:

$$F(t) = \int_{-\infty}^{t} f(\tau)d\tau \tag{10.6}$$

$$R(t) = 1 - F(t) = \int_{t}^{\infty} f(\tau)d\tau \tag{10.7}$$

$F(t)$ can be interpreted as the probability that the device fails before t, and $R(t)$ is the probability that the device does not fail before t. Note that $F(\infty)=1$, i.e., the device will eventually fail. Assuming we know $f(t)$, we can also express the probability of device failure between t_1 and t_2 as follows:

$$P(t_1, t_2) = \int_{t_1}^{t_2} f(\tau)d\tau = F(t_2) - F(t_1) = R(t_1) - R(t_2) \tag{10.8}$$

Another useful quantity is the *failure rate*. It is defined as the probability of failure between t_1 and t_2, assuming the device has not failed up to t_1, divided by $(t_1 - t_2)$. Failure rate is in general a function of time and, based on its definition, is given by

$$\lambda(t) = \frac{R(t_1) - R(t_2)}{R(t_1)(t_2 - t_1)} \tag{10.9}$$

Usually, we are interested in estimating the *average* time it takes for a device to fail. This parameter, known as *mean time to failure*, or MTTF, is the expectancy of the probability density function and can be obtained as

$$MTTF = \int_{0}^{\infty} t\, f(t)dt \tag{10.10}$$

In the next section we will discuss how these simple definitions are applied in simple statistical analysis of real parts. More advanced treatments can be found in other sources [2–5].

10.8.2 Failure rates and MTTF

As noted before, in order to apply statistical concepts to real devices and systems, we need to choose a model to describe the failure modes we are interested in. For many engineering systems, including electronic systems, observations have shown that the failure rate can often be approximated by the well-known bathtub curve, shown in Fig. 10.9.

Usually different failure types are distinguished. The first failure type known as infant mortality refers to failures that occur early during the life of the device. These failures are often quality related and result from poor components, inadequate screening, design failures, and lack of manufacturing process control. Random failures refer to unpredictable failures that occur completely by chance and more or less at a constant rate. In a random failure mode, the component or device shows no prior sign of failure, and there is no way to distinguish a part that is about to fail from one that will work much longer. The third phase, known as wear-out, includes failures that occur as a result of components aging and wear-out. Here the failure rate increases as a function of time. The sum of these different failure types result in the familiar bathtub curve.

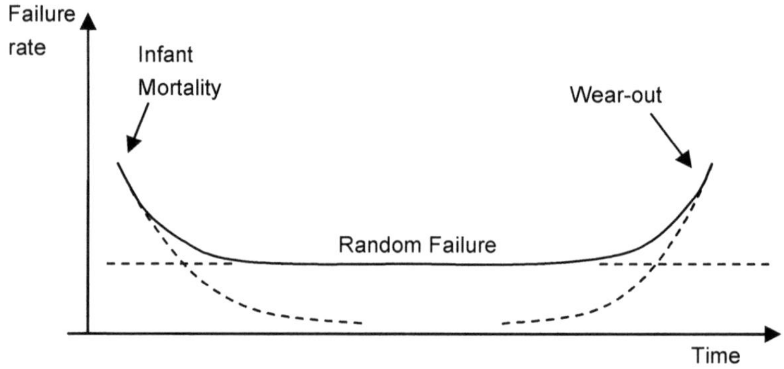

Fig. 10.9. The bathtub curve and different failure modes as a function of time

These three phases can be characterized by different probability density functions. For instance, infant mortality and wear-out can be described by gamma or normal distributions. In practice, oftentimes infant mortalities can be screened out by a well-designed screening process. Such a screening process typically involves temperature cycling the device in fast cycles in order to induce failures in weak parts. Here we will concentrate on the random failure phase, which is characterized by a constant failure rate.

The failure rate is given by Eq. (10.9). Assuming $\lambda(t)=\lambda$ (a constant), we can integrate Eq. (10.9) in the limit of $(t_1 - t_2)\to 0$ and obtain $R(t)$:

$$R(t) = e^{\lambda t} \tag{10.11}$$

This is the exponential distribution function that is usually taken to describe the random failures of electronic and optical components during their lifetime. In an exponential distribution, the MTTF, given by Eq. (10.10), is simply the inverse of failure rate:

$$MTTF = \frac{1}{\lambda} \tag{10.12}$$

Both MTTF and failure rate are commonly used parameters. Failure rates for common electronic and optoelectronic components are usually very small. For instance, the MTTF of a diode laser operating at normal room temperature may be in the range of hundreds of thousands of hours, corresponding to an exceedingly small number of failures in normal units of time (say, seconds or hours). As a result, device failure rates are usually characterized by a *failure in time* (FIT) number. By definition, a FIT number of 1 corresponds to one failure per 10^9 device-hour operation of the part:

$$1\,Fit = \frac{1\,Failure}{10^9\ hrs} \tag{10.13}$$

We should once again draw attention to the statistical nature and interpretation of failure rate, MTTF, and FIT numbers. For example, a FIT number of 10 does not mean that the part is going to fail after 10^8 hours of operation. For one thing, 10^8 is equal to 11,000 years! Nor is it a guarantee that it will work for 10^8. Instead, a more realistic interpretation is that in a sample size of 10^8 of these parts working under given conditions, we expect to observe, on average, one failure per hour.

A convenient advantage of FIT numbers is that they are additive. So for a system comprised of *n* components, if we can somehow characterize the FIT number of each component, we can estimate the FIT number for the system by adding the individual FIT numbers:

$$FIT_{SYS} = \sum_n FIT_n \tag{10.14}$$

The assumption behind Eq. (10.14) is that any failure in any of the components will result in a total system failure. In reality, this may not be true. For instance, if the design includes redundancies, the failure of one part may not result in the failure of the whole system. Moreover, not all components are equally critical to the function of the whole system, and it is possible for the system to continue to operate even if certain components of the system fail. Nevertheless, Eq. (10.14) highlights the usefulness of the concepts of failure rates and FIT numbers.

10.8. 3 Activation energy

One of the most important factors impacting reliability is temperature. Higher temperature implies that atoms and molecules oscillate with higher speeds, thus increasing the probability of collisions. This is especially important in chemical reactions. It is well known that the rate at which a wide variety of chemical reactions occur increases exponentially with increasing temperature. Likewise, the rate for almost all failure mechanisms that involve some form of chemical reaction (or other similar processes like diffusion and migration) also increases exponentially with temperature. This dependence is commonly modeled by the Arrhenius equation [38–40]:

$$K = A e^{-E_a/kT} \tag{10.15}$$

where K is the reaction rate, A is a fitting factor, k is Boltzmann's constant ($k = 8.62\times10^{-5}$ eV/K), E_a is the activation energy of the process, and T is the absolute temperature in Kelvin. The activation energy represents the energy it takes for a reaction to start.

In the previous section we discussed the concept of FIT number, but left open the question of actually measuring it. The Arrhenius equation provides a convenient way of measuring failure rates, and it is widely used as the basis for predicting failure rates for electronics and optoelectronic devices. Because of the exponential dependence of failures on temperature, rising temperature will significantly increase the failure rate of a device whose expected FIT number is too low to allow for real measurements. Thus, at high temperatures, the FIT numbers could be measured with acceptable confidence in a reasonable amount of time. Consequently, to relate the failure rate at two temperatures, Arrhenius equation can be written as

$$\frac{FIT(T_1)}{FIT(T_2)} = Ae^{-\frac{E_a}{k}\left(\frac{1}{T_1}-\frac{1}{T_2}\right)} \tag{10.16}$$

Equation (10.16) is the standard method of relating the results of accelerated aging tests at high temperatures to realistic operating temperatures. However, it should be noted that the proper value of activation energy E_a used in Eq. (10.15) is a function of device structure and the failure mode that we are trying to measure.

It is worth noting that a higher activation energy implies more temperature dependence for the associated failure mechanism. That is, if a failure mechanism has a higher activation energy, changes in temperature will cause a more dramatic change in failure rates. On the other hand, a lower activation energy means that the failure rate is relatively less sensitive to temperature change. Table 10.3 summarizes common values of activation energy for optoelectronic devices [14]. Note that the activation energy is generally lower for random failures compared to wear-out failures. This is because wear-out failures are expected to be more temperature dependent, whereas random failures are assumed to be mostly due to manufacturing deficiencies and thus less temperature dependent.

Table 10.3. Activation energies for random and wear-out failures for common optoelectronic devices

	E_a (eV)	
	Random failure	Wear-out failure
Lasers [1]	0.35	0.4
LEDs [1]	0.35	0.5
Receivers [2]	0.35	0.7
External modulators	0.35	0.7
Electro-absorption modulators	0.35	0.4

[1] Both device level and module level
[2] Includes photodiodes and receiver modules

10.9 Summary

Reliability is an important concern in all fields of engineering. Reliability is defined as the probability of a no-failure occurrence during the expected life time of a part under normal operating condition. Therefore, it is clear that embedded within the notion of reliability is the concept of probability.

Reliability discussions can be divided into three categories. One category has to do with design practices that will likely improve the reliability of a device or system. The design phase is the best time to address reliability risks, because at the design phase the risks are easier to identify and much cheaper to fix. The second category has to do with mathematical and statistical models of reliability. The third category deals with specific tests aimed at measuring the reliability of an actual device or system.

We started with a review of design practices aimed at improving reliability. A sound design approach requires dividing the system under question into smaller blocks. For fiber optic systems, a reasonable division includes optical transmitter, optical receiver, control circuits, optical subassemblies, electrical and optical interfaces, and power management. Each of these sub-systems can be designed and optimized individually over a sufficiently large range of parameters. It is easier to identify potential failure mechanisms in such smaller blocks than it is to study a system as a whole. Moreover, good design practice requires careful testing.

Without sufficient and well-planned testing, it is hard if not impossible to discover all design flaws purely on theoretical grounds.

We also divided the reliability issues into categories of electrical, mechanical, thermal, optical, and software. This categorization helps in focusing attention on the different aspects that a designer must keep in mind, because oftentimes focus on problems in one domain may increase the risk of creating reliability problems in other domains. This is especially true in cases where there is a trade-off between performance and reliability. Having the "big picture" in mind is always useful to achieve the optimal balance between reliability and performance.

The second and third categories involved quantification and testing of reliability. We gave a brief introduction to the definition of the main statistical functions used in reliability analysis. Using these functions, we defined several practical and commonly used reliability parameters, including MTTF and FIT number. Finally, we discussed the Arrhenius equation and the concept of activation energy. These expressions provide a way of estimating the failure rate for a given device by measuring the failure rate at higher temperatures and using the results to calculate the failure rate at normal operating conditions.

References

[1] R. Barlow, *Engineering Reliability*, American Statistical Association and the Society for industrial and Applied Mathematics, Philadelphia, PA, 1998

[2] A. Birolini, *Reliability Engineering*, Springer, Berlin, 2004

[3] P. D. T. O'Connor, D. Newton, and R. Bromley, *Practical Reliability Engineering*, John Wiley & Sons, New York, 2002

[4] B. S. Dhillon, *Design Reliability, Fundamentals and Applications*, CRC Press, Boca Raton, FL, 1999

[5] M. G. Stewart and R. E. Melchers, *Probabilistic Assessment of Engineering Systems*, Chapman and Hall, London, 1997

[6] L. A. Doty, *Reliability for the Technologies*, Industrial Press, New York, 1989

[7] A. Porter, *Accelerated Testing and Validation*, Elsevier Inc., Amsterdam, 2004

[8] MIL-HDBK-217, *Reliability Prediction of Electronic Equipment*, Department of Defense, Washington, DC

[9] MIL-HDBK-251, *Reliability/Design, Thermal Applications*, Department of Defense, Washington, DC

[10] MIL-HDBK-338, *Electronics Reliability Design Handbook*, Department of Defense, Washington, DC

[11] MIL-STD-785, *Reliability Program for Systems and Equipment, Development and Production*, Department of Defense, Washington, DC

[12] MIL-STD-790, *Reliability Assurance Program for Electronics Parts Specifications*, Department of Defense, Washington, DC

[13] GR-78, Generic Requirements for the Physical Design and Manufacture of Telecommunications Products and Equipment, Telcordia, 1997

[14] GR-468, Generic Reliability Assurance Requirements for Optoelectronic Devices Used in Telecommunications Equipment, Telcordia, 2004

[15] Institute for Interconnecting and Packaging Electronic Circuits (IPC), www.ipc.org

[16] G. Chen et al., "Characterizing diodes for RF ESD protection," IEEE Electron Device Letters, Vol. 25, pp. 323–325, 2004

[17] H. Ishizuka, "A study of ESD protection devices for input pins. Discharge characteristics of diode, lateral bipolar transistor, and thyristor under MM and HBM tests," IEEE Transactions on Components, Packaging, and Manufacturing Technology, Part C, Vol. 21, pp. 257–264, 1998

[18] J. Lee et al., "Design of ESD power protection with diode structures for mixed-power supply systems," IEEE Journal of Solid-State Circuits, Vol. 39, pp. 260–264, 2004

[19] J. S. Huang, "Temperature and current dependences of reliability degradation of buried heterostructure semiconductor lasers," IEEE Transactions on Device and Materials Reliability, Vol. 5, pp. 150–154, 2005

[20] M. Fukuda, "Laser and LED reliability update," Journal of Lightwave Technology, Vol. 6, pp. 1488–1495, 1988

[21] A. R. Goodwin, et al., "The design and realization of a high reliability semiconductor laser for single-mode fiber-optical communication links," Journal of Lightwave Technology, Vol. 6, pp. 1424–1434, 1988

[22] M. Ettenberg and H. Kressel, "The reliability of (AlGa)As CW laser diodes," IEEE Journal of Quantum Electronics, Vol. 16, pp. 186–196, 1980

[23] D. Kim et al., "Design and fabrication of a transmitter optical subassembly (TOSA) in 10-Gb/s small-form-factor pluggable (XFP) transceiver," IEEE Journal of Selected Topics in Quantum Electronics, Vol. 12, pp. 776–782, 2006

[24] M. S. Cohen et al., "Low-cost fabrication of optical subassemblies," Proceedings of the 46th Electronic Components and Technology Conference, pp. 1093–1100, 1996

[25] TIA-455-160-A, IEC-60793-1-50 (FOTP-160): Measurement Methods and Test Procedures-Damp Heat, TIA, 2003

[26] GR-326, Single mode Optical Connectors and Jumper Assemblies, Telcordia, 1999

[27] TIA-455-85-A (FOTP-85) Fiber Optic Cable Twist Test, TIA, 2005

[28] TIA/EIA-455-88 (FOTP-88) Fiber Optic Cable Bend Test, TIA, 2001

[29] TIA/EIA-455-158 (FOTP-158) Measurement of Breakaway Frictional Force in Fiber Optic Connector Alignment Sleeves, TIA, 1997

[30] D. Kececioglu, *Robust Engineering Design-by-Reliability with Emphasis on Mechanical Components and Structural Reliability*, DEStech Publications, Lancaster, PA, 2003

[31] O. Vinogradov, *Introduction to Mechanical Reliability, A Designer's Approach*, Hemisphere Publishing Corporation, New York, 1991

[32] G. M. Bubel and M. J. Matthewson, "Optical fiber reliability implications of uncertainty in the fatigue crack growth model," Optical Engineering, Vol. 30, 737–745, 1991

[33] A. Immonen and E. Niemela, "Survey of reliability and availability prediction methods from the viewpoint of software architecture," Software and System Modeling, Vol. 7, pp. 49–65, 2008

[34] M. Hecht, "The role of safety analyses in reducing products liability exposure in "Smart" consumer products containing software and firmware," Annual Reliability and Maintainability Symposium, pp. 153–158, Jan. 2003

[35] J. Musa, *Software Reliability Engineering: More Reliable Software, Faster and Cheaper*, AuthorHouse, Bloomington, IN, 2004

[36] D. Peled, *Software Reliability Methods*, Springer, Berlin, 2001

[37] H. Pham, *Software Reliability*, Springer, Berlin, 2000

[38] M. Tencer, J.S. Moss, and T. Zapach, "Arrhenius average temperature: the effective temperature for non-fatigue wearout and long term reliability in variable thermal conditions and climates," IEEE Transactions on Components and Packaging Technologies, Vol. 27, pp. 602–607, 2004

[39] M. S. Cooper, "Investigation of Arrhenius acceleration factor for integrated circuit early life failure region with several failure mechanisms," IEEE Transactions on Components and Packaging Technologies, Vol. 28, pp. 561–563, 2005

[40] G. Hartler, "Parameter estimation for the Arrhenius model," IEEE Transactions on Reliability, Vol. 35, pp. 414–418, 1986

Chapter 11
Test and Measurement

11.1 Introduction

This chapter deals with measurement and characterization of the critical parameters for fiber optic devices and links. These parameters include quantities such as optical power, extinction ratio, rise and fall time, bit error rate, wavelength, and spectral width. Measurement methods are crucial for evaluation of existing devices and systems as well as for validation and debugging of new designs. Without accurate test and measurement methods, the critical relationship between theory and real world breaks down, effectively rendering most engineering efforts meaningless.

We will start this chapter by an overview of the various elements that are required for a successful test or measurement experiment. Among these are a clear understanding of the physics involved in the test and a knowledge of the working principle and limitations of the test equipment used. Moreover, the set-up used in experiment must be designed carefully so that the parameter or quantity that is being tested is isolated from the influence of other factors that can disturb the measurement.

The reminder of the chapter deals with some of the most common tests that characterize fiber optic devices and links. These include optical power measurement, time domain analysis of optical waveform, optical and electrical spectrum analysis, sensitivity tests, as well analog tests for parameters such as modulation depth and harmonic distortion. In each section we will discuss both the quantities of interest that need to be tested or measured and the operating principles behind the common test instruments that are used for these measurements.

11.2 Test and measurement: general remarks

As noted above, test and measurement techniques are the crucial links that bridge the gap between theory and reality. Theories are always based on models that involve simplifications and approximations of reality, and it is through experiment, test, and eventually measurements that the relevance and success of these models and the theories that are based on these models can be determined. This is true both in pure science and in applied fields such as engineering. As a result, a great deal of effort has gone into the theories and techniques involved in test and measurements, including those related to fiber optics [1–6].

It is worthwhile to have a brief discussion about the different elements that are needed for a successful experiment design, be it a simple measurement or a complex system level test. Figure 11.1 summarizes some of these factors.

M. Azadeh, *Fiber Optics Engineering*, Optical Networks,
DOI 10.1007/978-1-4419-0304-4_11,

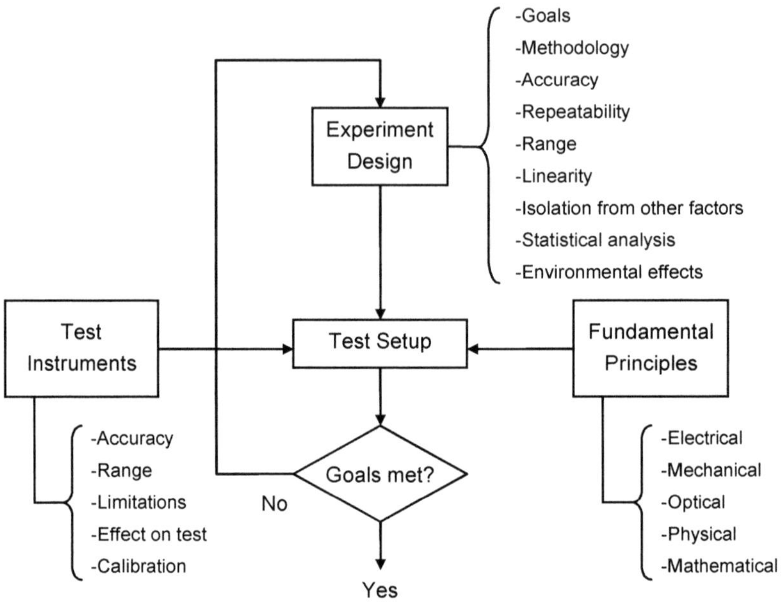

Fig. 11.1. Elements of a measurement experiment

An experiment typically involves a design step. First and foremost, the goal of the experiment must be defined clearly. This goal can be a simple quantity measurement or a complex system-level test involving many measurements. Once the goal is identified, a test methodology must be chosen that can clearly achieve the defined goal. We can think of test methodology as a detailed flow chart that spells out every intermediate step process, instrument, software, measurement, calibration, algorithm, etc., that will be used during the experiment. The methodology must take into account factors such as the target accuracy of the tests, the range and linearity of the parameters involved, and the effects of other system or environmental parameters on the experiment. Moreover, the experiment design should include applicable statistical provisions and analysis. For instance, oftentimes the accuracy of an experiment or measurement can be improved by averaging the results over a number of repeated tests or a larger sample size.

Any experiment is based on a set of theories, and knowledge of these theories is necessary both for the design of the experiment and for the interpretation of the results. For instance, fiber optic-related tests are often based upon a mixture of electrical, physical, and optical theories. Moreover, oftentimes an experiment is based on certain mathematical foundations, theorems, or principles. A frequently encountered example in engineering is the Fourier analysis that relates time and frequency domain quantities.

Another crucial element that plays a critical role is the test equipment and instruments. It is important to take into account the inherent shortcomings of all test equipment that show up in the form of limitations such as inadequate range, insufficient bandwidth, noise floor, and limited accuracy. Another important point to consider is that in order to conduct a measurement, the test equipment oftentimes has to disturb the very quantity that it is trying to measure. For example, the input impedance of an oscilloscope port or the parasitics of the oscilloscope probe can disturb the voltage of the node whose waveform is being measured.

The experiment design should also take into account the issue of standards. Standards of measurement are especially key in those areas where the quantity to be measured has a strong element of "definition" to it. This means the measured value strongly depends on how that quantity is defined or tested, or on how the test results are interpreted. There are a wide range of standards for fiber optic tests and measurements, for instance the FOTP series of documents from the Telecommunication Industry Association (TIA) [7]. We will return to the subject of standards in Chapter 12.

Maintenance and calibration of test instruments is also very important [8]. Without periodic calibration, a test instrument can deviate from its expected performance and yield erroneous measurements. As a result, and if possible, it is good practice to have some means of independently verifying the calibration of test instruments. For example, the accuracy of an optical power meter can be verified if a standard laser source is available. Alternatively, it is always possible to compare the measurement results obtained through different instruments if more than one piece of instrument for a particular measurement is available. Obviously if two instruments report different values for the same measurement, they both need to be calibrated and verified because at least one of them is out of calibration.

Figure 11.1 is intentionally organized in the form of a flow chart with a main loop. The reason is that a well-designed experiment should take into account the possibility of something going wrong and not meeting the goals of the experiment. In that case, some or all of the above-mentioned factors must be revisited, and the root cause of the problem should be identified. The problem can result from a simple reason, for instance a wrong setup on an instrument, or it could be due to deeper reasons, for example a flaw in the theory behind the experiment. Thus, it may take several iterations before the experiment is tailored and tuned so that the defined goal can be achieved.

In the reminder of this chapter, we will review the test procedures and instruments for measuring the most common parameters in fiber optic links.

11.3 Optical power

One of the most fundamental parameters in optical links is optical power. For instance, the average power of an optical transmitter is a critical measure of the physical distance its signal can reach. Similarly, the sensitivity of an optical receiver, which is the minimum optical power that the receiver can work with, is a figure of merit that determines the link budget that the receiver can support.

As noted in Chapter 3, optical power can have several meanings. The instantaneous power associated with the optical frequency is a quantity that oscillates at twice the optical frequency. Therefore it cannot be measured directly and it is almost never used in any practical application. More practical quantities related to power include average and instantaneous power. In fiber optics, oftentimes average power is the primary quantity of interest.

The average optical power is a convenient parameter that is widely used to characterize, among other things, light output of transmitters and sensitivity of receivers. The optical power is usually measured in units of decibel-milliwatt, which we introduced in Chapter 1, and which we repeat here for reference:

$$P_{\mathrm{dBm}} = 10\log_{10}\left(P_{\mathrm{mW}}\right) \tag{11.1}$$

Here P_{dBm} is power in units of dBm, and P_{mW} is power in mW. The convenience of average power lies in the fact that it can be measured easily with an *optical power meter* using a *broad area detector*, schematically shown in Fig. 11.2.

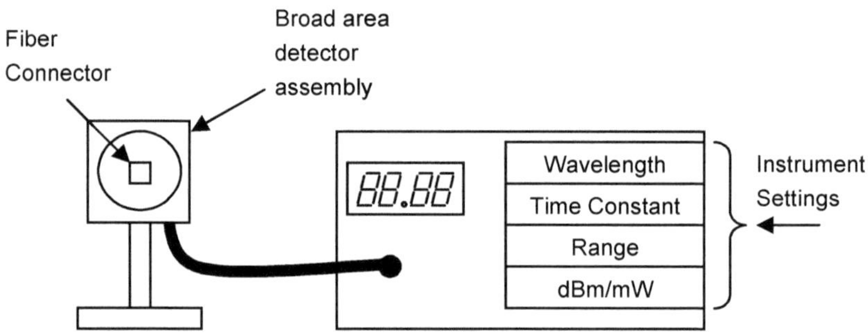

Fig. 11.2. Optical power meter based on a broad area detector

Using a large detector has the advantage of reducing the dependence of measurement on coupling efficiency, because all the optical power exiting from a fiber is collected by the detector. Naturally, a large detector has a slower time response, but this is not an issue when it comes to measuring average power. Most optical power meters have several instrument settings that need to be considered for accurate measurement.

• **Wavelength:** It is important to set the wavelength of the light that is to be measured correctly. This is because the responsivity of photodetectors is not constant across wavelengths, and the instrument uses the wavelength setting in order to calibrate the responsivity. Without correct wavelength settings, the power measurement may be off by as much as 1 dB.

• **Time Constant:** It is usually possible to set the time constant over which power measurements are averaged. The time constant settings can change from a fraction of second to seconds. Using a longer time constant setting is advantageous in terms of filtering noise effects and is preferred whenever possible.

• **Range:** Most modern equipment automatically select the correct range for a particular measurement. However, if the equipment does not include this feature or if this feature is optionally disabled (for instance when the device is controlled through a GPIB interface) the correct range must be selected. The typical dynamic range for most power meters is at least tens of decibels. When the magnitude of the power is completely unknown, it is a good practice to start from a higher range and select lower ranges afterward.

• **dBm/mW:** It is usually possible to select between logarithmic and linear measurements. As noted before, in most cases power measurements are carried out in logarithmic units. The dBm logarithmic scale is very convenient when it comes to measurements that involve comparing power levels, dynamic range, gain or loss, and power penalties.

11.4 Optical waveform measurements

A wide variety of signal parameters are related to the optical waveform. In general, these parameters can be divided into two categories. We discussed these parameters in Chapter 3 in terms of the optical signal itself. In this chapter, our focus is on measurement aspects, especially with respect to the instruments that are used for these measurements.

11.4.1 Electrical oscilloscopes with optical to electrical converter

One way to measure these parameters is to use an electrical oscilloscope in conjunction with an optical to electrical (OE) converter [9]. Naturally, the choice of the bandwidth of the scope, the optical to electrical converter, and the interconnecting signals is important. Lack of sufficient bandwidth on any of these components can cause an inaccurate representation of the time features of the signal. In general, the rise-time (or fall-time) value measured on a scope is given by

$$t_{measured}^2 = t_R^2 + t_{OE}^2 + t_{scope}^2 + t_{cable}^2 \tag{11.2}$$

where t_R is the actual rise-time (or fall-time) of the optical signal and t_{OE} , t_{scope} and t_{cable} are related to the nominal bandwidth of the OE converter, oscilloscope, and the interconnecting cables as[1]

$$t = 0.35 / BW \tag{11.3}$$

From Eq. (11.2), it can be seen that to capture high-frequency features of a signal accurately each of the components used in a setup must have a much wider bandwidth.

Using an OE converter requires additional calibration of the vertical axis, because in electrical oscilloscopes the vertical axis is calibrated in volts. One way to achieve this calibration is by measuring the average optical power with an optical power meter and correlating the results with the average voltage reading in the oscilloscope.

Modern oscilloscopes are often based on digital sampling of the signal. A high-speed analog to digital converter converts the signal into a series of digital data which is then stored in the memory and shown on the display. Working with such sampling scopes requires some caution. For instance, changing the time scale may cause some high-frequency details of the signal to be lost. Also, aliasing effects, resulting from under-sampling of signals, are a possibility [10–12].

11.4.2 Digital communication analyzer (DCA)

Measurement of high-speed optical signals is an active topic and both optical and electrical domain solutions have been proposed [13]. In most applications, digital sampling techniques are used to capture optical waveforms [14,15]. Specifically, manufacturers have come up with digital scopes specifically targeted for fiber optic applications. Sometimes called digital communication analyzer (DCA), these instruments are essentially a digital scope with a front-end optical-to-electrical converter and equipped with additional software to accomplish a variety of measurements.

From the stand point of capturing the optical waveform, in optical communication the challenge is that the involved frequencies may be too high, thus requiring extremely high-speed ADCs at the front end. For example, to represent a 10 Gbps signal accurately, we may need up to 20 GHz of analog bandwidth, which translates to at least 40, but preferably more, Giga samples of sampling per second. Achieving such sampling rates is difficult and expensive. For higher speed signals, it could simply be impractical.

DCAs solve this problem by using a sampling technique known as *equivalent time sampling* (ETS). [16] To do this, they take advantage of the fact that oftentimes we are not interested in the shape of each and every individual bit or wave-

---

[1] As noted in Chapter 3, the relationship between rise and fall times and bandwidth depends on the shape of the edges. Equation (11.3) is based on an exponential shape, which is more typical for RC time constants encountered in electrical components.

form in a signal. Instead, we would like to know certain properties of the signal as they relate collectively to all the bits or waveforms. In digital signals, this is typically done by the eye pattern, which is built from superposition of waveforms from many individual bits.[2] For instance, the rise-time measured in an eye pattern can be interpreted as a measurement averaged over many individual rise-times. However, the utility of such "averaged" measurements is not limited to digital signals and any repetitive pattern can be captured using ETS.

The principles of ETS are shown in Fig. 11.3, which shows a repetitive but otherwise arbitrary waveform. To capture the waveform, a trigger input that is synchronized with the repetition rate of the main signal is needed. Thus, each trigger event indicates the starting point of another window within which the waveform can be sampled. Next, the waveform is sampled once within each window, while the sampling point is moved forward in each subsequent window. In this way, the waveform can be scanned from beginning to the end after many windows. The sampled points can be stored in memory and used to reconstruct the shape of the waveform. It is evident from this description that the reconstructed waveform is not displayed as a function of real time, and that is why the time axis is called equivalent time.

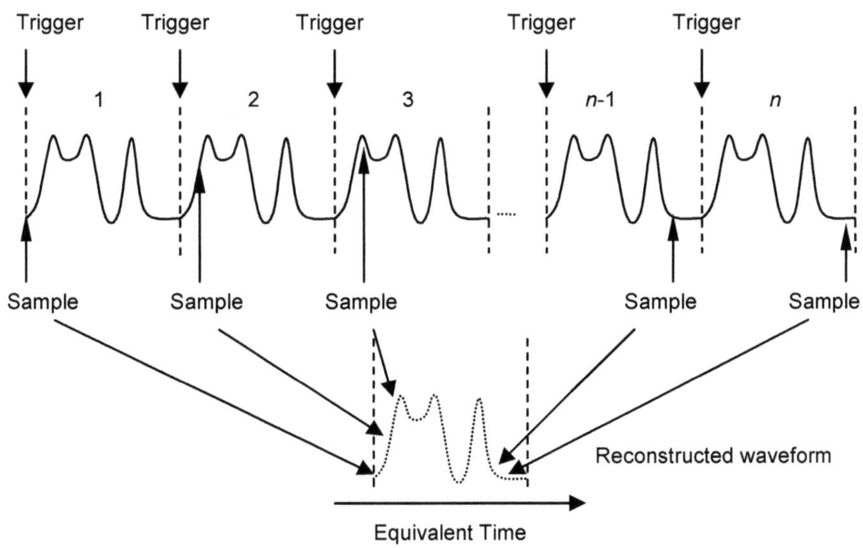

Fig. 11.3. Reconstruction of a fast waveform by equivalent time sampling

[2] For a more detailed discussion of eye patterns, see Chapter 3.

Typically, a DCA has two display modes. In the oscilloscope mode, the stored samples from the waveform are displayed only for a limited time before being discarded and replaced by new points. The display's *persistence time* can be set anywhere from a fraction of second to a few seconds. When the persistence time is short, the displayed waveform responds quickly to the changes in the actual waveform. When the persistence time is long, the displayed waveform maintains features of the physical waveform for a longer time. It is also possible to use an *infinite persistence* mode, where the display reflects a captured point forever (or until the display is erased). This mode is sometimes called the *eye-mask mode* and can be utilized to analyze the long-term behavior of the waveform. For example, the DCA may be left to run in this mode for days, after which the display contains a visual record of the behavior of the waveform. We will discuss this mode further in a later section.

As can be seen from Fig. 11.3, the waveform is reconstructed slowly and after many triggers, and this can only be done because an independent trigger input is (and should be) available. Thus, unlike traditional oscilloscopes, the instrument cannot trigger on the signal itself (i.e., edge trigger, level trigger, etc.) Moreover, only those features of the waveform are captured that are repeated in every waveform. For example, if a singular spike occurs in only one window, it is unlikely for it to coincide with the sampling point in that window, and therefore it will not be captured. However, the advantage of this technique is that the ADC at the front end of the scope needs only to sample once in a window. This would be a much lower rate compared to the required "real-time" sampling rate, dictated by the Nyquist theorem.[3] As a result, signals with much higher frequency components can be displayed and analyzed using this technique. Signals with bandwidths as high as 80 GHz can be handled by modern DCAs [17].

Another feature that makes DCAs particularly useful for fiber optic measurements is that most of them have standard "plug-in" modules that incorporate a calibrated high-speed optical to electrical converter. The instrument can display the waveform directly in the optical domain. Moreover, the manufacturers provide a variety of built-in software tools for measuring a wide range of parameters. In the next section, we will discuss some of these parameters briefly.

To increase the accuracy of waveform measurements, the signal-to-noise ratio at the input of the DCA must be maximized. This is because a DCA, like any other receiver, has an *input sensitivity* and a *noise floor*. If the amplitude of the signal gets close to the noise floor, the measurements will no longer be accurate. In general, the optical power of the signal should be kept close to the maximum level the DCA can accept. Consequently, the signal-to-noise ratio will be maximized, and measurements will be more accurate. Of course, this is only possible when the signal has enough optical power to begin with, so that by adding an optical attenuator the signal can be attenuated to the appropriate level for the DCA. On the other

[3] According to Nyquist theorem, in order to accurately reconstruct a signal, it should be sampled at a rate at least twice as high as the highest frequency components of that signal.

hand, if the optical power is already low compared to DCA's noise floor, an optical amplifier may have to be used to increase the signal-to-noise ratio.

Most DCAs have several low-pass filters built in that can be applied to the waveforms. These filters are defined by various standards and are generally used in eye mask measurements for digital signals (we will discuss eye mask tests later in this chapter). The effect of these filters is to remove high-frequency components of the signal. For instance, a typical OC48 filter is a low-pass Bessel–Thomson filter with a bandwidth of 1.87 GHz. Naturally, using a low-pass filter affects the optical eye. For example, rise and fall times will be slower, high-frequency laser relaxation oscillations will be removed, and in general the eye diagram will look cleaner and smoother.

An eye pattern, after passing through the appropriate filter, is effectively band limited according to the bandwidth requirements of the corresponding receiver. Thus, for instance, an eye pattern after going through an OC48 filter looks like what an OC48 receiver "sees" from that signal. Because of this band limiting effect, it should be ensured that the right filter is selected for each application.

11.4.3 Amplitude related parameters

Using a DCA, a number of parameters related to the amplitude (or instantaneous power) of an optical signal can be characterized. As noted in Chapter 3, these parameters mainly include optical "1" and "0" levels, overshoot, optical modulation amplitude (OMA), and extinction ratio. Alternatively, the built-in software provided with a DCA can be used. A point to notice in amplitude measurements is that they are referred to the average value of a distribution. Fortunately, most DCAs are equipped with built-in software that can capture a histogram of this distribution within any given window in an eye pattern. For example, Fig. 11.4 shows a typical histogram of the 0 and 1 levels within a predefined box in an optical eye diagram.

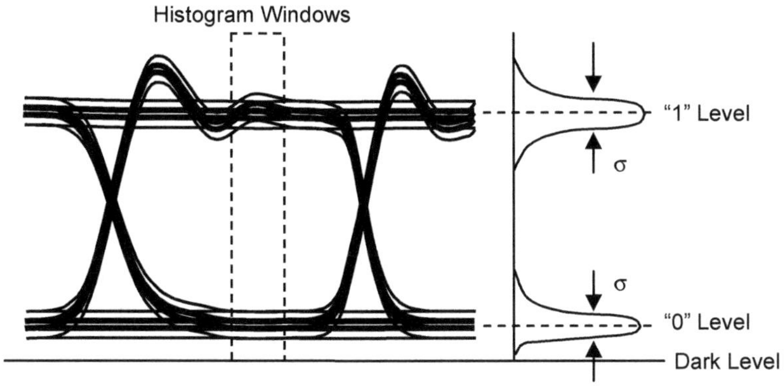

Fig. 11.4. Histogram of the distribution of 0 and 1 levels

The optical 1 and 0 levels correspond to the mean value of the distribution. By setting the corners of the histogram window, the statistical details of the signal can be examined more accurately.

An important parameter in digital signals is extinction ratio (ER), defined as the ratio of an optical "1" to an optical "0" level, in decibels. Most DCAs already have a predefined routine for measuring extinction ratio. However, in spite of the conceptual clarity of ER, accurate ER measurements are infamously difficult and affected by a number of uncertainties [18–20]. Among these is noise in the 0 and 1 levels, which makes it difficult to assign an exact value to the 0 and 1 levels. Also, signal oscillations especially at the 0 state can affect ER measurements, because they impact the exact value of the 0 state. Moreover, it becomes increasingly difficult for DCAs to resolve very high ERs, say, beyond 15 dB. This is because at such high ERs the 0 level gets very close to the actual dark level, or the noise floor of the DCA. Consequently, it becomes increasingly difficult to measure higher ERs.

11.4.4 Time-related parameters

Main time-related parameters include rise and fall time, eye width, and jitter. As noted before, it is important to ensure that when performing these measurements the equipment has sufficient bandwidth and that no additional filters are added to the signal path. In a DCA any time measurement, like amplitude measurement, is referenced to a distribution of events in time. Therefore, using time histogram features available in most DCAs is a valuable tool for signal analysis.

This is very obvious in jitter measurements. Figure 11.5a illustrates a typical jitter histogram for a representative eye pattern. If the distribution is Gaussian, it can be readily characterized by its rms and peak-to-peak value. The rms value refers to the standard deviation width of the distribution, and the peak-to-peak value, although theoretically unbounded, can be limited within a certain number of standard deviations in either side of the mean value. The number of σs used depends on the bit error rate.[4] Note, however, that jitter distribution is not always Gaussian. More often than not, jitter can be broken down into a deterministic and a random component. The random component can be characterized by a Gaussian, but the deterministic component is a function of the particular system under test.

For example, if the eye pattern shows a "double edge," the deterministic component can be attributed to the time delay between the two edges. This is shown in Fig. 11.5b. It can be seen that a time histogram gives a much more complete picture of the nature of jitter. The above discussion sets the basis for other time-related measurements as well. For example, two other important parameters in a digital signal are rise time and fall time (t_R and t_F). For optical signals, they reflect the speed of the modulating electrical circuit as well as the frequency response of the laser or the external modulator.

---

[4] For a more detailed discussion of jitter refer to Chapter 3.

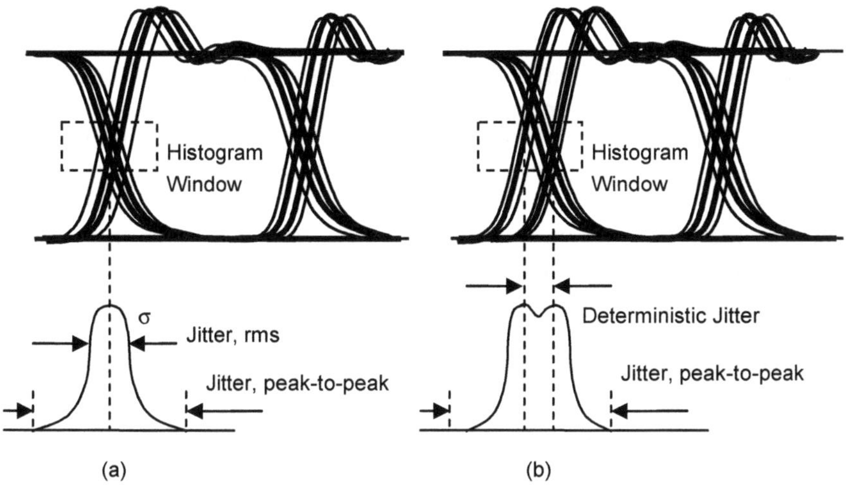

Fig. 11.5. Jitter histograms: (**a**) normal jitter distribution and (**b**) double-peak distribution

As shown in Fig. 11.6, the transition time for each of the rising and falling edges is a distribution and the mean value of the distribution should be used as the reference point for the edge. Both rise-time and fall-time can be measured either from 10 to 90% point or from 20 to 80% in a signal. Using the 20–80% levels typically provides more consistent results. On the other hand, using the 10–90% levels renders the measurement more sensitive to signal distortions and noise. For instance, in Fig. 11.6, t_R and t_F are based on the 90 and 10% levels. It can be seen that in this case the fall-time is disproportionately affected by a plateau in the falling edge. On the other hand, the 80–20% fall-time is much less sensitive to the same plateau.

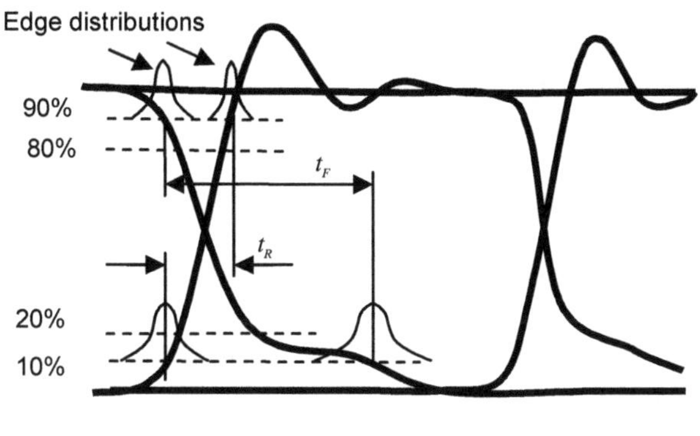

Fig. 11.6. Rise and fall-time measurement based on time histograms

11.4.5 Mask measurement

In a mask measurement, the shape of an eye pattern is laid against a pre-defined mask, which essentially verifies that the eye pattern is sufficiently open in time and amplitude [18]. It also ensures that the amount of noise in the 0 and 1 levels does not exceed the value set by the mask. In general, an eye mask determines the minimum acceptable "quality" for a particular digital signal, such that a non-conforming signal results in mask violations or hits. Figure 11.7 illustrates a mask test for an ideal eye diagram. Mask tests are typically performed in the infinite persistence mode of a DCA using the appropriate low-pass filter for that application. This will usually remove overshoot and laser relaxation oscillations that are normally out-of-band for many lower data rate applications, and in general will yield a "cleaner" eye pattern.

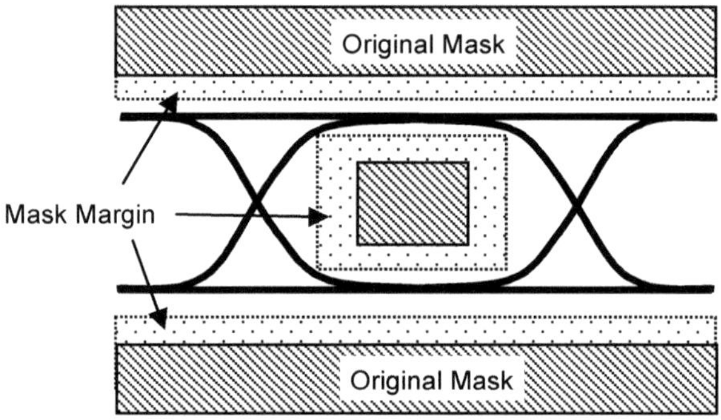

Fig. 11.7. Typical eye mask measurement

However, in-band distortions will pass through, and if they are large enough, they can cause mask hits. In cases where there is no mask hit, it is common to define a *mask margin* value. The mask margin can be measured by increasing the size of the mask and making it tighter around the eye pattern. The mask is made larger until it overlaps with the actual eye at some point. Mask margin is the percentage by which the mask has to increase in size before the first hit occurs. In Fig. 11.7, the original mask is shown by dashed lines and the extra margin is shown by the dotted area.

Certain factors can impact the accuracy of a mask test. As noted before, it is important to take into account the noise floor of the DCA itself. If the optical signal is weak and close to the noise floor, the mask margin can suffer. Moreover, any phase change between the data and the trigger signal will directly translate to a shift (and therefore "smearing") in the displayed eye. High-frequency phase shift, or jitter, will lower the mask margin. Low-frequency phase shift, also known as wander, can affect long-term measurements. In short, a mask test is a simultaneous reflection of both the signal itself and the noise floor of the equipment and the phase noise between the signal and the trigger.

11.5 Spectral measurements

The shape of a waveform in time domain does not provide any information about the spectrum of the signal. As discussed in Chapter 4, different laser structures have different spectral characteristics, resulting in differences in wavelength, spectral width, and spectral shape. The spectral characteristics of a signal have a direct impact on the amount of attenuation and dispersion the signal experiences as it propagates through the fiber. In this section we will discuss the principles of optical spectrum measurements.

11.5.1 Optical spectrum analyzer (OSA)

The main instrument used in the analysis of an optical spectrum is an optical spectrum analyzer (OSA). [21–24] In principle, an OSA works the same way as an electrical spectrum analyzer, with the difference that the frequencies involved are at much higher optical frequencies [25]. In both cases, the signal is passed through a variable narrowband filter, which only passes a very narrow range of frequencies from the signal. The filter is then swept across the frequencies of interest and its output is displayed as a function of the frequency. Figure 11.8 illustrates the concept behind a generic spectrum analyzer.

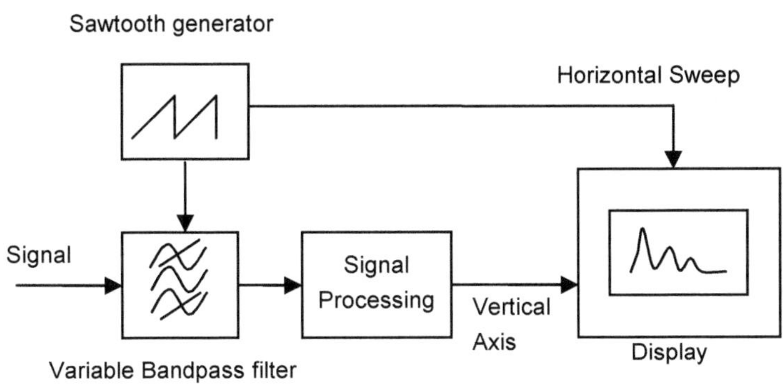

Fig. 11.8. Schematic diagram of a frequency spectrum analyzer

Because the phase information of the signal is lost, the equipment effectively displays the energy spectrum density of the signal, $S(f)$, given by

$$S(f) = |X(f)|^2 \tag{11.4}$$

where $X(f)$ is the Fourier transform of the time domain signal $x(t)$. In reality, the displayed signal is the convolution of the response of the filter and the spectrum of the signal. Thus, Eq. (11.4) holds if the bandwidth of the variable filter is very narrow compared to the spectral features of the signal. If this assumption does not hold, the displayed signal would be smoothed out as a result of the wider bandwidth of the filter and the spectral details of the signal will be lost.

In an optical spectrum analyzer, the variable filter must be capable of tuning its pass-band across the optical frequency range of interest. Many optical devices can act as an optical filter. For instance, we know that a Fabry–Perot resonator acts as a wavelength-selective device based on its length [26]. Thus, by changing the length of the cavity we can implement a variable optical filter. Another possibility is using a prism to split light into its constituent spectral components and then using a narrow slit to select only a specific band for further analysis. The advantage of this approach is that the prism maps the spectrum into the spatial domain, and therefore the filter has to be a spatial filter, i.e., a slit. However, a prism is not the best device to decompose the spectrum, because it has not enough dispersion in infrared. A better device for this purpose is a *grating*, which consists of a large number of closely spaced slits on a reflective or transmissive substrate. This is the basis of operation of most OSAs [27,28]. Figure 11.9 shows the block diagram of an OSA based on a reflective grating.

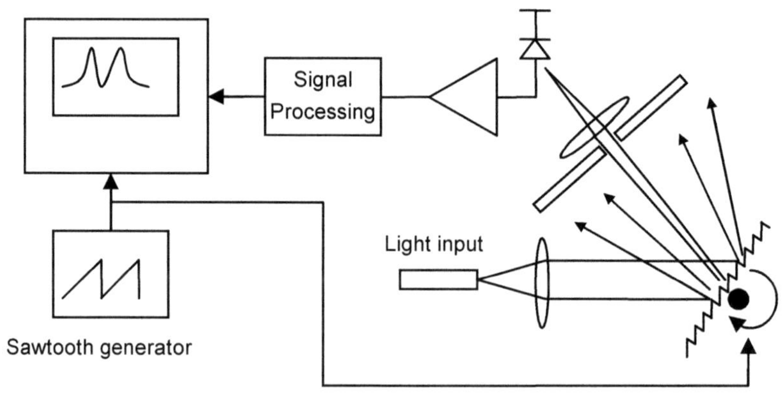

Fig. 11.9. Schematic of an optical spectrum analyzer based on reflection grating

Typically the light is brought to the equipment with a fiber where it is culminated and directed toward the grating. The grating separates the wavelengths, and a second slit acting as a spatial filter selects a narrow band. The light that passes through the slit is then focused on a detector, which converts the optical power to an electrical signal. After initial amplification the analog signal is digitized and goes through further signal processing. Eventually, the optical power is used as the vertical scale in a display. The horizontal scale is driven by a ramp generator which rotates the grating. Thus, the horizontal location of the trace in the display corresponds to the angular location of the grating, which in turn determines the range of wavelengths that go through the slit and detected by the detector. As a result, the display traces the optical power of the signal as a function of wavelength.

Figure 11.10 shows the typical settings and measurements available in an OSA. The left-hand side of the figure shows the settings that are available to the user. The wavelength settings control the horizontal scan that sets the start and stop wavelength points. Alternatively, a center wavelength and a wavelength span can be indicated. The level controls deal with the vertical axis of the display. The vertical axis can be changed between linear and logarithmic (dBm) scales. In most cases the logarithmic scale is used as it provides much more dynamic range in displaying the waveforms. In either case, the scale of the display grid can also be chosen (dB or mW per division), which determines the vertical span of the displayed waveform. The reference level selects the reference point for the vertical axis, so that changing it will shift the displayed waveform up or down. Several additional parameters related to the way the signal is scanned and displayed can also be set.

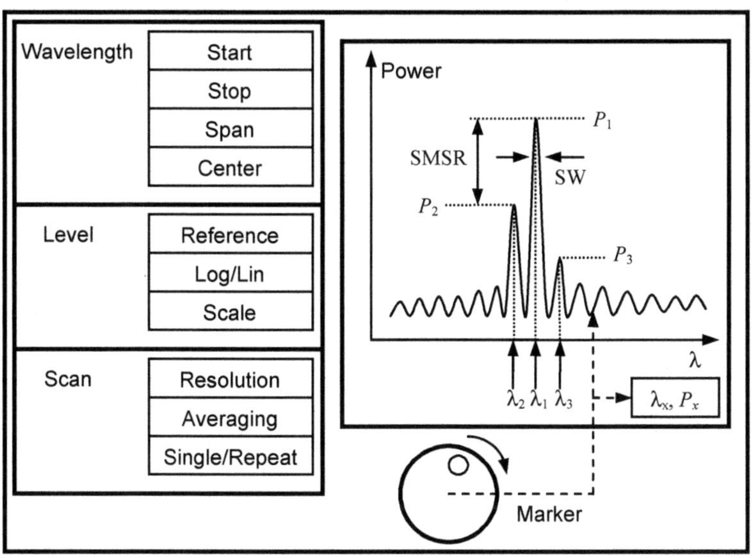

Fig. 11.10. Typical OSA controls and measurements

For instance, the OSA can be configured to scan the input signal continually or it can be set to perform a single scan. The resolution setting determines the bandwidth of the bandpass optical filter. Using higher resolution yields more details in the spectrum of the signal, but it will take longer to scan the waveform. Usually, resolutions of up to 0.1 nm can be achieved with standard OSAs. Using averaging enables the display to show the average of several scans. This is a useful feature when dealing with noisy signals or when measurements at very low optical powers are needed.

Figure 11.10 also shows some of the typical measurements that an OSA can perform on a waveform. Most OSAs are equipped with internal software that can automatically perform a number of key measurements. Among the most common measurements are optical power and wavelength for each of the peaks in the spectrum. Measurements of spectral width (SW) and side mode suppression ratio (SMSR) are also routine tests.[5]

In addition to automatic test routines, OSAs provide means of making direct measurements on the waveform. For instance, Fig. 11.10 shows a knob that can be used to move a marker along the waveform. At each point, the equipment displays the wavelength and the optical power. Using this feature, the user can perform direct measurements on various features of the waveform.

OSAs are versatile tools that can perform a host of spectral measurements on a signal, and their test capabilities suffice for most general applications. In some cases, however, the measurement accuracy of an OSA is not sufficient, in which case other test equipment specifically targeted for a particular measurement should be used.

11.5.2 Wavelength meters

One example where the accuracy of a typical OSA can be insufficient is when the wavelength of a signal must be measured with great precision, say, with ±0.001 nm accuracy. This is usually the case in WDM applications where wavelength spacing and stability of adjacent channels is of great importance. Wavelength meters are capable of measuring the wavelength of an optical signal with much more accuracy compared to an OSA.

Wavelength meters typically utilize interferometric principles to increase the measurement accuracy. For instance, in a Michelson interferometer, the optical signal is split into two beams in two branches and is then made to interfere with itself. When the length of one of the branches is varied, the output of the interferometer goes through successive minima and maxima, as a function of the length change. By counting these peaks, the wavelength of the light can be determined. Other methods are based on Fizeau, or Fabry–Perot interferometer effects, Doppler effect, and heterodyne principles [29–32].

---

[5] For a discussion of these and other spectral parameters refer to Chapter 3.

11.6 Link performance testing

Instruments like power meters or spectrum analyzers measure parameters associated with an individual device or test. However, it is also necessary to measure the performance of a system that consists of several individual elements. A prime example is a complete link, consisting of an optical driver, a laser, optical fiber, and an optical receiver. The ultimate measure of performance for a digital link is bit error rate (BER), which refers to the ratio of correctly received bits to the total number of bits sent [33]. Not only an entire link, but also many digital circuits and systems can be tested for BER, examples include receivers or clock and data recovery circuits. For the purpose of our current discussion, unless otherwise stated, we refer to the system being tested as device under test (DUT).

11.6.1 Bit error rate tester (BERT)

A typical BER test includes sending a known pattern of data through the DUT and comparing the data stream coming out of the DUT with the pattern that goes into the DUT. Figure 11.11 shows a schematic diagram of a commonly used bit error rate tester (BERT) for this purpose.

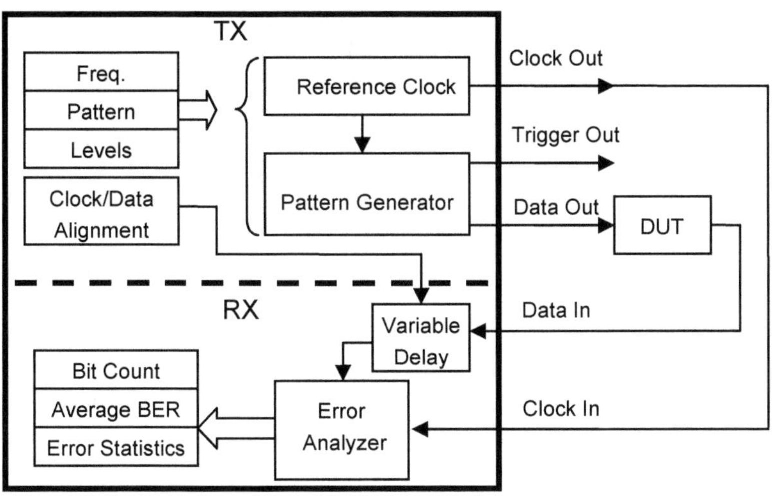

Fig. 11.11. Schematic diagram of a bit error rate tester (BERT)

The BERT is divided into two main blocks, a transmitter (TX) and a receiver (RX) unit. The user controls and interfaces are shown in the left side of the figure, while the instrument's signal inputs and outputs are shown on the right side. The BERT measures the DUT's BER as well as a variety of error statistics.

The user inputs include frequency, pattern, signal levels, and a delay between the receiver clock and data. The pattern and frequency inputs control a pattern generator that creates a test pattern simulating the real data traffic that the DUT is likely to encounter. The most common patterns for testing BER are pseudo-random (PRBS) patterns of different lengths, usually PRBS-7, PRBS-23, and PRBS-31. These patterns simulate random data traffic within their lengths, with longer patterns representing more random data. In most BERTs, there are options to define other user patterns too, usually to allow for stressing the DUT in some particular aspect.

The TX block has three main outputs. The data output includes a physical signal that represents the intended pattern. This pattern is usually applied to the DUT. There is also a clock output, which enables the RX block to synchronize with the TX block. Finally, the TX block provides a trigger output. This output is not used directly within the BERT. However, it is very useful if other tests are to be conducted. For example, it allows the use of a DCA for eye pattern analysis. The trigger output can be driven from two sources. It can either replicate the clock (or some divided version of the clock) or it can provide a trigger pulse every time the pattern generator has gone through the entire pattern and starts sending the pattern from the beginning again. As indicted in Fig. 11.11, the voltage level of these outputs can also be controlled by the user.

The RX section has two main inputs: a clock and a data input. The data output from the TX section is applied to the DUT and the output of the DUT goes to the data input of the RX section. On the other hand, the RX section uses the clock input to generate an exact replicate of the data pattern. An error analyzer compares this copy with the data it is receiving from the DUT bit by bit. As a result, the analyzer can determine the ratio of the bits that do not match the expected value (and therefore are considered errors) to the total number of received bits. This ratio, by definition, is the BER of the DUT.

Note that there is a time delay associated with the DUT and the electrical (or optical) interconnects in the setup. Therefore, the RX section must be able to change the delay of the received signal with respect to the pattern it generates internally to synchronize and match the two. This so-called clock and data alignment process is accomplished through a variable delay that can either be controlled by the user or optimized by the BERT. The alignment is optimized by changing the delay until the BER is minimized. This process is equivalent to changing the sampling point in time within an eye pattern. Placing the sampling point in the middle of an eye pattern will minimize the BER. On the other hand, by changing the delay, the sampling point can effectively be moved across the eye pattern, and by mapping the resulting BER vs. time delay the effective jitter at various BERs can be measured.

To accomplish the comparison between the two copies of the signal, the bit analyzer and the pattern generator need to run at exactly the same frequency. Usually, this is achieved by running both the pattern generator and the analyzer from the same clock source. This is the configuration shown in Fig. 11.11: the clock output from the TX section is connected directly to the clock input of the RX section. However, the clock can also be extracted from the output of the DUT using a clock and data recovery (CDR) circuit. Using a CDR becomes necessary whenever the phase difference between the clock and data is not constant. In such cases, a one time clock-data alignment will not work and a CDR provides continuous clock-data alignment. This is usually necessary in sensitivity testing, which is the topic of the next section.

11.6.2 Sensitivity measurement

A common use of a BERT system is in measuring sensitivity and sensitivity penalties associated with various phenomena that affect sensitivity. As we noted in Chapter 9, sensitivity is defined as the minimum optical power that a receiver can work with, while maintaining a given BER. Sensitivity is the most crucial parameter that characterizes a digital receiver, and therefore measuring both sensitivity and sensitivity penalties is very important. Figure 11.12 shows a setup for measuring sensitivity using a BERT.

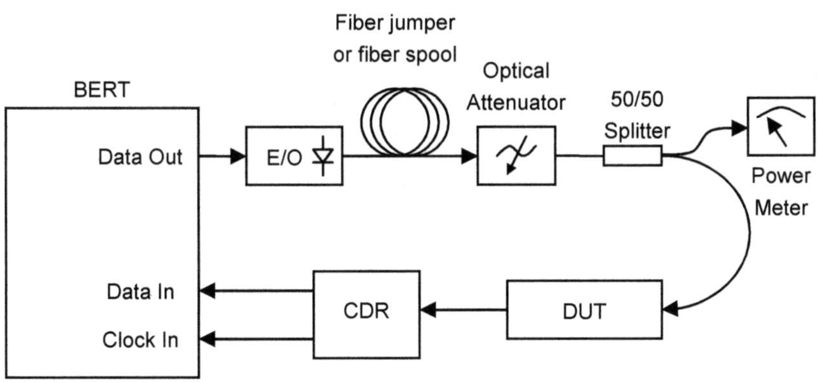

Fig. 11.12. Measurement setup for sensitivity tests

The electrical data output from the BERT drives an electrical to optical (EO) converter. This is essentially a "golden" transmitter that generates a well-controlled, high-quality optical signal. The optical signal is then sent through a given length of fiber. At the other side of the fiber, there is a variable optical attenuator. After the optical signal passes through the attenuator, it goes to a beam splitter which divides the optical power in half. One branch of the splitter goes to an optical power meter and the other branch goes to the DUT or the receiver whose sensitivity is being measured. Note that the optical power received by the DUT is the same as the power that goes to the power meter (assuming a 50/50 splitting ratio). Therefore, the power meter shows the optical power received by the DUT. The DUT converts the optical signal back to an electrical signal. The electrical signal is then applied to a clock and data recovery (CDR) unit, which extracts the clock out of the signal. The retimed data and the extracted clock are then sent back to the BERT's receiver for error analysis.

As noted in the previous section, the BERT's input clock can be driven directly from BERT's output clock. However, in this setup, because the optical signal travels through fiber, it may be necessary to use a CDR to remove random phase variations due to a long stretch of fiber. In this setup, by varying the optical attenuator, the level of optical power received by the DUT can be changed, while at each power level, the BERT can measure the bit error rate. Because the occurrence of errors is a random process, to get more accurate BER results for any given power level, it is necessary to run the test long enough to acquire a sufficient number of errors. Reasonable confidence levels may require as many as 100 errors, although this may be impractical for low BERs.

Using this setup, a number of measurements can be performed. The simplest test is to measure the receiver's sensitivity at zero dispersion. In this case, instead of a long fiber, a short fiber jumper with negligible dispersion is used and the attenuation is set to a nominal level. The link should be able to run error free at this point (BER=0). The attenuation is then increased until a relatively high error rate is observed. Next, the attenuation is reduced gradually until the BER falls below a desired level, say 10^{-12}. The power measured by the optical power meter at this point is the sensitivity of the receiver at zero dispersion.

The receiver's performance can also be characterized more accurately by measuring the waterfall curves.[6] This is especially useful in estimating the sensitivity at lower BERs, where it is impractical to run a test long enough to acquire sufficient number of errors for statistical confidence. Using this setup, the received power can be scanned across a number of equally spaced power levels, and at each point the BER can be measured. The plot of the BER vs. power level is the waterfall curve corresponding to the particular parameters used in the test.

11.6.3 Sensitivity penalty tests

The setup shown in Fig. 11.12 can also be used for a variety of sensitivity penalty measurements. Sensitivity penalty refers to the increase in optical power needed to compensate the degrading effects of a specific phenomenon on sensitivity, measured in decibels.

A common type of sensitivity penalty is associated with dispersion and is known as *dispersion penalty*. As discussed in Chapter 5, when optical signals travel in fiber they suffer from a variety of dispersion effects. Thus, dispersion penalty refers to the additional amount of optical power needed to compensate sensitivity degradation due to dispersion for a given length of fiber and at a given

---

[6] See Chapter 9 for a discussion of waterfall curves.

wavelength. To test dispersion penalty, the sensitivity of a receiver is first characterized with a minimum length of fiber, as discussed before. Then, the fiber jumper is replaced with the desired length of fiber, usually in the form of a fiber spool. Fiber spools are available in various lengths of fiber, typically from a few kilometers up to 100 km. Next, the optical attenuator is adjusted and the sensitivity of the receiver is characterized again. Obviously, as a result of the attenuation of fiber, the power level reaching the attenuator is going to be much lower, and therefore much less attenuation is needed to reach the sensitivity point. The power difference, in dB, to achieve the same BER level with and without the fiber spool is the dispersion penalty. For instance, a 1 dB dispersion penalty means that when a fiber spool is inserted in the link, the power at the receiver must be 1 dB higher to achieve the same target BER as before. Of course, this penalty is a function of several parameters, including bit rate, wavelength, length and design of fiber, and the type of optical transmitter used in the link.

This methodology can also be used to measure other types of power penalties. For example, we can measure the sensitivity penalty as a function of the extinction ratio (ER) of the transmitter. The process will be the same: the sensitivity is characterized when the transmitter (E/O converter) is set at a given reference ER. Then the ER is changed, and for every new value of ER sensitivity is re-measured. Thus, the sensitivity penalty associated with ER changes (specifically, lower ERs) can be characterized.

11.7 Analog modulation measurements

Although a majority of fiber optic links are based on digital transmission, fiber optic links based on analog formats are still in use in several applications, including CATV and microwave feeds. In a typical analog link, several carriers or channels at different frequencies are simultaneously modulated on the optical signal. It should be noted that frequency, in this context, refers to the electrical modulation frequency, and not to the optical frequency. Thus, a composite analog signal consisting of several sub-carriers modulate a single optical carrier. In this section, we discuss the lightwave signal analyzer (LSA), which is the primary tool for analog signal analysis and measurements. We will also discuss some of the main measurements that can be performed with an LSA.

11.7.1 Lightwave signal analyzer (LSA)

In general, signals can be analyzed in time or in frequency domain. For some applications, time domain analysis is very effective, as many important features of the signal, such as rise- and fall-time, or the shape of the waveform, are inherently time domain features. This is usually the case for baseband signals. In fiber optic communications, digital links are typically baseband, and therefore time domain analysis is the main tool in studying and characterizing them. The major tool for measurements in time domain is the oscilloscope or platforms based on an oscilloscope, such as the DCAs we considered previously.

On the other hand, for non-baseband signals where several frequency carriers exist simultaneously, usually frequency domain analysis provides a better insight into the nature of the composite signal. In RF applications, the electrical spectrum analyzer provides a convenient tool for analyzing signals in frequency domain. In principle, the same techniques can be used to analyze optical signals [34–36]. The lightwave signal analyzer (LSA) is an instrument that essentially integrates an RF spectrum analyzer with an OE converter front end. The LSA provides a convenient instrument for measurements of parameters such as modulation depth, relative intensity noise (RIN), inter-modulation interference, signal harmonics, laser relaxation frequency, and modulation frequency response [37]. These parameters are not only essential for characterization of analog signals, but are also useful tools in the analysis of digital signals in frequency domain.

.

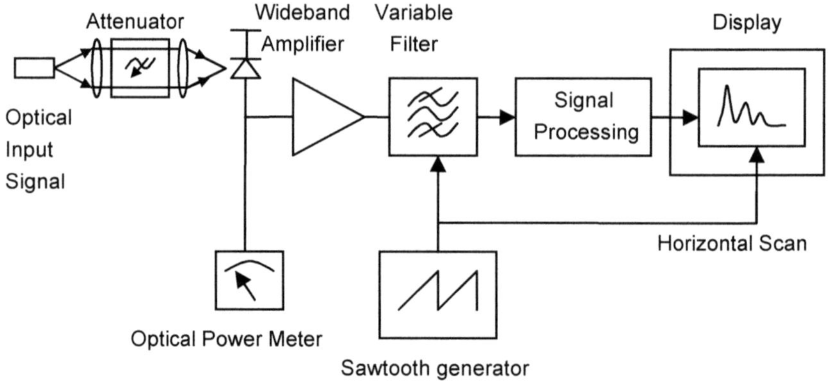

Fig. 11.13. Principals of operation of a lightwave signal analyzer (LSA)

Figure 11.13 shows the block diagram of an LSA. The input optical signal goes through an optical attenuator which brings down the optical power to a suitable level for the instruments detector. The light is focused on an optical detector which is followed by a wideband linear amplifier. The detector recovers the electrical signal from the optical carrier, and the linear amplifier provides additional gain and provides a suitable signal for further signal processing. An optical power meter is also provided that measures the average power in the signal. The output of the linear amplifier is fed to a variable narrow band filter whose bandpass is controlled by a sawtooth or ramp generator. The output of the filter, which represents the frequency contents of the signal in a narrow band, goes through further signal processing and ultimately drives the vertical axis of the display. The horizontal axis of the display is synchronously scanned with the bandpass of the filter, yielding the frequency spectrum of the signal.

A few points should be noted about the LSA. First, notice that the LSA is essentially a frequency spectrum analyzer shown in Fig. 11.8 preceded by an optical detector. Therefore, in principle, the same measurements that can be done with an LSA can also be done with an OE converter and a standard electrical spectrum analyzer. However, the advantage of an LSA is that it is already calibrated in terms of the optical power whereas such calibrations must be performed by the user if an OE converter along with a spectrum analyzer is used.

Another point to note is the apparent similarity between the schematic diagram of the LSA and that of the OSA shown in Fig. 11.9. However, it should be noted that in an OSA the filtering is done in the optical domain and at optical frequencies (wavelengths), as opposed to the LSA where the filtering is done in electrical domain and at much lower frequencies. Therefore, what an OSA displays should not be confused with what an LSA displays, as they represent totally different frequency domains.

11.7.2 Signal parameter measurements

Because the LSA essentially displays the signal in frequency domain, it shares many settings with ordinary electrical spectrum analyzers. Settings like start and stop frequency, and the bandwidth and reference level, are common. One difference is that in an LSA the units for the vertical axis can be chosen between optical dB or electrical dB. Because in optical to electrical conversion optical power is linearly converted to current, and because electrical power is proportional to the square of current, a distinction must be made between electrical and optical power changes. This distinction is reflected in the logarithmic units of optical dB (dB_{OPT}) and electrical dB (dB_{ELC}) where

$$\begin{aligned} dB_{OPT} &= 10\log\frac{P_1}{P_2} = 10\log\frac{I_1}{I_2} = \\ &\frac{1}{2}\times 20\log\frac{I_1}{I_2} = \frac{1}{2}\times 10\log\left(\frac{I_1}{I_2}\right)^2 = \frac{1}{2}dB_{ELC} \end{aligned} \tag{11.5}$$

Thus, for instance, a 10 dB change in optical power is equivalent to a 20 dB change in electrical power.

Figure 11.14 shows a representative spectrum on an LSA and the parameters that can be measured from the test. The average power of the signal is displayed on the average power bar in the left side of the screen. The modulated signal shows as a spike at the modulation frequency, f_0. From the display, the signal average power, modulation power, and modulation frequency can readily be measured. A particularly important parameter for analog systems is the *modulation depth*, MD, defined as

$$MD = 10\log\frac{P_{f0}}{P_{AVG}} \tag{11.6}$$

where P_{f0} is the power at the intended modulation frequency and P_{AVG} is the average power in the signal. From the LSA display, MD can be measured easily as the difference between the average power and the power in the main signal.

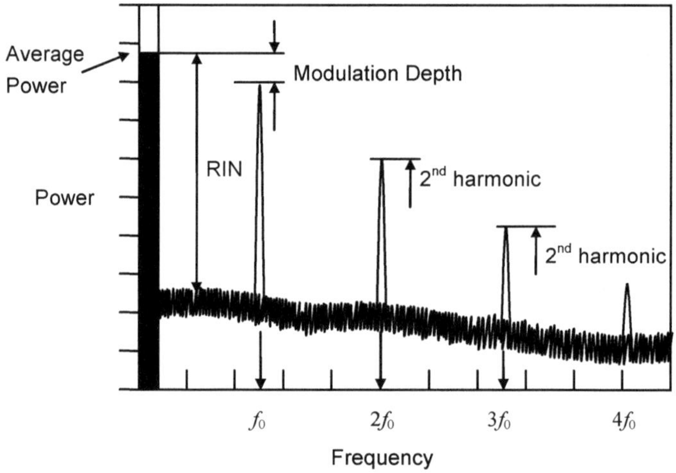

Fig. 11.14. Spectrum measurements using a lightwave signal analyzer (LSA)

It is also common to express MD as a linear percentage, which can be calculated directly from Eq. (11.6).

When a linear system is modulated at the frequency f_0, the output should only have a component at f_0. However, lasers and optical modulators are not completely linear devices, and therefore whenever they are modulated at a particular frequency, harmonics of that frequency will also show up in the spectrum. In Fig. 11.14, the second and third harmonics of the main signal, located at $2\times f_0$ and $3\times f_0$, are marked. The *harmonic distortion* associated with the *n*th harmonic, D_n, is defined as

$$D_n = 10\log\frac{p_{f_n}}{P_{f_0}} \tag{11.7}$$

Again, this distortion can easily be measured from the LSA's display. For example, the 2nd harmonic distortion in dBs is simply the difference between the peak powers at $2f_0$ and f_0. The *total harmonic distortion*, THD, is defined as

$$THD = 10\log\frac{\sum P_{f_n}}{P_{f_0}} \tag{11.8}$$

To obtain THD, the power in each harmonic (up to a reasonable number) must be measured from the display. These powers then need to be added and divided by the power in the main signal.

LSAs can also be used to characterize *inter-modulation distortion* (IMD) [38]. In a perfectly linear system several signals with different carrier frequencies can be present simultaneously without affecting each other. However, any nonlinearity causes the appearance of additional signals at sum or difference frequencies. Depending on the number of carriers present, the number of spurious frequencies multiplies. Thus, if the carriers are at frequencies $f_1, f_2, \ldots, f_n$, 2nd order IMD involves terms at $f_i \pm f_j$, 3rd order IMD involves terms at $f_i \pm f_j \pm f_k$, etc. IMD can be a serious problem in CATV systems where several channels are modulated on the same optical carrier. Using an LSA, both the frequencies and powers of these beat terms as well as their relative location to the main signals can easily be determined.

Figure 11.14 also shows another important quantity, the signal's *relative intensity noise* (*RIN*). RIN is a measure of the noise power caused by random optical intensity variations that always accompany any optical source [39–41]. Because noise is always a function of bandwidth, it is common to refer to intensity noise spectral density, in other words, the amount of noise present per hertz of bandwidth. Moreover, to remove the needed calibrations resulting from the responsivity of the detector, the noise is normalized to the average power present in the signal. In this way, any responsivity variations are cancelled out because they affect the average power and noise power in the same fashion. Thus, RIN is defined as

$$RIN(f) = 10\log\frac{P_N^2(f)}{P_{AVG}^2} \tag{11.9}$$

The units of RIN are dB/Hz. The reason the powers appear as squared is that noise power as measured by any receiver is proportional to the square of current, and the current is proportional to the optical power. Thus, the noise power in a receiver is proportional to the square of optical power. It can be seen from Eq. (11.9) that in general RIN is a function of frequency. For instance, the RIN in the spectrum of a laser typically reaches a maximum at the relaxation oscillation of the laser.

LSAs can be used to measure RIN [42]. As can be seen from Fig. 11.14, RIN can be measured at any given frequency as the ratio of the noise at that frequency to the average power. Note that RIN can also be integrated over a frequency range, yielding the total intensity noise within that range. If the frequency range is narrow enough, or if the RIN is flat over the range, the total noise power can be obtained by simply multiplying the RIN spectral density by the bandwidth.

11.8 Summary

Test and measurement methodologies constitute an important part of fiber optic design and validation. In this chapter we reviewed some of the most common test instruments and measurements that are used in fiber optic links.

One of the most common measurements is optical power testing. Power is the main parameter in an optical link, at both transmitting and receiving ends. Average power measurements are done using power meters that utilize broad area detectors. These detectors are typically thermally stabilized silicon or InGaAs PIN diodes. Silicon diodes are used for shorter wavelengths, while InGaAs is usually used for longer wavelengths in the 1310 and 1550 nm window. In a standard power meter the wavelength must be set so that the equipment can use the proper calibration factor to calibrate out the wavelength dependence of the detector and report an accurate result.

An optical signal is characterized by the instantaneous optical power, and signal measurements are part of time domain measurements which require an oscilloscope-type platform capable of displaying the optical waveform. While an OE converter and a standard oscilloscope can in principle accomplish this task, the most common tool used for this purpose is a digital communication analyzer, or DCA. Usually DCAs have an optical input port which allows for a direct connection to an optical signal. Moreover, DCAs are internally calibrated to optical power, and they include software utilities for testing a variety of additional parameters. Another advantage of a DCA is that by sampling in equivalent time, it can provide an effective bandwidth much higher than what is available in a standard digital scope. Thus, a DCA provides a convenient way of measuring a large number of time domain parameters on a given optical signal. The most important of these parameters include the shape of the signal, rise- and fall-time, optical modulation amplitude, overshoot, extinction ratio, jitter, and compliance mask tests. Time domain measurements are particularly useful for analysis of digital signals, where they are sometimes grouped together as eye pattern analysis.

Frequency domain analysis is another important area for measurements. Here one needs to distinguish between two separate frequency domains: optical frequencies and signal frequencies. Optical frequencies are related to the wavelength or spectrum of the light. The most common tool for optical frequency analysis is an optical spectrum analyzer, or OSA. An OSA can display the wavelength components of an optical signal. Thus, facts like whether an optical source is single mode or multimode can readily be determined by an OSA. Typical parameters an OSA can measure include peak wavelength(s), spectral width (SW), and side mode suppression ratio (SMSR).

The other frequency range for which a variety of parameters need to be measured is the signal or electrical frequency range. These measurements are specifically significant for analog applications, where oftentimes several channels are frequency multiplexed on a single optical carrier. In principle, a calibrated electrical spectrum analyzer preceded by an OE converter is capable of performing signal spectrum measurements. This is the operating basis of a light spectrum analyzer or LSA. An LSA has an optical input, with the convenience of providing calibration in both optical and electrical power units. Parameters that can be measured by an LSA include modulation depth (MD), distortion, total harmonic distortion (THD), inter-modulation distortion (IMD), and noise parameters such as relative intensity noise (RIN).

References

[1] J. S. Wilson et al., *Test and Measurement*, Elsevier, Amsterdam, 2008

[2] A. F. Van Putten, *Electronic Measurement Systems, Theory and Practice*, Taylor & Francis, Inc., London, 1996

[3] N. Kularatna, *Digital and Analogue Instrumentation Testing and Measurement*, Institution of Electrical Engineers, London, 2003

[4] P. D. Q. Campbell, *An Introduction to Measuration and Calibration*, Industrial Press Inc., New York, 1995

[5] G. Lawday et al., *A Signal Integrity Engineer's Companion: Real-Time Test and Measurement, and Design Simulation*, Prentice Hall, Englewood Cliffs, NJ, 2008

[6] D. Derickson, *Fiber Optic Test and Measurement*, Prentice Hall, Englewood Cliffs, NJ, 1997

[7] Telecommunication Industry Association (TIA), www.tiaonline.org

[8] A. S. Morris, *Measurement and Calibration Requirements for Quality Assurance to ISO 9000: For Quality Assurance to ISO 9000*, John Wiley & Sons, New York, 1997

[9] C. Foley, "Check optical data signals with an oscilloscope," Microwave & RF, Vol. 36, pp. 155–156, 1997

[10] A. Moschitta, F. Stefani, and D. Petri "Measurements of transient phenomena with digital oscilloscopes," Proceedings of the 20th IEEE Instrumentation and Measurement Technology Conference, 2003, Vol. 2, pp. 1345–1349, May 2003

[11] A. Ferrero et al., "Voltage Measurement," in *Measurement, Instrumentation, and Sensors*, Edited by J. Webster, Springer, Berlin, 1999

[12] Y. Lembeye, J. P. Keradec, and G. Cauffet, "Improvement in the linearity of fast digital oscilloscopes used in averaging mode," IEEE Transactions on Instrumentation and Measurement, Vol. 43, pp. 922–928, 1994

[13] C. Dorrer, "High-speed measurements for optical telecommunication systems," IEEE Journal of Selected Topics in Quantum Electronics, Vol. 12, pp. 84–858, 2006

[14] M. Kahrs, "50 years of RF and microwave sampling," IEEE Transactions on Microwave Theory and Techniques, Vol. 51, pp. 1787–1805, 2003

[15] R. L. Morrison et al., "Design and demonstration of a high-speed, multichannel, optical-sampling oscilloscope," Applied Optics, Vol. 35, pp. 1187–1194, 1996

[16] Product Note 86100-5, "Triggering wide-bandwidth sampling oscilloscopes for accurate displays of high-speed digital communications waveforms," Agilent Technologies, available from http://www.agilent.com

[17] For example, Agilent 86100C Infiniium DCA-J, datasheet available at www.agilent.com

[18] IEC 61280-2-2, "Fiber optic communication subsystem test procedures-Part 2-2: digital systems-optical eye pattern, waveform and extinction ratio measurement," International Electrotechnical Commission (IEC), 2008

[19] P. O. Anderson and K. Akermark, "Accurate optical extinction ratio measurements," IEEE Photonics Technology Letters, Vol. 6, pp. 1356–1358, 1994

[20] Technical Note, "Improving optical transceiver extinction ratio measurement accuracy by using reference receiver correction factors," Agilent, available at www.agilent.com

[21] C. Heras, et al., "High resolution light intensity spectrum analyzer (LISA) based on Brillouin optical filter," Optics Express, Vol. 15, pp. 3708–3714, 2007

[22] T. Saitoh et al., "Optical spectrum analyzer utilizing MEMS scanning mirror," IEEE Photonics Technology Letters, Vol. 18, pp. 767–769, 2006

[23] R. E. Saperstein, D. Panasenko, and Y. Fainman, "Demonstration of a microwave spectrum analyzer based on time-domain optical processing in fiber," Optics Letters, Vol. 29 pp. 501–503, 2004

[24] D. M. Baney, B. Szafraniec, and A. Motamedi, "Coherent optical spectrum analyzer," IEEE Photonics Technology Letters, Vol. 14, pp. 355–357, 2002

[25] A. D. Helfrick, *Electrical Spectrum and Network Analyzers: A Practical Approach*, Academic Press, New York, 1991

[26] A Rostami, "Full-optical spectrum analyzer design using EIT based Fabry-Perot interferometer," Telecommunications and Networking-ICT 2004, Vol. 3124, pp. 282–286, 2004

[27] J. Vobis and D. Derickson, "Optical spectrum analysis," in *Fiber Optic Test and Measurement*, Edited by D. Derickson, Prentice Hall, Englewood Cliffs, NJ, 1997

[28] AN 1550-4, "Optical spectrum analysis basics," Agilent Technologies, 2000, available from www.agilent.com

[29] M. B. Morris, T. J. McIlrath, and J. J. Snyder, "Fizeau wavemeter for pulsed laser wavelength measurement," Applied Optics, Vol. 23, pp. 3862–3868, 1984

[30] T. J. Scholl et al., "Broadband precision wavelength meter based on a stepping Fabry–Perot interferometer," Review of Scientific Instruments, Vol. 75, pp. 3318–3326, 2004

[31] M. Wakim et al., "Highly accurate laser wavelength meter based on Doppler effect," Optics Communications, Vol. 262, pp. 97–102, 2006

[32] X. Wang, Y. Li, and S. L. Zhang, "Heterodyne wavelength meter for continuous-wave lasers," Applied Optics, Vol. 46, pp. 5631–5634, 2007

[33] S. M. Berber, "An automated method for BER characteristics measurement," IEEE Transactions on Instrumentation and Measurement, Vo. 53, pp. 575–580, 2004

[34] C Dorrer and D. N. Maywar, "Ultra-high bandwidth RF spectrum analyzer for optical signals," Electronics Letters, Vol. 39, pp. 1004–1005, 2003

[35] H. Sobol, "The application of microwave techniques in lightwave systems," Journal of Lightwave Technology, Vol. 5, pp. 293–299, 1987

[36] D. M. Baney and W. V. Sorin, "Measurement of a modulated DFB laser spectrum using gated delayed self-homodyne technique," Electronics Letters, Vol. 24, 1988

[37] Application Note AN 371, "Agilent 71400 lightwave signal analyzer," Agilent Technologies, 2000, available form www.agilent.com

[38] M. R. Johnson, "Reviewing the basics of intermodulation distortion," Microwaves & RF, Vol. 46, pp. 74–82, 2007

[39] R. Zhang and F. G. Shi, "Manufacturing of laser diode modules: integration and automation of laser diode-fiber alignment and RIN characterization," IEEE Transactions on Advanced Packaging, Vol. 26, pp. 128 – 132, 2003

[40] S. L. Woodward, T. L. Koch, and U. Koren, "RIN in multisection MQW-DBR lasers," IEEE Photonics Technology Letters, Vol. 2, pp. 104–108, 1990

[41] K. Y. Lau and A. Yariv, "Ultra-high speed semiconductor lasers," IEEE Journal of Quantum Electronics, Vol. 21, pp. 121–136, 1985

[42] Product Note PN-71400-1, "Lightwave signal analyzers measure relative intensity noise," Agilent Technologies, 2000. Available from www.agilent.com

Chapter 12
Standards

12.1 Introduction

We started Chapter 1 with an example of spoken language as an intuitive model of a communication link. In language, a complex set of conventions and rules are at work to ensure that the sounds made by one side are interpreted correctly by the other side, so that the intended message is conveyed and communication is established. The role of standards in telecommunication is fundamentally not so different. However, fiber optic standards are not limited to those that cover communication aspects. There are also a wide range of standards related to components, systems, measurements, and test methodologies. Moreover, standards are constantly evolving as a result of new technologies, new needs, and new problems. This makes the topic of standards a vast, complex subject, especially in an area such as fiber optics which involves many different fields and technologies.

This chapter starts with a discussion of the main bodies that are active in defining fiber optics related standards. Next, we will divide the standards into several categories and provide a list of selected standards in each category. This classification, although by no means complete, can serve as a useful reference and provides a general perspective. Finally, we briefly discuss some standards that are commonly encountered in physical layer design. It goes without saying that our treatment of the topic is necessarily selective and brief. Therefore, for more details the reader can refer to the standards or references mentioned in each section.

12.2 Standards development bodies

There are numerous institutions that are active in defining standards, each from a specific perspective. For communication industries, the major bodies include the International Telecommunication Institute (ITU), the Telecommunication Industry Association (TIA), the International Electrotechnical Commission (IEC), the Institute of Electrical and Electronics Engineers (IEEE), Organization for Standardization (ISO), American National Standard Institute (ANSI), and Telcordia (also known as Bellcore). We will proceed by briefly introducting each of these organizations.

12.2.1 International Telecommunication Union (ITU)

ITU is an international agency within United Nations dedicated to information and communication technologies and is a major source of all kinds of standards for

M. Azadeh, *Fiber Optics Engineering*, Optical Networks,
DOI 10.1007/978-1-4419-0304-4_12, © Springer Science+Business Media, LLC 2009

telecommunication industries [1]. The international nature of ITU is a reflection of the necessity of having uniform standards for telecommunication to ensure compatibility between telecommunication infrastructures in different countries. ITU is headquartered in Geneva and most countries within the United Nations are its members. Moreover, it has numerous members from private sector, which include various communication companies, organizations, and agencies from all around the world. ITU's Telecommunication Standardization Sector, known as ITU-T, is one of the sectors within ITU whose role is directly related to defining telecommunication standards in the form of recommendations [2].

Many of ITU's recommendations related to communications and fiber optics are designated by the letter G, which is a class of recommendations related to digital systems and networks. For example, G.691 is a recommendation related to SDH systems operating at STM-64. Other categories of ITU-T's standards that contain fiber optics related documents include L (cables) and Y (information infrastructure). Most released standards from ITU-T are available to public free of charge [3].

12.2.2 International Electrotechnical Commission (IEC)

IEC is an international, non-governmental organization involved with defining standards related to all areas of electrical and electronics technologies [4]. IEC was founded in 1906 in London in recognition of the necessity of addressing questions regarding standardization of nomenclature and ratings for electrical devices and machines. Its central office is now located in Geneva, Switzerland. Examples of some units that IEC established include the hertz for frequency and the gauss for magnetic flux density. Another creation of IEC is the International Electrotechnical Vocabulary (IEV), now known as Electropedia [5]. IEC also proposed an international system for physical units which became known as the "Giorgi system." This system was later developed further and eventually became the International System of Units (SI) which is now used worldwide [6]. IEC works closely with other international bodies such as ISO and ITU. IEC covers a wide range of standards, including power, energy, semiconductor, and fiber optic industries.

The Standardization Management Board within IEC oversees the standards development work of IEC. IEC's standards are in the form of IEC-*x*, where *x* is the number of the document that starts from 60000. For example, laser safety issues are covered by the IEC-60825 series of documents. Documents developed jointly between the ISO and IEC are in the form of ISO/IEC-*x*.

12.2.3 Institute of Electrical and Electronics Engineers (IEEE)

IEEE is an international non-profit professional organization headquartered in New York City and is the largest professional organization in the world [7]. IEEE in its current form was established in 1984 through the merger of two older entities: American Institute of Electrical Engineers (AIEE) and Institute of Radio Engineers (IRE). IEEE is active in virtually all fields of electrical engineering and is involved in various scientific and educational activities. IEEE is responsible for the publication of many common journals and holds various conferences in all areas of electrical engineering. IEEE has regional sections all around the world, as well as specialized societies dedicated to specific fields, such as the IEEE Laser & Electro-Optics Society (LEOS).

IEEE is also a major source of standard development. These activities are typically carried out through IEEE Standard Association (IEEE-SA), which is a subset of IEEE and has relationships with other international organizations such as ITU and ISO [8]. IEEE standards are specified by numbers. For example, the 802.3 family are the main components defining the Ethernet standards.

12.2.4 Telecommunication Industry Association (TIA)

The Telecommunication Industry Association (TIA) is a trade association that represents numerous companies and industries in the field of telecommunications [9]. It was established in 1988 as a result of the merger of United States Telecommunications Suppliers Association (USTSA) and the Information and Telecommunications Technologies Group of Electronics Industries Alliance (EIA). In addition to representing the interests of its members before government agencies, TIA is active in certification, regulatory compliance, and standards development for a wide range of telecommunication applications. In terms of standard developments, TIA functions through several engineering committees. Issues related to fiber optics are mainly within the domain of User Premises Telecommunications Cabling Requirements Committee, or TR-42. The TR-42 committee itself consists of several subcommittees, each focusing on a certain area. Most of TIA's standards tend to have designations in the form of TIA/EIA-455-*X*, TIA-455-*x*, or TIA/EIA-*x*, where *x* is a designation number. Some of these documents are also known under the name of FOTP-*x*, where FOTP stands for Fiber Optic Test Procedures. TIA documents are mainly related to test and measurement methods for characterization of optical components. For example, the title of TIA/EIA-455-25-C, also known as FOTP-25, is Impact Testing of Optical Fiber Cables, and it is related to determining the tolerance of optical cables to impact loads.

12.2.5 ISO and ANSI

The International Organization for Standardization (ISO) is an international body comprised of national standard organizations from more than 140 countries [10]. Established in 1947, ISO is a non-governmental organization based in Geneva, Switzerland. ISO is active in the development of international standards in almost all fields of science and technology. Standard development activities are handled by technical committees that consist of representatives and experts from technology, industry, and business. ISO's standards documents are in the form of ISO-$x[y]$, where x is the document number, and y is the year the document was released. A well-known example of these documents is the ISO-9000 quality management systems family [11]. These documents define procedures that a company or organization must follow to ensure that the products and services that it offers satisfy certain consistency and quality metrics. The Open System Interconnection (OSI) model of networks we discussed in Chapter 2 is another well-known example in the area of communications.

American National Standard Institute (ANSI) is the US representative to ISO and is an active member and participant in many of ISO's standards development processes [12]. It also represents United States in many other international bodies such as ITU-T. ANSI itself is a private, not-for-profit organization founded in 1918 and is headquartered in Washington, DC. Its membership includes numerous companies, agencies, academic, and industrial organizations. Most of ANSI's fiber optics standards are common to those of TIA and IEC.

12.2.6 Telcordia (Bellcore)

In the 1980s, the then giant ATT was broken up into several entities. The regional companies that separated from ATT established Bellcore as a research institute, which later changed its name to Telcordia [13]. Telcordia is headquartered in Piscataway, New Jersey, and has been active both in systems development and in devising standards for telecommunications. For example, the toll-free number system or services such as caller ID and call waiting are among Telcordia's contributions. SONET is a primary example of Telcordia's contributions in fiber optics industry. Telcordia is currently active as a research company that provides support in areas of software, system design, R&D, wireless, cable networking, and other services to the telecommunication industry worldwide.

Many of the standards spelled out by Telcordia are in the form of General Requirement (GR) documents. Typically, there is a main document (sometimes called core document) that contains the general information about the subject, and a family of documents that cover details or various aspects of that topic. For example, the SONET standard is spelled out mainly in GR-253 (Synchronous Optical Network (SONET) Transport Systems: Common Generic Criteria). In addition to the GRs, Telcordia provides other documents such as Family of Requirements (FR), Special Reports (SR), and Technical Advisories (TA) [14].

12.2.7 Miscellaneous organizations

Although the organizations mentioned above provide standards in a wide variety of areas, sometimes the need arises for a relatively fast process to address a specific design or technology problem. In such cases, it is not uncommon for the involved industries to come together in the form of forums, organizations, and committees and to cooperate in defining protocols or standards related to their common area of interest.

An example of such an organization is the Optical Internetworking Forum (OIF) [15]. OIF is a member-driven non-profit organization consisting of more than 80 companies that acts as a forum for defining public domain technologies in the area of optical networking. The work of OIF is formalized as documents called Implementation Agreements (AIs). The existing AIs cover topics such as electrical interfaces, tunable lasers, and optical transponder interoperability.

Another example of such industry-driven efforts is the Small Form Factor (SFF) committee, which was established in 1990 as an ad hoc group by a number of companies with the narrow goal of defining mechanical dimensions for disk drives to ensure interoperability between various computers [16]. Later, SFF's goals were broadened to address other areas where for any reason industry needs could not be fulfilled by other standards bodies. SFF is clear that it does not want to replace other standards development activities, rather, it complements them. The SFF committee comprises a number of participating industries and their representatives.

Most documents produced by SFF are submitted to other standards bodies as well. Documents produced by SFF tend to be relatively short and focused on a narrow issue or problem. The typical numbering format for SFF recommendations is SFF-x or INF- x. An example of the SFF standards is the SFF-8472 (Diagnostic Monitoring Interface for Optical Transceivers), which we will discuss in more detail later in this chapter. The SFF committee also maintains documents such as the INF-8074i (SFP or Small Form-factor Pluggable Transceiver) and INF-8077i (XFP, or 10 Gbps Small Form Factor Pluggable Module) [17]. These documents can be accessed on SFF's FTP site [18]. Standards such as INF-8074i (SFP) and INF-8077i (XFP) are especially relevant to physical layer optical transceivers. These documents define the mechanical dimensions, electrical connection interface, and the general functionality of some of the most commonly used transceivers in fiber optics.

12.3 Standards classifications and selected lists

Standards cover a wide range of areas, and it is hard to classify them into well-defined categories. However, for the purpose of this chapter, we can roughly divide fiber optic related standards into four categories, as follows:

(1) Components and devices

(2) Test and measurement

(3) Reliability

(4) Networking standards

This classification is necessarily approximate, as there are some standards that do not clearly belong to any of these categories, while there are some that can be classified under more than one category. However, for the most part, this classification is sufficient as a starting point. In this section we provide a list of selected standards within each category. For simplicity, we do not distinguish between various categories of documents such as recommendations, requirements, technical reports, etc. Instead, we will treat all of these documents under the generic title of standards.

12.3.1 Standards related to components

Standards dealing with components typically define the parameters of common fiber optic components and devices [19]. These include passive components such as fibers or connectors, active components such as lasers or optical amplifiers, and even whole transceivers. These standards are generally used by component manufacturers who have to ensure their products meet the standard requirements. They are also useful for users, because compliant products from different manufacturers are (ideally) interoperable.

Broadly speaking, components can be divided into active and passive. Various kinds of optical fibers and cables can be considered important examples of passive components. For instance, G.651 to G.657 define the standard optical fibers. Another important class of passive components defined by standards is optical connectors, defined by the TIA/EIA-604-*x* family, also known as Fiber Optic Connector Intermateability Standard FOCIS-*x*. Common ST, SC, FC, and LC connectors are defined by FOCIS-2, FOCIS-3, FOCIS-4, and FOCIS-10, respectively. Active components typically include lasers and complete transceivers. The most common types of standard transceivers are those defined by multi-source agreements; examples include GBIC, SFP, and XFP. The following list includes some of the major standards related to component.

Optical fibers and cables

- **G.651.1:** Characteristics of a 50/125 µm Multimode Graded Index Optical Fiber Cable for the Optical Access Network, ITU-T, 2007
- **G.652:** Characteristics of a Single-Mode Optical Fiber and Cable, ITU-T, 2005
- **G.653:** Characteristics of a Dispersion-Shifted Single-Mode Optical Fiber and Cable, ITU-T, 2006
- **G.654:** Characteristics of a Cut-Off Shifted Single-Mode Optical Fiber and Cable, ITU-T, 2006
- **G.655:** Characteristics of a Non-zero Dispersion-Shifted Single-Mode Optical Fiber and Cable, ITU-T, 2006
- **G.656:** Characteristics of a Fiber and Cable with Non-zero Dispersion for Wide-band Optical Transport, ITU-T, 2006

- **G.657:** Characteristics of a Bending Loss Insensitive Single Mode Optical Fiber and Cable for the Access Network, ITU-T, 2006
- **G.971:** General Features of Optical Fiber Submarine Cable Systems, ITU-T, 2007
- **G.972:** Definition of Terms Relevant to Optical Fiber Submarine Cable Systems, ITU-T, 2008
- **G.973:** Characteristics of Repeaterless Optical Fiber Submarine Cable Systems, ITU-T, 2007
- **G.974:** Characteristics of Regenerative Optical Fiber Submarine Cable Systems, ITU-T, 2007
- **G.977:** Characteristics of Optically Amplified Optical Fiber Submarine Cable Systems, ITU-T, 2006
- **G.978:** Characteristics of Optical Fiber Submarine Cables, ITU-T, 2006
- **L.59:** Optical Fiber Cables for Indoor Applications, ITU-T, 2008
- **GR-20:** Optical Fiber and Optical Fiber Cable, Telcordia, 2008
- **GR-356:** Optical Cable Innerduct and Accessories, Telcordia, 1995
- **GR-2961:** Multi-Purpose Fiber Optic Cable, Telcordia, 1998
- **GR-3155:** Microducts for Fiber Optic Cables, Telcordia, 2007
- **TIA/EIA-472:** Family of Standards for Optical Fibers
- **TIA/EIA-492:** Family of Standards for Optical Fibers
- **TIA-568-C.3:** Optical Fiber Cabling Components Standard, 2008

Connectors

- **L.36:** Single Mode Fiber Optic Connectors, ITU-T, 2008
- **L.37:** Optical Branching Components (Non-Wavelength Selective), ITU-T, 2007
- **GR-1435:** Multi-Fiber Optical Connectors, Telcordia, 2008
- **GR-326:** Single Mode Optical Connectors and Jumper Assemblies, Telcordia, 1999
- **GR-1081:** Field-Mountable Optical Fiber Connectors, Telcordia, 1995
- **TIA/EIA-604:** Fiber Optic Connector Intermateability Standards, 1999
- **TIA/EIA-604-10-A (FOCIS-10):** Fiber Optic Intermateability Standard, 1999
- **TIA/EIA-604-12 (FOCIS-12):** Fiber Optic Connector Intermateability Standard, Type MT-RJ, 2002
- **TIA/EIA-604-2-A (FOIC-2):** Fiber Optic Connector Intermateability Standard, Type Fiber Jack Connector, 2003
- **TIA/EIA-604-2-B (FOIC-2):** Fiber optic connector Intermateability Standard, 2004
- **TIA/EIA-604-3-A (FOCIS-3):** Fiber Optic Connector Intermateability Standard, Type SC, 1997
- **TIA/EIA-604-3-B (FOCIS-3):** Fiber Optic Connector Intermateability Standard, Type SC and SC-APC, 2004
- **TIA/EIA-604-4-A (FOCIS-4):** Fiber Optic Connector Intermateability Standard, Type FC and FC-APC, 2000

- **TIA/EIA-604-4-B (FOCIS-4):** Fiber Optic Connector Intermateability Standard, Type FC and FC-APC, 2004
- **TIA/EIA-604-6 (FOCIS-6):** Fiber Optic Connector Intermateability Standard, Type Fiber Jack Connector, 2001

Lasers, amplifiers, and other devices

- **G.661:** Definition and Test Methods for the Relevant Generic Parameters of Optical Amplifier Devices and Subsystems, ITU-T, 2007
- **G.662:** Generic Characteristics of Optical Amplifier Devices and Subsystems, ITU-T, 2005
- **G.663:** Application Related Aspects of Optical Amplifier Devices and Subsystems, ITU-T, 2000
- **G.665:** Generic Characteristics of Raman Amplifiers and Raman Amplified Subsystems, ITU-T, 2005
- **L.31:** Optical Fiber Attenuators, ITU-T, 1996
- **G.666:** Characteristics of PMD Compensators and PMD Compensating Receivers, ITU-T, 2008
- **G.667:** Characteristics of Adaptive Chromatic Dispersion Compensators, ITU-T, 2006
- **G.671:** Transmission Characteristics of Optical Components and Subsystems, ITU-T, 2005
- **G.680:** Physical Transfer Functions of Optical Networks Elements, ITU-T, 2007
- **GR-910:** Fiber Optic Attenuators, Telcordia, 1998
- **GR-1209:** Passive Optical Components, Telcordia, 2001
- **GR-1009:** Fiber Optic Clip-on Test Sets, Telcordia, 1994
- **GR-1073:** Single-mode Fiber Optic Switches, Telcordia, 2001
- **GR-1095:** Multi-Fiber Splicing Systems for Single-Mode Optical Fibers, Telcordia, 1996
- **GR-2854:** Fiber Optic Dispersion Compensators, Telcordia, 1997
- **GR-2866:** Optical Fiber Fanouts, Telcordia, 2007
- **GR-2882:** Optical Isolators and Circulators, Telcordia, 1995
- **GR-2883:** Fiber Optic Filters, Telcordia, Telcordia, 1995
- **TIA-455-127 (FOTP-127):** Basic Spectral Characterization of Laser Diodes (2006)

Integrated transceivers

- **INF-8053i:** GBIC (Gigabit Interface Converter)
- **INF-8074i:** SFP Small Form-factor Pluggable Transceiver
- **INF-8077i:** XFP 10 Gbps Small Form Factor Pluggable Module
- **INF-8431:** Enhanced 8.5 and 10 Gigabit Small Form Factor Pluggable Module (SFP Plus)
- **SFF-8477:** Tunable XFP

- **INF-8475i:** XPAK Small Form-factor Pluggable Receiver

12.3.2 Standards related to measurements and procedures

As noted in Chapter 11, measurement methodologies and techniques constitute a critical part of any industry. One of the most active organizations in this respect is the TIA, which has produced measurement procedures for most aspects of fiber optics. Common examples include the Fiber Optic Test Procedures (FOTPs), some of which are listed below. These procedures can themselves be divided into several categories. For instance, some procedures define ways to measure fundamental physical quantities, such as length, optical power, or jitter. Others define standard ways of representing or categorizing physical measurements. An important example we will discuss in more detail later in this chapter is the SFF-8472, which defines a standard protocol for in situ measurement and reporting of critical physical layer parameters. There are also standards that define characteristics of test and measurement equipment used in fiber optics. The following is a selected list of common standards used in fiber optics.

Fibers and cables

- **G.650.1:** Definitions and Test Methods for Linear, Deterministic Attributes of Single-Mode Fiber and Cable, ITU-T, 2004
- **G.650.2** Definitions and Test Methods for Statistical and Non-linear Related Attributes of Single-Mode Fiber and Cable, ITU-T, 2007
- **G.650.3:** Test Methods for Installed Single-Mode Fiber Cable Sections, ITU-T, 2008
- **G.976:** Test Methods Applicable to Optical Fiber Submarine Cable Systems, ITU-T, 2007
- **GR-2923:** Generic Requirements for Optical Fiber Connector Cleaning Products, Telcordia, 1996
- **TIA/EIA-455-11-C (FOTP-11):** Vibration Test Procedure for Fiber Optic Components and Cables, 2002
- **TIA/EIA-455-25-C (FOTP-25):** Impact Testing of Optical Fiber Cables, 2002
- **TIA-455-78-B (FOTP-78)/IEC 60793-1-40:** Optical Fibers Part 1–40: Measurement Methods and Text Procedures – Attenuation, 2002
- **TIA-455-80-C (FOTP-80)/IEC 60793-1-44:** Optical Fibers Part 1–44: Measurement Methods and Test Procedures – Cut-off Wavelength, 2003
- **TIA-455-85-A (FOTP-85):** Fiber Optic Cable Twist Test, 2005
- **TIA/EIA-455-88 (FOTP-88):** Fiber Optic Cable Bend Test, 2001
- **TIA/EIA-455-95-A (FOTP-95):** Absolute Optical Power Test for Optical Fibers and Cables, 2000
- **TIA/EIA-455-132-A (FOTP-132):** Measurement of the Effective Area of Single-Mode Optical Fiber, 2001

- **TIA/EIA-455-158 (FOTP-158):** Measurement of Breakaway Frictional Force in Fiber Optic Connector Alignment Sleeves, 1997
- **TIA/EIA-455-162-A (FOTP-162):** Fiber Optic Cable Temperature-Humidity Cycling, 1999
- **TIA-455-177-B (FOTP-177)/IEC 60793-1-43:** Optical Fibers Part 1–43: Measurement Methods and Test Procedures – Numerical Aperture, 2003
- **TIA/EIA-455-204 (FOTP-204):** Measurement of Bandwidth on Multimode Fiber, 2000
- **TIA-455-218 (FOTP-218):** Measurement of Endface Geometry of Single Fiber Optical Connectors, 2002
- **TIA-455-220-A (FOTP-220):** Differential Mode Delay Measurement of Multimode Fiber in the Time Domain, 2003
- **TIA-526-7 (OFSTP-7):** Measurement of Optical Power Loss of Installed Single-Mode Fiber Cable Plant, 2003
- **TIA-526-14-A (OFSTP-14):** Optical Power Loss Measurement of Installed Multimode Fiber Cable Plant, 1998

Optical fiber amplifiers

- **TIA/EIA-455-206 (FOTP-206)/IEC 61290-1-1:** Optical Fiber Amplifiers – Basic Specification Part 1-1: Test Methods for Gain Parameters – Optical Spectrum Analyzer, 2000
- **TIA/EIA-455-207 (FOTP-207)/IEC 61290-1-2:** Optical Fiber Amplifiers – Basic Specification Part 1-2: Test Methods for Gain Parameters – Electrical Spectrum Analyzer, 2000
- **TIA/EIA-455-208 (FOTP-208)/IEC 61290-1-3:** Optical Fiber Amplifiers – Basic Specification Part 1-3: Test Methods for Gain Parameters – Optical Power Meter, 2000
- **TIA/EIA-455-209 (FOTP-209)/IEC 61290-2-1:** Optical Fiber Amplifiers – Basic Specification Part 2-1: Test Methods for Optical Power Parameters – Optical Spectrum Analyzer, 2000
- **TIA/EIA-455-210 (FOTP-210)/IEC 61290-2-2:** Optical Fiber Amplifiers – Basic Specification Part 2-2: Test Methods for Optical Power Parameters – Electrical Spectrum Analyzer , 2000
- **TIA/EIA-455-211 (FOTP-211)/IEC 61290-2-3:** Optical Fiber Amplifiers – Basic Specification Part 2-3: Test Methods for Optical Power Parameters – Optical Power Meter, 2000
- **TIA/EIA-455-212 (FOTP-212)/IEC 61290-6-1:** Optical Fiber Amplifiers – Basic Specification Part 6-1: Test Methods for Pump Leakage Parameters – Optical Demultiplexer, 2000
- **TIA/EIA-455-213 (FOTP-213)/IEC 61290-7-1:** Optical Fiber Amplifiers – Basic Specification Part 7-1: Test Methods for Out-of-Band Insertion Losses – Filtered Optical Power Meter, 2000

Component and link test procedures

- **G.8251:** The Control of Jitter and Wander Within the Optical Transport Network (OTN), ITU-T, 2001
- **TIA/EIA-455-34-A (FOTP-34):** Interconnection Device Insertion Loss Test, 1995
- **TIA/EIA-455-42-A (FOTP-42):** Optical Crosstalk in Fiber Optic Components, 1989
- **TIA/EIA-455-107-A (FOTP-107):** Determination of Component Reflectance or Link/System Return Loss Using a Loss Test Set, 2004
- **TIA/EIA-455-113 (FOTP-113):** Polarization-Mode Dispersion Measurement of Single-Mode Optical Fibers by the Fixed Analyzer Method, 1997
- **TIA-455-128 (FOTP-128):** Procedures for Determining Threshold Current of Semiconductor Lasers, 1996
- **TIA/EIA-455-157 (FOTP-157):** Measurement of Polarization Dependent (PDL) of Single-Mode Fiber Optic Components , 1995
- **TIA/EIA-455-180-A (FOTP-180):** Measurement of the Optical Transfer Coefficients of a Passive Branching Device, 1999
- **TIA/EIA-455-196 (FOTP-196):** Guideline for Polarization-Mode Measurement in Single-Mode Fiber Optic Components and Devices, 1999
- **TIA/EIA-455-197 (FOTP-197):** Differential Group Delay Measurement of Single-Mode Components and Devices by the Differential Phase Shift Method, 2000
- **TIA/EIA-455-200 (FOTP-200):** Insertion Loss of Connectorized Polarization-Maintaining Fiber or Polarizing Fiber Pigtailed Devices and Cable Assemblies, 2001
- **TIA/EIA-455-201 (FOTP-201):** Return Loss of Commercial Polarization-Maintaining Fiber or Polarizing Fiber Pigtailed Devices and Cable Assemblies, 2001
- **TIA/EIA-455-203 (FOTP-203):** Launched Power Distribution Measurement Procedure for Graded-Index Multimode Fiber Transmitters, 2000
- **TIA-455-228 (FOTP-228):** Relative Group Delay and Chromatic Dispersion Measurement of Single-Mode Components and Devices by the Phase Shift Method, 2002
- **TIA-455-229 (FOTP-229):** Optical Power Handling and Damage Threshold Characterization, 2002
- **TIA-455-235 (FOTP-235):** Fiber Optic Communication Subsystem Test Procedures Part 2-8: Digital Systems – Determination of Low BER Using Q-Factor Measurements, 2004
- **TIA-455-236 (FOTP-236):** Fiber Optic Communication Subsystem Test Procedures Part 2-9: Digital Systems – Optical Signal-to-Noise Ratio Measurement for Dense Wavelength-Division Multiplexed Systems, 2004
- **TIA-526-4-A (OFSTP-4)**: Optical Eye Pattern Measurement Procedure, 1997
- **TIA-526-10 (OFSTP-10):** Measurement of Dispersion Power Penalty in Digital Single-Mode Systems, 1998

- **TIA-526-11 (OFSTP-11):** Measurement of Single-Reflection Power Penalty for Fiber Optic Terminal Equipment, 1998
- **TIA-526-15 (OFSTP-15):** Jitter Tolerance Measurement, 1998
- **TIA-526-16 (OFSTP-16):** Jitter Transfer Function Measurement, 1998
- **TIA-526-17 (OFSTP-17):** Output Jitter Measurement, 1998
- **TIA-526-18 (OFSTP-18):** Systematic Jitter Generation Measurement, 1998
- **SFF-8472:** Diagnostic Monitoring Interface for Optical Transceivers, SFF Committee, 2007

Test equipment

- **GR-196:** Generic Requirements for Optical Time Domain Reflectometer (OTDR) Type Equipment, Telcordia, 1996
- **GR-198:** Generic Requirements for Optical Loss Test Sets, Telcordia, 1996
- **GR-2947:** Generic Requirements for Portable Polarization Mode Dispersion Test Sets, Telcordia, 1997
- **O.172:** Jitter and Wander Measuring Equipment for Digital Systems Which Are Based on the Synchronous Digital Hierarchy (SDH), 2005
- **O.173:** Jitter Measuring Equipment for Digital Systems Which Are Based on the Optical Transport Network (OTN) , ITU-T, 2007
- **O.181:** Equipment to Assess Error Performance on STM-N Interfaces, ITU-T, 2002
- **O.182:** Equipment to Assess Error Performance on OTN Interfaces, ITU-T, 2007
- **O.201:** Q-factor Test Equipment to Estimate the Transmission Performance of Optical Channels, ITU-T, 2003
- **TR-NWT-001137:** Generic Requirements for Hand-Held Optical Power Meters, Telcordia, 1991
- **TIA/EIA-455-224 (FOTP-224)** Calibration of Fiber Optic Chromatic Dispersion Test Sets, TIA, 2002
- **TIA-455-231 (FOTP-231)** Calibration of Fiber Optic Power Meters, TIA, 2003
- **TR-TSY-000886:** Generic Criteria for Optical Power Meters, Telcordia, 1990
- **TR-TSY-001028:** Generic Criteria for Optical Continuous Wave Reflectometers, Telcordia, 1990
- **TSB-142:** Optical Return Loss Meters Measurement and Application Issues, TIA, 2005
- **TSB-143:** Fiber Optic Power Meters: Measurement and Application Issues, TIA, 2008

12.3.3 Reliability and safety standards

Reliability issues deal with the long-term performance of devices and systems. Standards related to reliability can be used for two closely related purposes. On the one hand, they can be used as a source of information regarding design practices that will improve the long-term performance or reliability of a device or system. On the other hand, they define procedures and tests that attempt to quantify and predict the reliability of an existing device or system. This is usually done by stressing the device or the system in question in certain ways and then attempting to use the resulting failures as a basis for reliability predictions about similar devices or systems.

One of the main standards used for characterization of reliability of fiber optic components and systems is GR-468, which we will discuss in more detail later. However, GR-468 itself draws on a variety of standards from various bodies. The following is a list of selected standards from different bodies related to reliability.

Safety

- **G.664:** Optical Safety Procedures and Requirements for Optical Transport Systems, ITU-T, 2006
- **IEC-60825-1:** Safety of Laser Products – Part 1: Equipment Classification and Requirements, 2007
- **IEC-60825-2:** Safety of Laser Products – Part 2: Safety of Optical Fiber Communication Systems, 2007

Component reliability, aging, and stress tests

- **L.14:** Measurement Method to Determine the Tensile Performance of Optical Fiber Cables Under Load, ITU-T, 1992
- **L.45:** Minimizing the Effect on the Environment from the Outside Plant in Telecommunication Networks, ITU-T, 2000
- **SR-NWT-002855:** Optical Isolators: Reliability Issues and Proposed Tests, Telcordia, 1993
- **SR-TSY-001136:** Handbook for Digital Cross-Connect System Quality and Reliability Analyses, Telcordia, 1989
- **SR-3244:** Reliability Concerns with Lightwave Components, Telcordia, 1994
- **SR-332:** Reliability Prediction Procedure for Electronic Equipment, Telcordia, 2006
- **SR-TSY-001369:** Introduction to Reliability of Laser Diodes and Modules, Telcordia, 1989
- **TIA/EIA-455-2-C (FOTP-2):** Impact Test Measurements for Fiber Optic Devices, 1998
- **TIA/EIA-455-5-C (FOTP-5):** Humidity Test Procedure for Fiber Optic Components, 2002

- **TIA-526-27 (OFSTP-27):** Procedure for System-Level Temperature Cycle Endurance Test, 1998
- **TIA/EIA-455-28-C (FOTP-28):** Measuring Dynamic Strength and Fatigue Parameters of Optical Fibers by Tension, 2004
- **TIA/EIA-455-31-C (FOTP-31):** Proof Testing Optical Fibers by Tension, 2004
- **TIA/EIA-455-69-A (FOTP-69):** Test Procedure for Evaluating the Effect of Minimum and Maximum Exposure Temperature on the Optical Performance of Optical Fibers, 1991
- **TIA/EIA-455-71-A (FOTP-71):** Procedure to Measure Temperature-Shock Effects on Fiber Optic Components,1999
- **TIA/EIA-455-72 (FOTP-72):** Procedure for Assessing Temperature and Humidity Cycling Aging Effects on Optical Characteristics of Optical Fibers,1997
- **TIA/EIA-455-73 (FOTP-73):** Procedure for Assessing Temperature and Humidity Cycling Aging Effects on Mechanical Characteristics of Optical Fibers, 1997
- **TIA-455-129 (FOTP-129):** Procedures for Applying Human Body Model Electrostatic Discharge Stress to Package Optoelectronic Components, 1996
- **TIA/EIA-455-130 (FOTP-130):** Elevated Temperature Life Test for Laser Diodes, 2001
- **TIA-455-160-A, IEC-60793-1-50 (FOTP-160):** Optical Fibers – Part 1–50: Measurement Methods and Test Procedures – Damp Heat, 2003
- **TIA-1048 (IEC 62005-7):** Reliability of Fiber Optic Interconnecting Devices and Passive Components – Part 7: Life Stress Modeling, 2005
- **TSB-62-13 (ITM-13):** Measuring Dynamic Strength and Fatigue Parameters of Optical Fibers by Two-Point Bending, TIA, 2000
- **TSB-145:** Reliability of Passive Fiber Optic Components: Failure Modes and Mechanisms of Fiber Optic Connectors, TIA, 2003
- **TSB-149:** Generic Workmanship Guidelines for Fiber Optic Connector Interoperability, TIA, 2004

General reliability concerns, practices, methodologies, and tests

- **G.911:** Parameters and Calculation Methodologies for Reliability and Availability of Fiber Optic Systems, ITU-T, 1997
- **GR-78** Generic Requirements for the Physical Design and Manufacture of Telecommunications Products and Equipment, Telcordia, 2007
- **GR-357** Generic Requirements for Assuring the Reliability of Components Used in Telecommunications Equipment, Telcordia, 2001
- **GR-418:** Generic Reliability Assurance Requirements for Fiber Optic Transport Systems, Telcordia, 1999
- **GR-468:** Generic Reliability Assurance Requirements for Optoelectronic Devices Used in Telecommunications Equipment, Telcordia, 2004

- **GR-874:** An Introduction to the Reliability and Quality Generic Requirements (RQGR), Telcordia, 2003
- **GR-929:** Reliability and Quality Measurements for Telecommunications Systems (RQMS-Wireline), Telcordia, 2002
- **GR-1221:** Generic Reliability Assurance Requirements for Passive Optical Components, Telcordia, 1999
- **GR-1339:** Generic Reliability Requirements for Digital Cross-Connect Systems, Telcordia, 1997
- **GR-1421:** Generic Requirements for ESD-Protective Circuit Packed Containers, Telcordia, 1995
- **GR-1929:** Reliability and Quality Measurements for Telecommunications Systems (RQMS-Wireless), Telcordia, 2005
- **GR-2903:** Reliability Assurance Practices for Fiber Optic Data Links, Telcordia, 1995
- **MIL-HDBK-217**: Military Handbook, Reliability Prediction of Electronic Equipment, Department of Defense, Washington DC
- **MIL-HDBK-338:** Electronic Reliability Design Handbook, Department of Defense, Washington DC
- **SR-1171:** Methods and Procedures for System Reliability Analysis, Telcordia, 2007
- **SR-TSY-000385:** Bell Communications Research Reliability Manual, Telcordia, 1986
- **SR-TSY-001130:** Reliability and System Architecture Testing, Telcordia, 1989

Software reliability

- **ANSI/AIAA R-013:** Recommended Practice for Software Reliability, 1992
- **GR-282:** Software Reliability and Quality Acceptance Criteria (SRQAC) , Telcordia, 2006
- **SR-1547:** The Analysis and Use of Software Reliability and Quality Data, Telcordia, 2000
- **SR-2785:** Software Fault Insertion Testing (SFIT) Methodology, Telcordia, 2000
- **GR-2813:** Generic Requirements for Software Reliability Prediction, Telcordia, 1994
- **SR-3299:** Broadband Software Robustness Guidelines, Telcordia, 1994

12.3.4 Networking and system standards

Within the past few decades, optical networking has come a long way, and this is reflected in continuous updates to the existing standards as well as new standards related to and covering the new technologies as they become available [20,21].

In networking standards, the industry has witnessed a trend from lower speed, single-wavelength optical links to higher speeds, multi-wavelength, longer reach, complex networking interfaces. Moreover, there have been attempts to define various pieces of the networks in as much detail as possible, so that *transverse compatibility* can be achieved [22]. This means equipment from different vendors can be used at either side of an optical link.

The following list includes some of the major standards covering optical networks. Among these, G.957, G.691, and GR-253 are the main documents covering SONET/SDH networks up to 10 Gbps. DWDM networks are mostly covered by G.692, G.694.1, G.696.1, and G.698.1. CWDM applications are covered by G.694.2 and G.695. Optical transport networks (OTNs), which incorporate the next generation of optical networks beyond SONET, are mostly covered by G.709, G.872, G.693, G.959.1. Passive optical networks (PONs) are defined by G.983.1-5 and G.984.1-6. Ethernet applications are mainly covered by IEEE 802.3, which in its 2005 revision incorporates subfamilies such as 802.3ae. We will discuss the physical layer parameters of SONET, Ethernet, and PON networks briefly later in this chapter. The following list includes the main standards related to optical networks and systems.

SONET/SDH

- **G.691** Optical Interfaces for Single Channel STM-64 and Other SDH Systems with Optical Amplifiers, ITU-T, 2006
- **G.707** Network Node Interface for the Synchronous Digital Hierarchy (SDH), ITU-T, 2007
- **G.957** Optical Interfaces for Equipments and Systems Relating to the Synchronous Digital Hierarchy, ITU-T, 2006
- **GR-253:** Synchronous Optical Network (SONET) Transport Systems: Common Generic Criteria, Telcordia, 2005
- **GR-496:** SONET Add-Drop Multiplexer (SONET ADM) Generic Criteria, Telcordia, 2007
- **GR-1042:** Generic Requirements for Operations Interfaces Using OSI Tools – Information Model Overview: Synchronous Optical Network (SONET) Transport Information Model, Telcordia, 1998
- **GR-1230:** SONET Bi-directional Line-Switched Ring Equipment Generic Criteria, Telcordia, 1998
- **GR-1244:** Clocks for the Synchronized Network: Common Generic Criteria, Telcordia, 2005
- **GR-1250:** Generic Requirements for Synchronous Optical Network (SONET) File Transfer, Telcordia, 1999
- **GR-2891:** SONET ATM Virtual Path Digital Cross-Connect Systems – Generic Criteria, Telcordia, 1998
- **GR-2899:** Generic Criteria for SONET Two-Channel (1310/1550-NM) Wavelength Division Multiplexed Systems, Telcordia, 1995

- **GR-2900:** SONET Asymmetric Multiples Functional Criteria, Telcordia, (1995)
- **GR-2950:** Information Model for SONET Digital Cross-Connect Systems (DCSS), Telcordia, 1999
- **GR-2955:** Generic Requirements for Hybrid SONET/ATM Element Management Systems (EMSS), Telcordia, 1998

WDM networks

- **G.692** Optical Interfaces for Multi-channel Systems with Optical Amplifiers, ITU-T, 2005
- **G.694.1** Spectral Grids for WDM Applications: DWDM Frequency Grid, ITU-T, 2002
- **G.694.2:** Spectral Grids for WDM Applications: CWDM Wavelength Grid, ITU-T, 2003
- **G.695:** Optical Interfaces for Coarse Wavelength Division Multiplexing Applications, ITU-T, 2006
- **G.696.1:** Longitudinally Compatible Intra-domain DWDM Applications, ITU-T, 2005
- **G.697** Optical Monitoring for DWDM Systems, ITU-T, 2004
- **G.698.1** Multi-channel DWDM Applications with Single Channel Optical Interfaces, ITU-T, 2006
- **G.698.2** Amplified Multi-channel DWDM Applications with Single Channel Optical Interfaces, ITU-T, 2007
- **GR-2918:** DWDM Network Transport Systems with Digital Tributaries for Use in Metropolitan Area Applications: Common Generic Criteria, Telcordia, 2003

Ethernet

- **802.3:** IEEE Standards for Information Technology, Carrier Sense Multiple Access with Collision Detection (CSMA/CD) Access Method and Physical Layer Specifications, IEEE, 2005
- **802.3aq:** 10GBASE-LRM 10 Gbps (1,250 MB/s) Ethernet over Multimode Fiber, IEEE, 2006
- **802.3ba:** 40 Gbps and 100 Gbps Ethernet Task Force
- **G.985:** 100 Mbit/s Point-to-Point Ethernet Based Optical Access System, ITU-T, 2003
- **G.8010/Y.1306:** Architecture of Ethernet Layer Networks, ITU-T, 2004
- **TIA-785-1:** 100 Mb/s Physical Layer Medium Dependent Sublayer and 10 Mb/s and 100 Mb/s Auto-Negotiation on 850nm Fiber Optics, 2002

Passive optical networks (PON)

- **802.3av:** 10Gb/s PHY for EPON Task Force, IEEE task force
- **G.983.1** L**:** Broadband Optical Access Systems Based on Passive Optical Networks (PON), ITU-T, 2005
- **G.983.2:** ONT Management and Control Interface Specification for B-PON, ITU-T, 2005
- **G.983.3:** A Broadband Optical Access System with Increased Service Capability by Wavelength Allocation, ITU-T, 2001
- **G.983.4:** A Broadband Optical Access System with Increased Service Capability Using Dynamic Bandwidth Assignment, ITU-T, 2001
- **G.983.5:** A Broadband Optical Access System with Enhanced Survivability, ITU-T, 2002
- **G.984.1:** Gigabit-Capable Passive Optical Networks (GPON): General Characteristics, ITU-T, 2008
- **G.984.2:** Gigabit-Capable Passive Optical Networks (GPON): Physical Media Dependent (PMD) Layer Specification, ITU-T, 2003
- **G.984.3:** Gigabit-Capable Passive Optical Networks (G-PON): Transmission Convergence Layer Specification, ITU-T, 2008
- **G.984.4:** Gigabit-Capable Passive Optical Networks (G-PON): ONT Management and Control Interface Specification, ITU-T, 2008
- **G.984.5:** Enhancement Band for Gigabit Capable Optical Access Networks, ITU-T, 2007
- **G.984.6:** Gigabit-Capable Passive Optical Networks (GPON): Reach Extension, ITU-T, 2008

Optical transport networks (OTN)

- **G.693** Optical Interfaces for Intra-office Systems, ITU-T, 2006
- **G.709 (Y1331):** Interfaces for the Optical Transport Network (OTN), ITU-T, 2003
- **G.798** Characteristics of Optical Transport Network Hierarchy Equipment Functional Blocks, ITU-T, 2006
- **G.870 (Y.1352):** Terms and Definitions for Optical Transport Networks, ITU-T, 2008
- **G.871** Framework for Optical Transport Network Recommendations, ITU-T, 2000
- **G.872** Architecture of Optical Transport Networks, ITU-T, 2001
- **G.874** Management Aspects of the Optical Transport Network Element, ITU-T, 2008
- **G.959.1** Optical Transport Network Physical Layer Interfaces, ITU-T, 2008

12.4 Fiber standards

In the previous sections standards related to various aspects of fiber optics were listed. In this and the following sections, we discuss a few standards in more detail. The selected standards include those that are directly related to physical layer design and test procedures. We will start by discussing standards that define the optical fiber. Optical fibers are mainly covered by ITU-T's G.651 to G.657 series. G.651 defines the graded index multimode fiber, while the rest of them define various single mode fibers. The graded index fiber has a core and cladding diameters of 50 and 125 μm. Attenuation is 4 dB/km at 850 nm and 2 dB/km at 1310 nm [23,24].

The most commonly used single-mode fibers (SMF) are defined by G.652 [25]. These are non-dispersion shifted fibers, intended primarily for use in the 1310 nm window, although they can be used in the 1550 nm window too. They are divided into four classes designated as A, B, C, and D. Their zero chromatic dispersion point is in the 1310 nm window, while their chromatic dispersion is around 17 ps/nm·km in the 1550 nm window. The maximum attenuation is around 0.4 dB/km at 1310 nm and 0.3 dB/km at 1550 nm. Cladding diameter for these fibers is 125 μm, and the diameter of the mode field is approximately 9 μm. The standard also defines various parameters such as cladding concentricity, bend loss, and polarization mode dispersion. Class B is intended for higher data rates compared to class A, because its attenuation and polarization dispersion are lower. Classes C and D are similar to A and B, except that in class C and D the water absorption peak is reduced.

Table 12.1. Characteristics of standard single-mode fiber

Standard	Max attenuation (dB/km)	Dispersion (ps/nm/km)
G.652(A)	0.5@1310 nm 0.4@1550 nm	17@1550 nm 0@1310 nm
G.652(B)[1]	0.4@1310 nm 0.35@1550 nm 0.4@1625 nm	17@1550 nm 0@1310 nm
G.652(C)[2]	0.4@1310–1625 nm 0.3@1550 nm	17@1550 nm 0@1310 nm
G.652(D)[3]	0.4@1310–1625 nm 0.3@1550 nm	17@1550 nm 0@1310 nm

[1] Similar to G.652(A) but with reduced PMD
[2] Similar to G.652(A) but with reduced water absorption peak
[3] Similar to G.652(B) but with reduced water absorption peak

Therefore, wavelengths between the 1310 and 1550 window, otherwise unusable due to high absorption coefficient, can be used. Table 12.1 summarizes the properties of these fibers.

Dispersion-shifted fibers are defined by G.653 and in two classes of A and B [26]. In Class A, the zero dispersion wavelength must fall somewhere between 1500 and 1600 nm window. Class B is essentially similar to Class A, except that the dispersion limits are more accurately defined in terms of both minimum and maximum, by "mask lines."

Another type of widely used fibers is defined by G.655, under classes A through E. These are non-zero dispersion-shifted fibers with some minimum dispersion throughout the wavelength range of 1530–1565 nm. These fibers are primarily defined for the 1550 window of operation in WDM applications. Other classes of fibers include G.654 (cut-off shifted) [27], G.656 (non-zero dispersion in the range of 1460–1625 nm) [28], and G.657 [29] which defines a class of fibers less sensitive to bending loss, intended for access network applications.

The G.651-7 series are ITU-T standards. Fibers with equivalent characteristics are also defined in International Electrotechnical Commission standard IEC 60793-2-10, multi-mode fiber class A1, and 60793-2-50, single-mode fibers classifications B1 to B5 [19].

12.5 Laser safety

Almost all optical sources used in fiber optics utilize a semiconductor laser or LED as the source of optical power. Laser radiation is essentially coherent, low divergent light, and because human eye is an excellent optical instrument specialized at focusing light, laser radiation can cause harm to the eye. Also, for intense laser radiation, skin damage is a possibility. As a result, optical sources used in fiber optics must comply with safety standards. Laser safety standards are primarily covered by IEC-60825.

These standards classify lasers into classes ranging from 1 to 4:

- Class 1 lasers are those with sufficiently low power that are deemed harmless. Devices that incorporate high-power lasers in an enclosure in a way that no light is radiated out of the enclosure can also be classified as Class 1. Class 1M lasers are also harmless unless the light is viewed through certain magnifying or light-concentrating optics, for instance a lens.
- Class 2 lasers are visible light lasers (400–700 nm) with powers less than 1 mW. These are classified as low-risk lasers due to the natural response of the eye to light. Thus, accidental viewing of a Class 2 laser, as long as exposure is short and within the reaction time of the eye, is considered low risk.
- Class 2M lasers are also devices with visible light output that are safe because of the natural reaction of the eye. However, they can be hazardous if viewed through magnifying optics.

- Class 3R lasers have a power range of 1–5 mW in the 400–700 nm range. Maximum allowed power may be lower for shorter wavelengths. They pose low levels of risk to eye and skin, but they should be handled with care.
- Class 3B lasers have an output power range of 5–0.5 W. They are dangerous when viewed directly, even with the natural reaction of the eye. However, viewing their reflection off a scattering (non-shiny) surface is safe.
- Class 4 laser are lasers with output powers above 0.5 W. They are hazardous for both eye and skin, and they may also pose fire risks.

Optical transmitters used in fiber optics are generally expected to use class 1 lasers. Because the optical power from a laser is a function of the bias current, optical transmitters are expected to have current limiting circuitry to ensure that optical power stays at a safe level under all conditions, including those resulting from a fault or short circuit.

12.6 SFF-8472 digital monitoring interface

The purpose of the Digital diagnostic and monitoring interface (DDMI) is to provide a set of information regarding the curial link parameters of transceivers that perform the optical to electrical and electrical to optical conversion at the physical layer. By monitoring this information, the systems can identify the transceivers that are deployed at the physical layer, determine their compliance, oversee the performance of the optical link, determine the link margin, and identify or even predict certain failure modes. Thus, the probability of catastrophic link failures or link down time can be reduced. In a way, these functions behave like the dials on a dashboard in a car that provide the driver (the system) with some of the vital information about the physical performance of the car.

The SFF-8472 is the most common standard that defines the types of information available for optical transceivers, as well as their format [30]. It is a Multi Source Agreement (MSA) developed by a committee consisting of a variety of companies whose products are somehow related to optical transceivers. It uses the 2-wire interface (commonly known as I2C) to establish communication with the transceiver. Two serial bus addresses 10100000 (A0h) and 10100010 (A2h) are defined. We will consider each space separately.

12.6.1 Identification data (A0h)

The A0 address space was in fact originally defined as part of the Small Form Pluggable (SFP) MSA in the INF-8074 document [31]. The A0 space is primarily static data, i.e., it contains data that is not changed during system operation. It consists of two pages. The lower page, which is 128 bytes long, includes information that identifies the type of transceiver, the manufacturer, date of manufacturing, the part number, and the serial number. It also includes information about details such as the optical connector type, the operating wavelength, and transceiver's intended data rate. The last 32 bytes of the first page are vendor specific. The second page of A0 space, which covers addresses 128–255, is reserved.

12.6.2 Diagnostic data (A2h)

The A2 space is primarily defined as a means of providing a range of live information about certain critical parameters measured inside the transceiver. These parameters typically include supply voltage, temperature, optical source bias current, transmitter optical power, and received optical power. An 8472 compliant transceiver is expected to have analog to digital converters on board that constantly convert the corresponding analog signals to digital values and update predefined addresses in the memory map with the new values in the format(s) specified by 8472. In addition to these live data, if the transceiver is *externally calibrated*, the transceiver manufacturer must also provide calibration coefficients that will enable the user to convert the live digital values to meaningful physical measurements. On the other hand, if the transceiver is *internally calibrated*, the transceiver must report the real physical values directly, again according to the format specified by the 8472 MSA.

For each of the measured quantities, four additional values need to be available in the transceiver: a low-value warning and alarm threshold, and a high-value warning and alarm threshold. In this manner, the user can determine if the quantity that is being measured is within the normal operating range, or it is approaching a potentially unacceptable level.

The reporting format for transceiver temperature, supply voltage, laser bias current, and optical output power follows a linear fit:

$$[\mathrm{X}] = [\mathrm{SLOPE}] \times [\mathrm{X}_{\mathrm{ADC}}] + [\mathrm{OFFSET}] \tag{12.1}$$

Here [X] is the measured quantity as a 16 bit wide digital number in units defined by SFF-8472, $[\mathrm{X}_{\mathrm{ADC}}]$ is the "raw digital value" resulting from analog to digital conversion inside the transceiver, and [SLOPE] and [OFFSET] are static 16 bit calibration coefficients that are needed to convert the raw value of $[\mathrm{X}_{\mathrm{ADC}}]$ to a meaningful physical measurement of [X].

An externally calibrated module reports the live value of $[\mathrm{X}_{\mathrm{ADC}}]$ directly. The slope and offsets are also stored by the manufacturer in predefined addresses. The user can then read all the quantities on the right-hand side of Eq. (12.1) from the module through the I2C bus and calculate the value of the measured quantity. On the other hand, an internally calibrated module reports the final calibrated value directly, thus no further calculations are needed on the part of the user.

The first-order linear calibration defined by Eq. (12.1) is sufficiently accurate for most measurements. An exception is the received power, which usually needs a higher order of calibration. The reason is that the received power can vary by several orders of magnitude, and it is usually not possible to maintain reasonable measurement accuracy without including higher order correction terms. As a result, SFF-8742 defines a fourth-order polynomial calibration for received power, instead of the linear fit of Eq. (12.1).

12.7 Reliability standards

Reliability standards attempt to address the long-term operation and performance of the device under test (DUT) during its normal life time. They do so by organizing risk factors into specific categories and by defining methods of measuring as well as stressing various parameters associated with those categories.

For fiber optic components and devices, one of the most widely used standards for test and reliability is GR-468, which is a base document from Telcordia [32]. It describes a host of tests designed to qualify and/or quantify the major reliability risk factors associated with optoelectronic devices used in telecommunication systems. In order to assess a part or a product from a reliability perspective, a reasonable sample size is selected and subjected to a set of reliability tests. If the sample parts pass these tests, it can be assumed that they are representative of other similar parts that have gone through the same manufacturing process.

GR-468 highlights the main tests that need to be carried out for reliability purposes as related to optoelectronic devices. However, it does not define all the specifics of the methods used in each of those tests. For that, it draws on a variety of other relevant standards used throughout different industries. What follows is a brief discussion of the major reliability tests that a device needs to pass, primarily based on GR-468. Familiarity with these tests is not only necessary for assessing reliability, but also valuable from a design standpoint, as they highlight risk factors that can affect the long-term performance of a design.

- **Hermeticity:** Typically, optoelectronic devices such as lasers and photodetectors are expected to be hermetically sealed. To test hermeticity, the DUT is placed in a high-pressure helium chamber after which it is placed in a vacuum chamber. Any potential helium leakage from the DUT is then picked up by a leak detector and could indicate a lack of hermeticity. GR-468 recommends the hermeticity test procedure described in MIL-STD-883E as a reference for this test [33].

- **Electro-static discharge (ESD):** To test the susceptibility of a device to ESD, generally a human body model is used. As the name indicates, the human body model attempts to simulate the effects of physical contact between the human body and the device. A common equivalent circuit for human body is a 100 pF capacitor charged to the target voltage, which then is discharged into the device through a 1500 Ω resistor. After that, the device must be tested for functionality, and if no adverse effects are found, the device has passed that ESD test.

A device is normally expected to tolerate 500 V pulses, although usually a higher rating is expected. The tests methodology is to apply ESD pulses with increasing amplitude, until the device fails. For integrated modules, discharges of ±8 and ±15 kV are additionally tested. The reference documents for ESD testing are FOTP-129 and GR-78.

- **Fiber mechanical integrity tests:** There are a number of tests designed to measure reliability of the fiber. These include fiber integrity, twist, side pull, retention, durability of connectors, pull, mating, and wiggle tests. The basis for a lot of these tests is GR-326 [34].

- **Mechanical shock and vibration:** Mechanical shock is a test to verify the mechanical integrity of a device. The test involves applying a series of acceleration pulses to the device. The pulses are to be applied five times in each of the six spatial directions (±x, ±y, ±z). The amount of acceleration for small optoelectronic devices should be 500 g (1 g=9.8 m/s). The reference for this test is MIL-STD-883E [33]. Vibration involves much less acceleration compared to mechanical shock (20 g compared to 500 g), but the acceleration is applied over a long period of time. The test includes applying vibrations with a range of frequencies over a specified period of time. The vibrations must be applied along each of the three spatial axis (x, y, z), over a time cycle at least 4 min long, and then each cycle must be repeated at least four times along each of the axis. The reference document for this test is MIL-STD-883E [33].

- **Temperature cycling:** The purpose of temperature cycling test is to expose the DUT to the stresses associated with changing temperature. Such changes may affect the mechanical integrity of the DUT due to thermal expansion and contraction, or they may affect DUT's operation, for instance, change the optical alignment in the module. In this test the DUT must be subjected to cycles of fast temperature change from −40° C to +85° C. Also the device must stay at each extreme temperature for at least 10 min. The minimum number of cycles is 50 for laser diodes and detectors, 100 for modules rated as commercial temperature, and 500 for modules rated as industrial or extended temperature parts. This test is based on MIL-STD-883E [33].

- **Damp heat:** This is a test designed to measure the reliability of the part against the combination of heat and humidity. The DUTs must be exposed to 85°C and 85% humidity for 500 h, after which they should be tested for functionality. The basis for this test is MIL-STD-202G [35] and IEC 60068-2-3 [36].

- **Powered high-temperature operation:** Generally, increasing the operating temperature of a device will increase the probability of failure. A high-temperature test is designed to expose such potential failures. For optoelectronic devices, the DUT must be tested for 2000 h at the maximum temperature that it is rated for, while it is powered up and is subjected to its worse case operating conditions. A minimum of 11 samples must be used for the test.

- **Cyclic moisture operation test:** This test is designed to expose potential weaknesses against environmental conditions that include high humidity and extreme temperatures. This test requires at least 20 (in some special cases 10) temperature cycles, where each cycle involves two temperature ramps to 65°C, two soaking periods at 65° C, two temperature ramps down to room temperature, and two soaking periods at room temp, all at high humidity. The basis for this test is MIL-STD-883E [33].

- **Power damp heat test:** For non-hermetic optoelectronic devices, a powered damp heat test is also required. This test involves running the DUTs for 1000 h at 85° C with 85% humidity. The minimum number of parts needed is 11. In some cases if it can be shown that the powered and non-powered damp heat tests stimulate the same failure modes then only one of them can be performed.

- **Accelerated aging:** GR468 refers to accelerated aging as a "quantitative" test, because unlike the previous tests, the result is not a mere "pass/fail." These tests are deigned to accelerate the failure mechanisms that occur either randomly during the life of a device or as a result of wear out. To perform the test, a number of DUTs are exposed to extreme operating conditions, generally involving high temperature. The time is 5000 h for most optoelectronic modules and 10,000 h for devices such as laser diodes. The test temperature is 85°C for modules designed to work in uncontrolled environment and 70°C for devices designed for controlled temperature environments. The number of units required to go through the test is between 5 and 10. During the test, critical parameters need to be tested periodically and compared against levels deemed to represent end of life conditions. At the end of the tests, if critical parameters have not degraded to the end of life threshold, the degradation can be extrapolated to the end of life threshold. The results of accelerated aging can be used in reliability calculations to estimate the performance degradation of the device with time.

In addition to the tests outlined above, a number of other tests are performed to evaluate the reliability of elements such as connectors and fibers. For more details, reliability standards such as GR-468, G.911, and GR-357 should be consulted.

12.8 Networking standards

In Chapter 2 we reviewed some of the fundamental concepts in networking. Because of the nature of the industry, networking is a technology heavily involved with standards. There are countless protocols and standards already in place for all layers of networks, while standardization bodies are constantly working on new standards in response to the evolving technologies and market needs. Because the focus of this book is physical layer, we limit our discussion to major physical layer link parameters for SONET, Ethernet, and PONs.

12.8.1 SONET/SDH

SONET (synchronous optical network) is a popular standard for digital data transmission over optical fibers, based on time division multiplexing of various digital signals in a hierarchical way [37]. It is mainly defined in Telcordia's GR-253 and in T1.105 from ANSI [38]. The ITU-T's equivalent to SONET is the synchronous digital hierarchy, or SDH, defined in several recommendations, including G.957 and G.691 [39, 40].

SONET's hierarchy is based on interleaving lower rate signals to form higher data rates. Because the clocks used throughout the system are extremely accurate, the lower rate signals can be stacked perfectly to generate higher levels.

Table 12.2. Physical layer specifications for some SONET links[1]

Class	Optical signal	Source	Nominal λ (nm)	Budget (dB)	Optical PWR (dBm)	ER Min (dB)	RX PWR (dBm)
SR-1	OC-3	MLM/LED	1310	0 to 7	−15 to −8	8.2	−23 to −8
SR-1	OC-48	MLM/LED	1310	0 to 7	−10 to −3	8.2	−18 to −3
IR-1	OC-48	SLM	1310	0 to 12	−5 to 0	8.2	−18 to 0
IR-2	OC-48	SLM	1550	0 to 12	−5 to 0	8.2	−18 to 0
IR-1	OC-192	DM/EM	1310	6 to 11	1 to 5	6	−11 to −1
IR-2	OC-192	EM	1550	3 to 11	−1 to 2	8.2	−14 to −1
LR-1	OC-48	SLM	1310	10 to 24	−2 to +3	8.2	−27 to −9
LR-2	OC-48	SLM	1550	10 to 24	−2 to +3	8.2	−28 to −9
LR-1	OC-192	DM/EM	1310	17 to 22	+4 to +7	6	−19 to −10
LR-2(a,c)	OC-192	EM	1550	11 to 22	−2 to +2	10	−26 to −9
LR-2b	OC-192	EM	1550	16 to 22	+10 to +13	8.2	−14 to −3
VR-1	OC-192	EM	1310	22 to 33	+10 to +13	6	−24 to −9
VR-2	OC-192	EM	1550	22 to 33	+10 to +13	10	−25 to −9

[1] MLM: Multi-longitudinal mode laser, SLM: single-longitudinal mode, DM: directly modulated laser, EM: externally modulated laser

The fundamental data rate defined in SONET is 51.84 Mb/s, which is known as synchronous transport signal (STS-1) in electrical domain, or optical carrier-1 (OC-1) signal in optical domain. Higher data rates are then multiple integers of this base data rate. Signals for OC-3, OC-12, OC-48, OC-192, and OC-768 rates are defined. Table 2.1 in Chapter 2 includes the data rates for SONET signals.

The digital format used in SONET is NRZ. The SONET packets include the main information bits, called the payload, along with overhead bits added for various information management purposes as defined within the SONET standards. These extra functions make SONET very flexible.

At the physical layer, SONET defines different link categories based on the link budget and data rate. Generally, they are categorized as short reach (SR), intermediate reach (IR), long reach (LR), and very long reach (VR). The suffix 1 refers to the wavelength window of 1310 nm, and higher suffixes (usually 2) refer to the 1550 nm window. Similar link categories are defined in G.957 and G.691 [39, 40]. Table 12.2 includes a summary of the physical layer specifications for some SONET links ranging from OC-3 to OC-192.

12.8.2 Ethernet

Ethernet is a popular standard for data communication over various networks, including optical networks. It was originally developed for LAN applications where several computers shared a common coax cable medium for communication. However, it has since expanded into a complex technology covering point-to-point links, through both coax cables and optical fibers. The operating speeds that started with 10 Mbps have since been expanded into 100 Mbps, 1000 Mbps, and 10 Gbps. Currently, work is under way to study the possibility of 100 Gbps links [41]. Like much of the communication networks, higher data rates require increasing reliance on fiber optics as opposed to copper links. Although copper links are defined even at 10 Gbps, the physical distances they are defined for are significantly shorter compared to fiber-based links.

Originally, Ethernet was based on a broadcast concept, meaning when a computer on a network sent its information, all other computers received that information and had to determine whether the packet was destined for them by looking at the destination address within the packet. If two (or more) computers attempted to use the shared medium at the same time, a collision would occur. To resolve the conflict, both would back away and wait for a random amount of time before attempting to send information again. This scheme is known as carrier sense multiple access with collision detection (CSMA/CD). However, current Ethernet networks utilize a variety of additional networking concepts to improve the communication efficiency. In this section we are mostly concerned with Ethernet's physical layer implementations in optical fiber. It is defined through the IEEE 802.3 family of standards, which is a family of documents that defines data rates of 10 Mbps, 100 Mbps, 1000 Mbps, and 10 Gbps [42].

The designation for Ethernet physical layer implementations is typically in the form of X-Base-Y where X is a reference to the data rate and Y designates some specific aspect such as the medium. The standards defined over copper cable typically end with T, while fiber-based links usually end with suffixes that include X. Tables 12.3 and 12.4 include main physical layer parameters for Ethernet optical links.

Table 12.3. Gigabit Ethernet physical layer specification

Optical signal	Fiber	Nominal λ (nm)	Distance (m)	Budget (dB)	PWR (dBm)	ER Min (dB)	RX PWR (dBm)
1000Base-SX	62.5MMF	770–860	220–275	7.5	−9.5	9	−17 to 0
1000Base-SX	50MMF	770–860	500–550	7.5	−9.5	9	−17 to 0
1000Base-LX	62.5MMF	1270–1355	550	7.5	−11.5	9	−19 to −3
1000Base-LX	50MMF	1270–1355	550	7.5	−11.5	9	−19 to −3
1000Base-LX	10SMF	1270–1355	5000	8	−11	9	−19 to −3

Table 12.4. 10G Ethernet physical layer specification

Optical Signal	Data rate (Gbps)	Nominal λ (nm)	Distance (m)	Budget (dB)	PWR (dBm)	ER Min (dB)	RX PWR (dBm)
10GBase-SR	10.31	850	26–300	7.5	−1	3	−12* to −1
10GBase-SW	9.95	850	26–300	7.5	−1	3	−12* to −1
10GBase-LR	10.31	1310	2–10000	9.4	−4.1* to 0.5	4	−13.2* to −0.5
10GBase-LW	9.95	1310	2–10000	9.4	−4.1* to 0.5	4	−13.2* to −0.5
10GBase-ER	9.95	1550	2–40000	18	−1.39* to 4	3	−15.39* to −3
10GBase-EW	10.31	1550	2–40000	18	−1.39* to 4	3	−15.39* to −3

* Specified in terms of optical modulation amplitude (OMA)

Within the 10G specifications, 10G-SR and 10G-SW are intended for MMF, and the rest are intended for SMF. Also note that some of the minimum powers and sensitivities are expressed for optical modulation amplitude. This is because the minimum required extinction ratio is rather low. At such low ERs, the transmitter is required to generate a minimum amount of modulation, and the receiver is required to be able to receive a minimum amount of modulation, regardless of how high the average power in the signal is.

12.8.3 Passive optical networks (PON)

Passive optical networks are at the front line of advancement of fiber optics toward end users[43].[2] In a typical PON, many users are passively connected to a central office. The central office is generically called optical line terminal (OLT) and the users are called optical network unit (ONU). An ONU may be located in a home, in which case a fiber link directly goes to home. In other cases, an ONU may be a point of optical to electrical conversion, which brings the high-speed signal to the premise, and after that the signal is distributed between a number of users, say, in an apartment complex. There are three main families of PON in common use: broadband PON (BPON), Ethernet PON (EPON), and gigabit PON (GPON). Table 12.5 gives a summary of these three standards.

Table 12.5. Common PON standards

Technology	Standard	Distance (km)	Splitting ratio (max)	Downstream (Mbps)	Upstream (Mbps)
BPON	G.983	20	32:1	155, 622, 1244	155, 622
EPON	802.3	10	32:1	1244	1244
GPON	G.984	20	64:1	1244, 2488	155 to 2488

BPON is the original standard used in PON networks and is based on the well-established ATM protocol. It is defined in G.983.1 to G.983.5. A key feature of ATM is that it defines different classes of service that makes it suitable for carrying various types of traffic, including voice, video, and data. The downstream data rate defined in BPON is 622 or 1244 Mbps, and it is transmitted on the 1490 nm wavelength. The upstream data rate is defined at 155 or 622 Mbps and is carried back over a 1310 nm wavelength. BPON also allows a video channel on the 1550 nm wavelength to be added to the downstream data [44]. In the downstream

[2] See Chapter 2 for a discussion about PON concepts.

direction, the OLT transmits in a CW mode to all the ONUs using ATM format. Each ONU can distinguish the cells that are intended for it by looking at the headers in each packet. In the upstream direction, only one ONU can transmit at a time.

This requires the ONUs to work synchronously with each other, so that no two ONUs "talk" at the same time. From a network perspective, a *ranging* process has to be performed through which the OLT can calibrate the various delays between different ONUs, because in general, ONUs are located at different distances from the OLT and therefore the time it takes for their output to reach the ONU varies. From a physical layer perspective, the upstream traffic is burst mode in nature, because depending on the distance of the particular ONU, data will be received at the OLT as a mixture of high and low power bursts.

Gigabit PON or GPON are based on BPON concepts and therefore are very similar to BPON. The GPON standards are covered by the ITU-T family of G.984 documents, which include G.984.1 to G.984.6. The purpose of developing the gigabit PON was to extend the speed into the Gbps range. In GPON the downstream data rate can be 1.24 or 2.48 Gbps, and the upstream rate can be from 155 Mbps up to 1.24 Gbps. GPON also uses the 1490 nm wavelength for downstream and the 1310 nm for upstream [45, 46].

Both BOPN and GPON are based on the ATM signaling technology. However, the pervasive use of Ethernet especially in local area networks makes it an alternative method for transmission of signals. Technologies that use Ethernet for access networks are generally known as ethernet in the first mile (EFM) [47]. EPON (Ethernet PON) is a primary example of EFM. EPON also uses the 1310 nm window for upstream and 1490 nm window for downstream transmission. The 1550 nm window is open and can be used independently for applications such as video. The nominal bit rate, however, is 1 Gbps, which as a result of 8B10B encoding translates to a line rate of 1.25 Gbps. EPON utilizes this bit rate symmetrically for both downstream and upstream directions. The splitting ratio can be either 32 or 64. The physical distance reach for EPON is either 10 or 20 km. EPON is mainly covered by the 802.3ah standard, which is now part of the 802.3 standard.

12.9 Summary

Standards are an essential component of any industry. In fiber optics, standards define the specifications and performance of optical components. Standards also define methodologies for measuring physical quantities encountered in fiber optics. For reliability purposes, standards define methods of testing and stressing devices to characterize their long-term performance. For networks, standards define a common language that enables different systems and equipment from different manufacturers to work and communicate with each other.

We started this chapter by introducing the major bodies active in defining the standards that are used in the fiber optic industry. These include ITU-T, IEC, ISO,

IEEE, TIA, and Telcordia, among others. We then proceeded by presenting a list of selected standards in each of the four different categories of (1) components, (2) test and measurement, (3) reliability, and (4) networks and systems.

The next sections in this chapter were dedicated to a brief review of selected standards that are more frequently needed and encountered in physical layer design. In terms of components, we provided a brief review of optical fiber standards. We also considered standards of laser safety especially with regards to optical transceivers.

In terms of measurement standards, we went over the SFF-8472 protocol, which defines an interface between physical layer and higher network layers. Modern transceivers are commonly expected to provide live information about critical link parameters, and 8472 is the default protocol used for this purpose.

In terms of reliability standards, we focused on GR-468, which is an accepted basis for testing the reliability of optical components and systems. We noted that GR-468 is a base document, which in turn draws on a variety of other standards for various aspects of reliability. Knowledge of these standards is not only necessary for reliability testing, but also provides a useful background for design engineers in terms of expected levels of reliability from optoelectronic components.

Finally, in terms of networks standards, we reviewed key physical layer parameters of SONET/SDH, Ethernet, and PON networks. SONET and Ethernet are the main protocols for data and telecom applications. PON networks, especially BPON, GPON, and EPON, are at the forefront of the wave that is bringing the advantages of fiber optics to the end users. Thus, these networks, along with SONET and Ethernet, constitute the basis of optical networking infrastructure that is driving the information technology presently and in the years to come.

References

[1]International Telecommunication Union (ITU), www.itu.int

[2] Telecommunication Standardization Sector (ITU-T), http://www.itu.int/ITU-T

[3] For a list of ITU-T's standards see http://www.itu.int/publ/T-REC/en

[4] International Electrotechnical Commission (IEC), www.iec.ch

[5] http://www.electropedia.org

[6] For a review of SI system of units, see National Institute of Standards and Technology's (NIST) website at http://physics.nist.gov/cuu/Units

[7] Institute of Electrical and Electronics Engineers (IEEE), www.ieee.org

[8] IEEE Standards Association (IEEE-SA), http://standards.ieee.org

[9] Telecommunication Industry Association (TIA), www.tiaonline.org

[10] International Organization for Standardization (ISO), www.iso.org

[11] See, for instance, http://www.iso9000council.org

[12] American National Standard Institute (ANSI), www.ansi.org

[13] Telcordia, www.telcordia.com

[14] Telcordia, "Telcordia Roadmap to Fiber and Optical Technologies Documents," Issue 4, Aug 2008, available from http://telecom-info.telcordia.com/site-cgi/ido/index.html

[15] www.oiforum.com

[16] www.sffcommittee.org

[17] www.xfpmsa.org

[18] ftp://ftp.seagate.com/sff/

[19] G.Sup.40, ITU-T Supplement 40, Optical fiber and cable recommendations and standards guideline, ITU-T, 2006

[20] K. Kazi, *Optical Networking Standards*, Springer, New York, 2006

[21] G.SUP.42, ITU-T Supplement 42, Guide on the use of the ITU-T Recommendations related to Optical Technology, ITU-T, 2008

[22] G.SUP.39, ITU-T Supplement 39, Optical system design and engineering considerations, 2006

[23] ITU-T, Recommendation G.651, Characteristics of a 50/125 μm multimode graded index optical fiber cable, 1998

[24] ITU-T, Recommendation G.651.1, Characteristics of a 50/125 μm multimode graded index optical fiber cable for the optical access network, ITU-T, 2007

[25] ITU-T, Recommendation G.652, Characteristics of a single-mode optical fiber and cable, 2005

[26] ITU-T, Recommendation G.653, Characteristics of a dispersion-shifted single-mode optical fiber and cable, 2006

[27] ITU-T, Recommendation G.654, Characteristics of a cut-off shifted single-mode optical fiber and cable, 2006

[28] ITU-T, Recommendation G.656, Characteristics of a fiber and cable with non-zero dispersion for wideband optical transport, 2006

[29] ITU-T, Recommendation G.657, Characteristics of a bending loss insensitive single mode optical fiber and cable for the access network, 2006

[30] SFF-8472, Diagnostic Monitoring Interface for Optical Transceivers, SFF Committee, 2007

[31] INF-8074i, Specification for SFP (Small Formfactor Pluggable) Transceiver, SFF Committee, 2001

[32] GR-468-Core, Generic Reliability Assurance Requirements for Optoelectronic Devices, Telcordia, 2004

[33] MIL-STD-883E, *Test Method Standard, Microcircuits*, Department of Defense, Washington, DC, 1996

[34] GR-326, Generic Requirements for Single mode Optical Connectors and Jumper Assemblies, Telcordia, 1999

[35] MIL-STD-202G, *Test Method Standard, Electronic and Electrical Component Parts*, Department of Defense, Washington, DC, 2002

[36] IEC-60068-2-30, Environmental testing – Part 2-30: Tests – Test Db: Damp heat, cyclic (12 h + 12 h cycle), International Electrotechnical Commission, 2005

[37] W. Goralski, *SONET/SDH*, McGraw-Hill, New York, 2002

[38] GR-253, Synchronous Optical Network (SONET) Transport Systems: Common Generic Criteria, Telcordia, 2005

[39] G.957, Optical interfaces for equipments and systems relating to the synchronous digital hierarchy, ITU-T, 2006

[40] G.691, Optical interfaces for single channel STM-64 and other SDH systems with optical amplifiers, ITU-T, 2006

[41] IEEE 802.3ba, 40Gb/s and 100Gb/s Ethernet Task Force, http://www.ieee802.org/3/ba/index.html

[42] IEEE, 802.3 Standard, 2005, available from www.ieee802.org

[43] G. Keiser, *FTTX Concepts and Applications*, John Wiley & Sons, Hoboken, NJ, 2006

[44] G.983.1, Broadband optical access systems based on Passive Optical Networks (PON), ITU-T, 2005

[45] G.984.1, Gigabit-capable passive optical networks (GPON): General characteristics, ITU-T, 2008

[46] G.984.2: Gigabit-capable Passive Optical Networks (GPON): Physical Media Dependent (PMD) layer specification, ITU-T, 2008

[47] Michael Beck, *Ethernet in the First Mile*, McGraw-Hill, New York, 2005

Appendix A
Common Acronyms

ADC	Analog to Digital Converter
ADM	Add/Drop Multiplexer
AGC	Automatic Gain Control
ANSI	American National Standard Institute
APC	Angled Physical Contact
APC	Average Power Control
APD	Avalanche Photo Diode
ASE	Amplified Spontaneous Emission
ATM	Asynchronous Transfer Mode
AWG	Arrayed Waveguide Grating
BER	Bit Error Rate
BERT	Bit Error Rate Tester
BMR	Burst Mode Receiver
BPON	Broadband PON
CATV	Community Access TV
CDR	Clock and Data Recovery
CID	Consecutive Identical Digits
CML	Current Mode Logic
CMRR	Common Mode Rejection Ratio
CWDM	Coarse Wavelength Division Multiplexing
DAC	Digital to Analog Converter
dB	Decibel
DCA	Digital Communication Analyzer
DCF	Dispersion Compensating Fiber
DDMI	Digital Diagnostic Monitoring Interface
DFA	Doped Fiber Amplifier
DFB	Distributed Feedback Laser
DSL	Digital Subscriber Line
DUT	Device Under Test
DWDM	Dense Wavelength Division Multiplexing
EDFA	Erbium-Doped Fiber Amplifier
EPON	Ethernet PON
ER	Extinction Ratio
ETS	Equivalent Sampling Time
FC	Fixed Connector
FIT	Failure in Time
FOCIS	Fiber Optic Connector Intermateability Standard
FOTP	Fiber Optic Test Procedure
FP	Fabry–Perot
FTTC	Fiber to the Curb
FTTH	Fiber to the Home
FTTP	Fiber to the Premise
FTTX	Fiber to the X
FWM	Four Wave Mixing
Gbps	Giga Bits Per Second
GHz	Giga Hertz
GPON	Gigabit PON
GRIN	Graded Index
HDTV	High-Definition TV
IEC	International Electrotechnical Commission
IMD	Inter-modulation Distortion
IP	Internet Protocol
ISO	International Organization for Standardization
ITU	International Telecommunication Union
Kbps	Kilo Bits Per Second
LAN	Local Area Network
LED	Light Emitting Diode
LIA	Limiting Amplifier
LLC	Logical Link Control
LPF	Low-Pass Filter
LSA	Lightwave Signal Analyzer
LVCMOS	Low-Volatge CMOS
LVDS	Low-Voltage Differential Signaling

M. Azadeh, *Fiber Optics Engineering*, Optical Networks,
DOI 10.1007/978-1-4419-0304-4,

LVTTL	Low-Voltage TTL
MAC	Media Access Control
MAN	Metropolitan Area Network
Mbps	Mega Bits Per Second
MD	Modulation Depth
MHz	Mega Hertz
MMF	Multimode Fiber
MMTF	Mean Time to Failure
MZ	Mach–Zehnder
NRZ	Non-Return to Zero
NTC	Negative Temperature Coefficient
OADM	Optical Add/Drop Multiplexer
OC-N	Optical Carrier N
OEO	Optical–Electrical–Optical
OIF	Optical Internetworking
OLT	Optical Line Terminal
OMA	Optical Modulation Amplitude
ONT	Optical Network Terminal
ONU	Optical Network Unit
OSA	Optical Spectrum Analyzer
OSI	Open System Interconnect
OTN	Optical Transport Network
OXC	Optical Crossconnect
PC	Physical Contact
PCB	Printed Circuit Board
PECL	Positive Emitter Coupled Logic
PLL	Phase Lock Loop
PMD	Polarization Mode Dispersion
POF	Plastic Optical Fiber
PON	Passive Optical Network
PRBS	Pseudo Random Bit Sequence
RIN	Relative Intensity Noise
rms	Root Mean Square
ROSA	Receiver Optical Subassembly
RX	Receiver
SAP	Service Access Point
SBS	Stimulated Brillouin Scattering
SC	Subscriber Connector
SDH	Synchronous Digital Hierarchy
SFF	Small Form Factor
SFP	Small Form Pluggable
SMF	Single-Mode Fiber
SMSR	Side Mode Suppression Ratio
SNR	Signal to Noise Ratio
SOA	Semiconductor Optical Amplifier
SONET	Synchronous Optical Network
SPM	Self-Phase Modulation
SRS	Stimulated Raman Scattering
ST	Straight Tip
STM-N	Synchronous Transport Module-N
SW	Spectral Width
Tbps	Terra Bits Per Second
TEC	Thermoelectric Cooler
TEM	Transverse Electromagnetic
THD	Total Harmonic Distortion
THz	Terra Hertz
TIA	Transimpedance Amplifier
TIA	Telecommunication Industry Association
TOSA	Transmitter Optical Subassembly
TX	Transmitter
UHF	Ultra High Frequency
UI	Unit Interval
VCO	Voltage-Controlled Oscillator
VCSEL	Vertical Cavity Surface Emitting Laser
VHF	Very High Frequency
WAN	Wide Area Network
WDM	Wavelength Division Multiplexing
XPM	Cross-Phase Modulation

Appendix B
Physical Constants

Planck's constant	h	6.625×10^{-34} J.s
Speed of light in vacuum	c	2.998×10^{8} m/s
Electron charge	e	-1.602×10^{-19} C
Mass of electron	m	9.109×10^{-31} kg
Electron-Volt	eV	1.602×10^{-19} J
Boltzmann constant	k	1.381×10^{-23} J/°K
Permittivity of vacuum	ε_0	8.85×10^{-12} F/m
Permeability of free space	μ_0	1.257×10^{-6} H/m
Angstrom	Å	10^{-10} m
Nanometer	nm	10^{-9} m
Micron	μm	10^{-6} m
Silica fiber core index of refraction	n_0	1.48
Speed of electrical signals in a 50 Ω FR4 microstrip	v	$\sim8\times10^{7}$ m/s

Index

M

MIX
Papier aus verantwortungsvollen Quellen
Paper from responsible sources
FSC® C105338

If you have any concerns about our products,
you can contact us on
ProductSafety@springernature.com

In case Publisher is established outside the EU,
the EU authorized representative is:
Springer Nature Customer Service Center GmbH
Europaplatz 3, 69115 Heidelberg, Germany

Printed by Libri Plureos GmbH
in Hamburg, Germany